T0210966

Communications in Computer and Information Science   **1120**

Commenced Publication in 2007
Founding and Former Series Editors:
Phoebe Chen, Alfredo Cuzzocrea, Xiaoyong Du, Orhun Kara, Ting Liu,
Krishna M. Sivalingam, Dominik Ślęzak, Takashi Washio, Xiaokang Yang,
and Junsong Yuan

More information about this series at http://www.springer.com/series/7899

Hai Jin · Xuemin Lin · Xueqi Cheng ·
Xuanhua Shi · Nong Xiao ·
Yihua Huang (Eds.)

Big Data

7th CCF Conference, BigData 2019
Wuhan, China, September 26–28, 2019
Proceedings

 Springer

Editors
Hai Jin
Huazhong University of Science
and Technology
Wuhan, China

Xueqi Cheng
Chinese Academy of Sciences
Beijing, China

Nong Xiao
National University of Defense Technology
Changsha, China

Xuemin Lin
East China Normal University
Shanghai, China

University of New South Wales
Sydney, Australia

Xuanhua Shi
Huazhong University of Science
and Technology
Wuhan, China

Yihua Huang
Nanjing University
Nanjing, China

ISSN 1865-0929 ISSN 1865-0937 (electronic)
Communications in Computer and Information Science
ISBN 978-981-15-1898-0 ISBN 978-981-15-1899-7 (eBook)
https://doi.org/10.1007/978-981-15-1899-7

This Springer imprint is published by the registered company Springer Nature Singapore Pte Ltd.
The registered company address is: 152 Beach Road, #21-01/04 Gateway East, Singapore 189721, Singapore

Preface

With the rapid development of technology and industry, big data is a national strategy now, and it has become a link for different disciplines. The CCF Academic Conferences on BigData (CCF BigData) series are the premier forums for interdisciplinary big data research in China. The 7th CCF Academic Conference on BigData (CCF BigData 2019) was held during September 26–28, 2019, at a specific place called Optics Valley, in Wuhan, China. This volume constitutes the proceedings of it, with the theme "Gathering the Data, Intelligent Computing for the Future."

CCF BigData 2019 was initiated by the China Computer Federation (CCF), and was co-organized by the China Computer Federation Task Force on Big Data and Huazhong University of Science and Technology. There were 5 keynote speeches, 2 invited talks, 188 invited workshop talks, 13 best-paper candidates presentations, 141 posters, and 10 exhibitions at the conference. There were 324 submissions (in English and Chinese) from 700 authors. Each submission was reviewed by at least three Program Committee members. Finally, 30 high-quality papers written in English were selected for inclusion in this proceedings volume. The acceptance rate was 9.26%.

We sincerely thank the keynote speakers, invited speakers, authors, and the workshop organizer for creating the wonderful academic activities at CCF BigData 2019, and we sincerely thank the Program Committee members for the paper reviewing. We also want to thank the China Computer Federation Task Force on Big Data and the Huazhong University of Science and Technology for the great support and for organizing the conferences. We appreciate Springer for publishing the proceedings in CCIS (*Communications in Computer and Information Science*). Moreover, we want to thank the sponsors for their contributions.

September 2019

Hai Jin
Xuemin Lin
Xueqi Cheng
Xuanhua Shi
Nong Xiao
Yihua Huang

Organization

CCF BigData 2019 was organized by the China Computer Federation and co-organized by the China Computer Federation Task Force on Big Data and Huazhong University of Science and Technology.

Conference Honorary Chairs

Guojie Li	Institute of Computing Technology, Chinese Academy of Sciences, Academician of Chinese Academy of Engineering, China
Hong Mei	Academy of Military Sciences PLA China, Academician of Chinese Academy of Sciences, China

Steering Committee Chair

Hong Mei	Academy of Military Sciences PLA China, Academician of Chinese Academy of Sciences, China

General Chairs

Hai Jin	Huazhong University of Science and Technology, China
Xuemin Lin	The University of New South Wales, Australia, and East China Normal University, China
Xueqi Cheng	Institute of Computing Technology, Chinese Academy of Sciences, China

Program Chairs

Xuanhua Shi	Huazhong University of Science and Technology, China
Nong Xiao	National University of Defense Technology, China
Yihua Huang	Nanjing University, China

Publicity Chairs

Jiang Xiao	Huazhong University of Science and Technology, China
Chao Gao	Southwest University, China

Publication Chairs

Li Wang	Taiyuan University of Technology, China
Yang Gao	Nanjing University, China
Ran Zheng	Huazhong University of Science and Technology, China
Zeguang Lu	Zhongke Guoding Institute of Data Science, China

Finance Chairs

Li Tao	Huazhong University of Science and Technology, China
Jin Yang	Institute of Computing Technology, Chinese Academy of Sciences, China

Sponsorship Chairs

Yunquan Zhang	Institute of Computing Technology, Chinese Academy of Sciences, China
Laizhong Cui	Shenzhen University, China

Local Arrangement Chairs

Yingshu Liu	Huazhong University of Science and Technology, China
Tao Jia	Southwest University, China

Workshop Chairs

Tong Ruan	East China University of Science and Technology, China
Rui Mao	Shenzhen University, China

Best Paper Award Chairs

Jiajun Bu	Zhejiang University, China
Shaoliang Peng	Hunan University, China

Program Committee

Peng Bao	Beijing Jiaotong University, China
Jiajun Bu	Zhejiang University, China
Yi Cai	South China University of Technology, China
Ruichu Cai	Guangdong University of Technology, China
Guoyong Cai	Guilin University of Electronic Technology, China
Fuyuan Cao	Shanxi University, China

Hanhua Chen	Huazhong University of Science and Technology, China
Hong Chen	Renmin University of China, China
Jun Chen	Institute of Applied Physics and Computational Mathematics, China
Jun Chen	China Defense Science and Technology Information Center, China
Shimin Chen	Institute of Computing Technology, Chinese Academy of Sciences, China
Xuebin Chen	North China University of Science and Technology, China
Yueguo Chen	Renmin University of China, China
Gong Cheng	Nanjing University, China
Peng Cui	Tsinghua University, China
Laizhong Cui	Shenzhen University, China
Lizhen Cui	Shandong University, China
Cheng Deng	Xidian University, China
Zhaohong Deng	Jiangnan University, China
Shifei Ding	China University of Mining and Technology, China
Zhiming Ding	Institute of Software Chinese Academy of Sciences, China
MingGang Dong	Guilin University of Technology, China
Junping Du	Beijing University of Posts and Telecommunications, China
Jianzhou Feng	Yanshan University, China
Shaojing Fu	National University of Defense Technology, China
Chao Gao	Southwest University, China
Sheng Gao	Beijing University of Posts and Telecommunications, China
Xiaofeng Gao	Shanghai Jiao Tong University, China
Xinbo Gao	Xidian University, China
Yanhong Guo	Dalian University of Technology, China
Kun Guo	Fuzhou University, China
Bin Guo	Northwestern Polytechnical University, China
Daojing He	East China Normal University, China
Guoliang He	Wuhan University, China
Jieyue He	Southeast University, China
Liwen He	Jiangsu Industrial Technology Research Institute, China
Zhenying He	Fudan University, China
Qiangsheng Hua	Huazhong University of Science and Technology, China
Lan Huang	Jilin University, China
Zhexue Huang	Shenzhen University, China
Bin Jiang	Hunan University, China
Yichuan Jiang	Southeast University, China

Bo Jin	The Third Research Institute of The Ministry of Public Security, China
Peiquan Jin	University of Science and Technology of China, China
Xin Jin	Central University of Finance and Economics, China
Xiaolong Jin	Chinese Academy of Sciences, China
Jianhui Li	Chinese Academy of Sciences, China
Cuiping Li	Renmin University of China, China
Dongsheng Li	National University of Defense Technology, China
Jianxin Li	Beihang University, China
Kai Li	University of Science and Technology of China, China
Ru Li	Inner Mongolia University, China
Ruixuan Li	Huazhong University of Science and Technology, China
Shanshan Li	National University of Defense Technology, China
Shijun Li	Wuhan University, China
Tianrui Li	Southwest Jiaotong University, China
Wu-Jun Li	Nanjing University, China
Yidong Li	Beijing Jiaotong University, China
Yufeng Li	Nanjing University, China
Wangqun Lin	Academy of Military Sciences PLA China, China
Jie Liu	Nankai University, China
Qi Liu	University of Science and Technology of China, China
Qingshan Liu	Nanjing University of Information Science and Technology, China
Shijun Liu	Shandong University, China
Yang Liu	Shandong University, China
Yubao Liu	Sun Yat-sen University, China
Tun Lu	Fudan University, China
Jiawei Luo	Hunan university, China
Jun Ma	Shandong University, China
Rui Mao	Shenzhen University, China
Wenji Mao	Institute of Automation, Chinese Academy of Sciences, China
Qiguang Miao	Xidian University, China
Baoning Niu	Taiyuan University of Technology, China
Jianquan Ouyang	Xiangtan University, China
Shaoliang Peng	Hunan University, China
Zhiyong Peng	Wuhan University, China
Yuhua Qian	Shanxi University, China
Shaojie Qiao	Chengdu University of Information Technology, China
Lu Qin	University of Technology Sydney, Australia
Zheng Qin	Hunan University, China
Jiadong Ren	Yanshan University, China
Lei Ren	Beihang University, China
Tong Ruan	East China University of Science and Technology, China

Jin Zhang	Institute of Computing Technology, Chinese Academy of Sciences, China
Zili Zhang	Southwest University, China
Le Zhang	Sichuan University, China
Dongyan Zhao	Peking University, China
Gansen Zhao	South China Normal University, China
Pengpeng Zhao	Soochow University, China
Xiang Zhao	National University of Defense Technology, China
Xiaolong Zheng	Institute of Automation, Chinese Academy of Sciences, China
Bin Zhou	National University of Defense Technology, China
Fengfeng Zhou	Shenzhen Institutes of Advanced Technology, Chinese Academy of Sciences, China
Tao Zhou	Alibaba Group, China
Yangfan Zhou	Fudan University, China
Yuanchun Zhou	Chinese Academy of Sciences, China
Feida Zhu	Singapore Management University, Singapore
Xiaofei Zhu	Chongqing University of Technology, China
Fuhao Zou	Huazhong University of Science and Technology, China
Lei Zou	Peking University, China

Sponsoring Institutions

Contents

Big Data Modelling and Methodology

A Constrained Self-adaptive Sparse Combination Representation Method
for Abnormal Event Detection   3
    Huiyu Mu, Ruizhi Sun, Li Li, Saihua Cai, and Qianqian Zhang

A Distributed Scheduling Framework of Service Based ETL Process   16
    DongJu Yang and ChenYang Xu

A Probabilistic Soft Logic Reasoning Model with Automatic
Rule Learning   33
    Jia Zhang, Hui Zhang, Bo Li, Chunming Yang, and Xujian Zhao

Inferring How Novice Students Learn to Code: Integrating Automated
Program Repair with Cognitive Model   46
    Yu Liang, Wenjun Wu, Lisha Wu, and Meng Wang

Predicting Friendship Using a Unified Probability Model   57
    Zhijuan Kou, Hua Wang, Pingpeng Yuan, Hai Jin, and Xia Xie

Product Feature Extraction via Topic Model and Synonym
Recognition Approach   73
    Jun Feng, Wen Yang, Cheng Gong, Xiaodong Li, and Rongrong Bo

Vietnamese Noun Phrase Chunking Based on BiLSTM-CRF Model
and Constraint Rules   89
    Hua Lai, Chen Zhao, Zhengtao Yu, Shengxiang Gao, and Yu Xu

Big Data Support and Architecture

A Distributed Big Data Discretization Algorithm Under Spark   107
    Yeung Chan, Xia Jie Zhang, and Jing Hua Zhu

A Novel Distributed Duration-Aware LSTM for Large Scale Sequential
Data Analysis   120
    *Dejiao Niu, Yawen Liu, Tao Cai, Xia Zheng, Tianquan Liu,
and Shijie Zhou*

Considering User Distribution and Cost Awareness to Optimize
Server Deployment   135
    Yanling Shao and Wenyong Dong

Convolutional Neural Networks on EEG-Based Emotion Recognition   148
    Chunbin Li, Xiao Sun, Yindong Dong, and Fuji Ren

Distributed Subgraph Matching Privacy Preserving Method for Dynamic
Social Network.   159
    Xiao-Lin Zhang, Hao-chen Yuan, Zhuo-lin Li, Huan-xiang Zhang,
    and Jian Li

Big Data Processing

Clustering-Anonymization-Based Differential Location Privacy
Preserving Protocol in WSN.   177
    Ren-ji Huang, Qing Ye, and Mo-Ci Li

Distributed Graph Perturbation Algorithm on Social Networks
with Reachability Preservation   194
    Xiaolin Zhang, Jian Li, Xiaoyu He, and Jiao Liu

Minimum Spanning Tree Clustering Based on Density Filtering   209
    Ke Wang, Xia Xie, Jiayu Sun, and Wenzhi Cao

Research of CouchDB Storage Plugin for Big Data Query Engine
Apache Drill.   224
    Yulei Liao and Liang Tan

Visual Saliency Based on Two-Dimensional Fractional Fourier Transform . . .   240
    Haibo Xu and Chengshun Jiang

Big Data Analysis

An Information Sensitivity Inference Method for Big Data Aggregation
Based on Granular Analysis.   255
    Lifeng Cao, Xin Lu, Zhensheng Gao, and Zhanbing Zhu

Attentional Transformer Networks for Target-Oriented
Sentiment Classification.   271
    Jianing Tong, Wei Chen, and Zhihua Wei

Distributed Logistic Regression for Separated Massive Data.   285
    Peishen Shi, Puyu Wang, and Hai Zhang

Similarity Evaluation on Labeled Graphs via Hierarchical
Core Decomposition   297
    Deming Chu, Fan Zhang, and Jingjing Lin

Weighted Multi-label Learning with Rank Preservation   312
  Chong Sun, Weiyu Zhou, Zhongshan Song, Fan Yin, Lei Zhang,
  and Jianquan Bi

Orthogonal Graph Regularized Nonnegative Matrix Factorization
for Image Clustering   325
  Jinrong He, Dongjian He, Bin Liu, and Wenfa Wang

Big Data Application

Characteristics of Patient Arrivals and Service Utilization
in Outpatient Departments   341
  Yonghou He, Bo Chen, Yuanxi Li, Chunqing Wang, Zili Zhang,
  and Li Tao

College Academic Achievement Early Warning Prediction Based
on Decision Tree Model   351
  Jianlin Zhu, Yilin Kang, Rongbo Zhu, Dajiang Wei, Yao Chen, Zhi Li,
  Zhentong Wang, and Jin Huang

Intelligent Detection of Large-Scale KPI Streams Anomaly Based
on Transfer Learning   366
  XiaoYan Duan, NingJiang Chen, and YongSheng Xie

Latent Feature Representation for Cohesive Community Detection Based
on Convolutional Auto-Encoder   380
  Chun Li, Wenfeng Shi, and Lin Shang

Research on Monitoring Method of Ethylene Oxide Process
by Improving C4.5 Algorithm.   395
  Xuehui Jing, Hongwei Zhao, Shuai Zhang, and Ying Ruan

Web API Recommendation with Features Ensemble
and Learning-to-Rank   406
  Hua Zhao, Jing Wang, Qimin Zhou, Xin Wang, and Hao Wu

EagleQR: An Application in Accessing Printed Text for the Elderly
and Low Vision People   420
  Zhi Yu, Jiajun Bu, Chunbin Gu, Shuyi Song, and Liangcheng Li

Author Index   431

Big Data Modelling and Methodology

A Constrained Self-adaptive Sparse Combination Representation Method for Abnormal Event Detection

Huiyu Mu[1], Ruizhi Sun[1,2(✉)], Li Li[1], Saihua Cai[1], and Qianqian Zhang[1]

[1] College of Information and Electrical Engineering,
China Agricultural University, Beijing 100083, China
{b20183080630,sunruizhi,lili_2018,caisaih,qqzhang}@cau.edu.cn
[2] Scientific Research Base for Integrated Technologies of Precision Agriculture
(Animal Husbandry), The Ministry of Agriculture, Beijing 100083, China

Abstract. Automated abnormal detection system meets the need of society for detecting and locating anomalies and alerting the operators. In this paper, we proposed a constrained self-adaptive sparse combination representation (CSCR). The spatio-temporal video volumes low-level features, which be stacked with multi-scale pyramid, can extract features effectively. The CSCR strategy is robust to learn dictionary and detect abnormal behaviors. Experiments on the published dataset and the comparison to other existing methods demonstrate the certain advantages of our method.

Keywords: Visual surveillance · Abnormal event detection · Sparse representation · Structure information

1 Introduction

Abnormal event detection capture the abnormal behavior with alert and minimize the loss for different application scenarios. It has fatal research significance and practical value for the intelligent video monitoring system that can automatically detect abnormal events. The definition of abnormal event detection have not form a standardized outlook. Current research is mainly focused on the limited category of behavior recognition or detection in specific situations with the idea of classification.

Depending on the different detection tasks in different scenes, there are three broad categories, which are behavioral relationship analysis [1–4], motion trajectory analysis [5–7], and low-level feature analysis [8–12]. Specifically, behavioral relationship methods learn the normal and abnormal with probability model, and the abnormal detection is defined with the statistical probability of behavior analysis, e.g., Hidden Markov model [1], Markov random fields [2,3] and Gaussian model, et al. However, those methods are sensitive to parameters [4]. Another line is extracted the motion trajectory from the object-of-interested [5].

© Springer Nature Singapore Pte Ltd. 2019
H. Jin et al. (Eds.): BigData 2019, CCIS 1120, pp. 3–15, 2019.
https://doi.org/10.1007/978-981-15-1899-7_1

These methods commonly use the trajectory information of speed and direction for feature modeling, such as spatiotemporal trajectory detection methods [7]. Tracking-based approaches can obtain satisfied results with a few people scenes. However, it is difficult in the crowded scene and background should be static. Lastly, low-level feature achieves excellent performance for the event description. Such as spatial-temporal energy [8], topic model [9], neural network [10,11] and sparse representation [12,13], etc. Especially, sparse representation for anomaly event detection makes full use of the information of video datasets which can achieve efficient representation of data. Nevertheless, it is difficult to find the suitable basis vectors from the large-scale dictionary, and not suitable for multiple scenarios with poor generalization ability. Furthermore, the method with the fixed dictionary commonly causes more false alarms and shows the incapacity for adjustment to the varying scene, leads to a higher error rates [12].

In order to sufficiently utilize the prior motion information in video sequences, we propose a constrained self-adaptive sparse representation (CSCR) method for solving the problem of detecting abnormal behavior. We adopt a sparse combination representation method, instead of using fixed dictionary, to meet the need of deployment practically for the real-time process. We also add the local geometric structure for the data in the sparse coding stage to improve the detection accuracy.

2   Related Works

In general, abnormal event detection involves three basic steps, including feature extracting, feature depicting, and scoring about these features for abnormal event detection. Specifically, feature extracting is usually described such as the shape, texture and motor information [14–16]. Feature description is used to describe predefined region in the frame motion with the basis vectors of the low-level features. Several previous works are focused in the feature description, for example, shape feature in [17], trajectory feature in [18], texture feature in [19], social force feature in [20]. Abnormal event detection level provides semantic information about the event and judges whether the event is abnormal or not. Event detection can be categorized into supervised (bag of words [21]) and unsupervised methods (space and support vector data description [22], sparse representation [12]).

We notice that sparse representation obtains state-of-the-art performance which makes no prior knowledge for abnormal event definition. Sparse representation theory are mainly divided into fixed base dictionary methods (e.g., wavelet wavelets [23], discrete cosine transform [24], discrete Fourier transform) and learning dictionary methods (e.g., generalized principal component analysis, the MOD [25] and K-SVD [26], etc.). Those methods with fixed base dictionary can not make a good match through analyzing the structure features of image signal. The latter with learning dictionary methods generate different dictionaries for different types of signals, have strong adaptive ability, which can better suit for image datasets. For all this, the search space for finding the suitable basis vectors from the dictionary remains larger which cost high computational complexity in learning dictionary methods. Meanwhile, detecting abnormal events require to meet the real-time process. So the efficiency problem determines whether or not these methods can be deployed practically.

3 Proposed Approach

In view of internal redundancy and structure information of video datasets, we propose a constrained self-adaptive sparse combination representation method. The flowchart of our method is shown in Fig. 1.

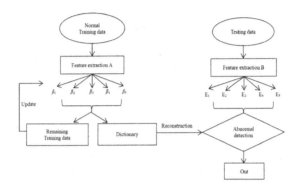

Fig. 1. The flowchart of CSCR algorithm.

Spatio-temporal cube are extracted using multi-scale pyramid from surveillance videos. Then we compute spatio-temporal video volumes (STVVs) gradients features on non-overlapping patches following [27]. For training, the improved sparse combination representation is used for training dictionary. For testing, we also using the same method in training for feature selection. Consequently, whether the test data is normal or not is judged by the reconstruction error.

3.1 Feature Extraction

In order to describe the video data, we adopt a multi-scale STVVs low-level feature. Here, we resize each frame into 3-scale and corresponding subdomain in 5 sequence frames are accumulated to constitute a STVVs as shown in Fig. 2.

With consideration of the target morphological features and motion features at different specific application, we compute gradient features for each STVVs. The size of STVVs are denoted as $V_x * V_y * V_t$, V_x and V_y are the spatial size, V_t is the temporal dimensions. Then the gradient features G_x, G_y and G_t are calculated as one order differential approximation corresponding and transformed into polar coordinate presentation with r, θ and φ in [27]. Before gradient calculation, each frame is cleaned with Gaussian kernel to de-noising. Normalization is performed to the feature vector.

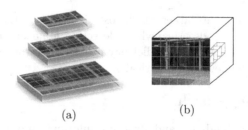

(a)   (b)

Fig. 2. The figure of spatio-temporal features: (a) Multi-scale pyramid. (b) Spatio-temporal video volumes.

3.2 Constrained Self-adaptive Sparse Representation Learning

In our method, abnormal event detection is performed as a sparse representation problem. The objective function of sparse representation is expressed as

$$Y = \arg\min ||x - B\beta||_2^2 + \lambda||\beta|| \tag{1}$$

where x is observed data, B is dictionary, β is weight vectors, λ is regularization threshold. $\arg\min ||*||$ is reconstruction error. It is small for normal pattern. $\lambda||*||$ is sparsity regularization. It should be sparsity for reconstructing normal events and results a dense reconstruction weight vector for abnormal events. In this function, it is intended for finding the perfect basis vectors from the dictionary B to represent data x. The dictionary space is vast with a high testing cost and it cannot be deployed practically. So we use the sparse combination learning for detection. Based on this strategy, we use the sparse combination learning other than searching large space from the dictionary B.

Specifically, the video data is represented as $X = [x_1, ..., x_n] \in R^{m \times n}$, sparse basis combination set $B = [B_1, ..., B_k]$, $B_i \in R^{m \times s}$. The sparse combination learning function can expressed as

$$\min_{B,r,\beta} r_j^i(||x_j - B_i\beta_j^i||_2^2 - \lambda_1 + \lambda_2 \sum_{i=1}^{n} |\beta_j^i|) \tag{2}$$

where $r_j^i = \{0, 1\}$ for the purpose of only one combination is selected for data x_j. λ_1 is the reconstruction error upper bound for getting the combinations of k. λ_2 is regularization parameter. β_j^i is the corresponding coefficient.

However, the objective function above cannot maintain the local structure information for the input data. We use the reconstruction coefficient to expression the geometric structure of data. Specifically, the sample x_j is represented with the sparse coding vector of p nearest neighbor. Therefore, for maintaining the local geometric structure for the data in the sparse coding stage, we add the following constraint function

$$\min_{\beta} \sum_{i=1}^{n} ||\beta_j^i - \sum_{j=1}^{p} w_{jk}\beta_k^i||^2 \tag{3}$$

where w_{jk} is reconstruction coefficient threshold. It can be solved as

$$\min_{w} \sum_{j=1}^{n} ||x_j - \sum_{L \in N_p(x_j)} w_{jL} x_L||^2$$
$$s.t. \sum_{L \in N_p(x_j)} w_{jL} = 1 \tag{4}$$

where $N_p(x_j)$ is sample x_j with p nearest neighbor sample.

It can be solved with the optimal solution

$$\min_{\beta} \sum_{j=1}^{n} ||\beta_j^i - \sum_{j=1}^{p} w_{jk} \beta_k^i||^2 = tr(\beta Z \beta^T) \tag{5}$$

where $Z = (I - W)(I - W)^T$.

In order to facilitate calculation, the above formula is rewritten into the vector form

$$tr(\beta Z \beta^T) = \sum_{j,k=1}^{n} Z_{jk} \beta_j^T \beta_k \tag{6}$$

Combined with the Eq. (6), the improved Eq. (2) is

$$\min_{B,r,\beta} r_j^i (||x_j - B_i \beta_j^i||_2^2 - \lambda_1 + \alpha \sum_{j,k=1}^{n} Z_{jk} \beta_j^T \beta_k + \lambda_2 \sum_{i=1}^{n} |\beta_j^i|)$$
$$s.t. \sum_{i=1}^{k} r_j^i = \{0,1\} \tag{7}$$

We needs to solve three variables B, r and β. In each round i, solving the Eq. (7) with two steps to iteratively update r and B, β.

(1) Update r Fixed B, β

For each x_j, the objective function becomes

$$\min_{r_j^i} r_j^i (||x_j - B_i \beta_j^i||_2^2 - \lambda_1 + \alpha \sum_{j,k=1}^{n} Z_{jk} \beta_j^T \beta_k + \lambda_2 \sum_{i=1}^{n} |\beta_j^i|)$$
$$s.t. \sum_{i=1}^{k} r_j^i = \{0,1\} \tag{8}$$

r_j^i can be solved through the following closed-form

$$r_j^i = \begin{cases} 1 & \text{if} ||x_j - B_i \beta_j^i||_2^2 + \alpha \sum_{j,k=1}^{n} Z_{jk} \beta_j^T \beta_k + \lambda_2 \sum_{i=1}^{n} |\beta_j^i| \leq \lambda_1 \\ 0 & \text{otherwise} \end{cases} \tag{9}$$

(2) Update B, β Fixed r

The Eq. (7) becomes

$$\min_{B,r,\beta} ||x_j - B_i \beta_j^i||_2^2 + \alpha \sum_{j,k=1}^{n} Z_{jk} \beta_j^T \beta_k + \lambda_2 \sum_{i=1}^{n} |\beta_j^i| \tag{10}$$

We optimize B while fixing β, then exchange the computation order β and B. For solving B by block-coordinate descent method

$$B_i = \prod [B_i - \delta_t \nabla_{B_i} L(\beta, B_i)] \tag{11}$$

where $\prod$ is product and it denotes projecting the basis to a unit column. For solving β, the objective function becomes

$$\min_{\beta} \sum_{i=1}^{n} ||x_j - B_i \beta_j^i||_2^2 + \alpha \sum_{j,k=1}^{n} Z_{jk} \beta_j^T \beta_k + \lambda_2 \sum_{i=1}^{n} |\beta_j^i| \tag{12}$$

Then stepwise refinement β, the function becomes

$$\min_{\beta} f(\beta_j) = ||x_j - B_i \beta_j^i||^2 + \alpha Z_{jj} \beta_j^T \beta_j + \beta_j^T G_j + \lambda_2 \sum_{i=1}^{n} |\beta_j^i| \tag{13}$$

where $G_j = 2\alpha(\sum_{j \neq k} Z_{jk} \beta_k)$. We use feature-sign search method [28] to solve this function. Define $g(\beta_j) = ||x_j - B_i \beta_j^i||^2 + \alpha Z_{jj} \beta_j^T \beta_j + \beta_j^T G_j$, so $f(\beta_j) = g(\beta_j) + \lambda_2 \sum_{i=1}^{n} |\beta_j^i|$. Set $\nabla_j^i |\beta_j|$ is the i partial derivative with β_j. When $|\beta_j^i| > 0$, $\nabla_j^i |\beta_j| = sign(\beta_j^i)$. Or else, when $|\beta_j^i| = 0$, $\nabla_j^i |\beta_j|$ is not differentiable, $\nabla_j^i |\beta_j| = \{0, 1\}$. The optimization condition of the minimum $f(\beta_j)$ can be expressed as

$$\begin{cases} \nabla_j^i g(\beta_j) + \lambda sign(\beta_j^i) = 0, & if |\beta_j^i| \neq 0 \\ |\nabla_j^i g(\beta_j)| \leq \lambda_2, & \text{otherwise} \end{cases} \tag{14}$$

However, when $\beta_j^i = 0$ and $|\nabla_j^i g(\beta_j)| > \lambda_2$, the gradient direction $\nabla_j^i f(\beta_j)$ should be discussed separately.

(1) Assumption $\nabla_j^i g(\beta_j) > \lambda_2$, shows whatever the value $sign(\beta_j^i)$ is, $\nabla_j^i f(\beta_j) > 0$ invariably, we can decrease β_j^i to decrease $f(s_i)$. We set $sign(\beta_j^i) = -1$.
(2) Assumption $\nabla_j^i g(\beta_j) < -\lambda_2$, when $sign(\beta_j^i) = 1$. Equation (13) can be solved by quadratic optimization without constraints. The improved sparse combination method description as shown in Algorithm 1.

3.3   Abnormal Event Detection

For detecting anomalies, we achieve it by checking the least square error for each B_i. The large reconstruction error is represented as anomalous events. It is expressed as

$$\min_{\beta^i} ||x - B_i \beta^i||_2^2 \tag{15}$$

It can solve as

$$\widehat{\beta^i} = (B_i^T B_i)^{-1} B_i^T x \tag{16}$$

Algorithm 1. The Improved Sparse Combination Learning Method

Input: Dataset X, current data $X_r \in X$, nearest neighbors p, regularization parameter λ_1, λ_2 and α

Output: Corresponding coefficient β and sparse combinations B

Step1: Initialize $B = \emptyset$ and $i = 1$

Step2: Calculate weight W according to the Eq. (5)

Step3: Compute matrix Z with $Z = (I - W)(I - W)^T$

Step4: Optimize B using Eq. (11) and combine with block-coordinate descent method

Step5: Calculate β using Eq. (13) and combine with feature-sign search method

Step6: Until Eq. (10) converges

      break;

Step7: Remove trained data from X_r

      $i = i + 1$

Step8: Until $X_r = 0$

Step9: End

The reconstruction error in B_i is

$$||x - B_i \widehat{\beta^i}||_2^2 = ||(B_i(B_i^T B_i)^{-1} B_i^T - I_m)x||_2^2 \tag{17}$$

where I_m is a $m * m$ identity matrix.

We can estimate the abnormal sample by calculating the reconstruction cost. A high reconstruction cost represents the most probable outcomes of being an abnormal event.

4 Experiments

We test our method on two commonly used datasets. The UCSD Ped1 datasets [8] has 34 training samples and 36 testing samples. Avenue Datasets [13] contain 16 training and 21 testing video sequences.

As mentioned in the previous section, each frame is resized into 3 scales with $20 * 20$, $30 * 40$, $120 * 160$ sizes respectively for multi-scale pyramid. The size of each STVVs is $10 * 10 * 5$. Then we use PCA to dimensionality reduction, the parameter settings are similar to that of [13]. Nearest neighbors p is 11, regularization parameter λ_1 is 0.04, λ_2 is 0.2, and α is 0.1.

4.1 Experimental Results on the UCSD Ped1 Dataset

To evaluate the property of the improved sparse combination representation method, we compare the classical methods with sparse combination representation (SCR) [13], sparse [37], MOD [25], K-SVD [26], mixture of dynamic textures (MDT) [8], optical flow (OF) [29], bag of words (BoW) [30], robust PCA (RPCA) [31] methods. The result as shown in Fig. 3.

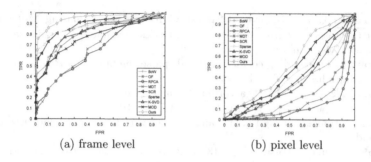

(a) frame level   (b) pixel level

Fig. 3. The output of the proposed method.

As shown in Fig. 3, our ROC curve performance is better than the other methods. It is even more remarkable that the detection results can be achieved the aim of expected function and performance. That is because our new sparse objective function add the reconstruction coefficient for maintaining the local geometric structure for the data in the sparse coding stage. Furthermore, the MOD and K-SVD methods performance is not good.

To assess the capability between our method and other recently proposed representations – unsupervised kernel learning with clustering constraint (CCUKL) [22], dynamic patch grouping (DPG) [32], dominant sets (DS) [33], hybrid histogram of optical flow (HHOF) [34], fuzzy weighted c-means (FWCM) [27], multi-scale histogram of optical flow (MHOF) [35] and earth mover's distance (EMD) [36] methods. Since code for these methods are not available, the results presented are obtained from the original papers in Table 1.

Table 1. The performance between different methods. AUFC: Arean under frame-level ROC; AUPC: Area under pixel-level ROC; EER: Equal error rate; RD: Rate of detection; RT: Running time; EE: Experiment environment.

Method	AUFC↑	AUPC↑	EER(%)↓	RD(%)↑	RT(s/frame)↓	EE(RAM/CPU)
CCUKL [22]	0.98	0.50	19	51	–	–
DPG [32]	0.94	0.47	23	47	1.2	4 GB/3 GHZ
DS [33]	–	–	26	–	–	–
HHOF [34]	0.85	–	17.4	–	2	–
FWCM [27]	0.87	–	21	63	3.4	2 GB/3 GHZ
MHOF [35]	–	0.48	19	46	3.8	2 GB/2.6 GHZ
EMD [36]	0.86	–	22	–	–	2 GB/3 GHZ
Ours	0.95	0.51	17.16	53.03	0.0067	8 GB/2.7 GHZ

Our method achieves higher AUFC value and AUPC value (except CCUKL method in AUFC) in Table 1. Meanwhile, our method has lower EER comparing with these latest algorithms (except HHOF method). In addition, we get high values for RD (except FWCM method). The running time of our is 0.0067 s/frame in Ped1 dataset though our working environment in 8 GB RAM. Therefore, it demonstrates that our algorithm develop a perfect abnormal detection solution. Some image results are shown in Fig. 4 for UCSD Ped1 datasets.

Fig. 4. Examples of experiment result for UCSD Ped1 datasets.

4.2 Experimental Results on Avenue Dataset

As we mentioned in Sect. 3.2, normal events are easily to obtain sparse reconstruction coefficient. Some video sequence and results are shown in Fig. 5 for Avenue dataset.

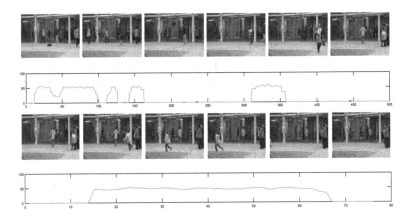

Fig. 5. Examples of two video sequences with the reconstruction cost. The bottom row under each video sequence provides the reconstruction cost respectively, the x value indexes frame and y-index donates reconstruction cost.

The experimental results which conducted on Avenue dataset of the CSCR method in Fig. 6, where the abnormal event regions are marked for a few frames. The corresponding ROC curve and the accuracy distribution for 36 test video sequences are seen in Fig. 7.

In Fig. 7(a), the area under frame-level ROC is 0.82 which prove the effectiveness of our task. In Fig. 7(b), the accuracy distribution of 36 video sequence in

Fig. 6. Examples of abnormal video detection results on the Avenue dataset.

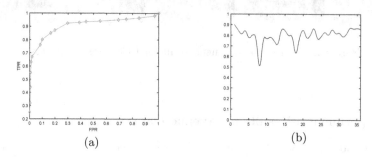

Fig. 7. (a) The corresponding ROC curve (left). (b) The accuracy distribution for 36 test video sequences (right).

75% float up or down, which show that the test result has reaches the anticipated performance. However, we also observe a curious phenomenon that the 8 and 18 video sequences achieve low accuracy. By research, the less frames of the 8 and 18 videos which contain less information is the main reason that draw down this problem. We use the object-level measurement for quantitative comparison [38]. An overlap threshold for detecting abnormal can be represented as follows:

Table 2. Average accuracies on the Avenue dataset. K-SVD [26]: K-singular value decomposition; SCR [13]: Sparse combination representation method; SHD [38]: Statistical hypothesis detector ROC method; LASC [39]: Locality-constrained affine subspace coding method.

ϑ	K-SVD(%)↑	SCR(%)↑	SHD(%)↑	LASC(%)↑	Ours(%)↑
0.2	61.3	70.0	74.2	72.4	75.1
0.3	58.8	67.3	70.3	70.1	70.6
0.4	54.2	63.3	66.5	65.8	66.1
0.5	51.4	59.3	64.0	63.4	63.7
0.6	49.2	57.5	62.1	61.7	62.4
0.7	47.1	55.7	61.2	59.8	61.5
0.8	46.7	54.4	60.8	58.1	60.9

$$\frac{\text{Detected Abnormality} \cap \text{True Abnormality}}{\text{Detected Abnormality} \cup \text{True Abnormality}} \geq \vartheta \qquad (18)$$

where ϑ is a parameter. Average accuracies are shown in Table 2.

The value of ϑ are all outperforms the SCR [13] method, which indicated that our new sparse objective function add the reconstruction coefficient for maintaining the local geometric structure for the data in the sparse coding stage, can improve the detection accuracy effectively. In short, our method can improve the abnormal detection adaptability and achieves superior performance compared with SHD [38] and LASC [39] in all situations (except the result of SHD method in $\vartheta = 0.4$ and $\vartheta = 0.5$).

5 Conclusion

In this work, we have presented a novel algorithm via constrained self-adaptive sparse combination representation for abnormal event detection. Firstly, STVVs low-level feature are stacked with multi-scale pyramid for feature extraction. Then, the constrained sparse combination representation strategy is used for efficient feature description which adopted dynamically update dictionary learning method. Furthermore, our new sparse objective function add the reconstruction coefficient for maintaining the local geometric structure for the data. Finally, abnormal events are judged by the reconstruction error. Experiments result on the published datasets confirm the method get higher accuracy and effectiveness.

References

1. Kratz, L., Nishino, K.: Anomaly detection in extremely crowded scenes using spatio-temporal motion pattern models. In: 2009 IEEE Conference on Computer Vision and Pattern Recognition, pp. 1446–1453 (2009)
2. Kim, J., Grauman, K.: Observe locally, infer globally: a space-time MRF for detecting abnormal activities with incremental updates. In: IEEE Conference on Computer Vision and Pattern Recognition, pp. 1–8 (2009)
3. Benezeth, Y., Jodoin, P.M., Saligrama, V.: Abnormality detection using low-level co-occurring events. Pattern Recogn. Lett. **32**(3), 423–431 (2011)
4. Hospedales, T.M., Li, J., Gong, S., et al.: Identifying rare and subtle behaviors: a weakly supervised joint topic model. IEEE Trans. Pattern Anal. Mach. Intell. **33**(12), 2451–2464 (2011)
5. Wang, X., Hou, Z.Q., et al.: Target scale adaptive robust tracking based on fusion of multilayer convolutional features. Acta Optica Sin. **11**, 232–243 (2017)
6. Xu, J.: Unusual event detection in crowded scenes. In: IEEE International Conference on Acoustics (2015)
7. Tran, D., Yuan, J., Forsyth, D.: Video event detection: from subvolume localization to spatiotemporal path search. IEEE Trans. Pattern Anal. Mach. Intell. **36**(2), 404–416 (2014)
8. Mahadevan, V., Li, W., Bhalodia, V., et al.: Anomaly detection in crowded scenes. In: 2010 IEEE Computer Society Conference on Computer Vision and Pattern Recognition, pp. 1975–1981 (2010)

9. Xia, L.M., Hu, X.J., Wang, J.: Anomaly detection in traffic surveillance with sparse topic model. J. Cent. South Univ. **25**(9), 2245–2257 (2018)
10. Cai, R.C., Xie, W.H., Hao, Z.F., et al.: Abnormal crowd detection based on multi-scale recurrent neural network. J. Softw. **26**(11), 2884–2896 (2015)
11. Sultani, W., Chen, C., Shah, M.: Real-world anomaly detection in surveillance videos. In: 2018 IEEE Computer Society Conference on Computer Vision and Pattern Recognition, pp. 6479–6488 (2018)
12. Zhao, B., Li, F.F., Xing, E.P.: Online detection of unusual events in videos via dynamic sparse coding. In: 2011 IEEE Conference on Computer Vision and Pattern Recognition, pp. 3313–3320 (2011)
13. Lu, C., Shi, J., Jia, J.: Abnormal event detection at 150 FPS in MATLAB. In: Proceedings of the IEEE International Conference on Computer Vision, pp. 2720–2727 (2013)
14. Zhao, Y., Yu, Q., Jie, Y., Nikola, K.: Abnormal activity detection using spatio-temporal feature and Laplacian sparse representation. In: International Conference on Neural Information Processing, pp. 410–418 (2015)
15. Tao, Z., Wenjing, J., Baoqing, Y., et al.: MoWLD: a robust motion image descriptor for violence detection. Multimed. Tools Appl. **76**(1), 1419–1438 (2017)
16. Mabrouk, A.B., Zagrouba, E.: Spatio-temporal feature using optical flow based distribution for violence detection. Pattern Recogn. Lett. **92**, 62–67 (2017)
17. Aslan, M., Sengur, A., Xiao, Y., et al.: Shape feature encoding via fisher vector for efficient fall detection in depth-videos. Appl. Soft Comput. **37**, 1023–1028 (2015)
18. Kim, H., Lee, S., et al.: Weighted joint-based human behavior recognition algorithm using only depth information for lowcost intelligent video-surveillance system. Expert Syst. Appl. **45**, 131–141 (2016)
19. Wang, J., Xu, Z.: Crowd anomaly detection for automated video surveillance. In: The 6th International Conference on Imaging for Crime Detection and Prevention, pp. 4–6 (2015)
20. Mehran, R., Oyama, A., Shah, M.: Abnormal crowd behavior detection using social force model. In: 2009 IEEE Computer Society Conference on Computer Vision and Pattern Recognition, pp. 20–25 (2009)
21. Rajkumar, S., Arif, A., Prosad, D.D., Pratim, R.P.: Surveillance scene segmentation based on trajectory classification using supervised learning. In: Proceedings of International Conference on Computer Vision and Image Processing, pp. 261–271 (2016)
22. Ren, W.Y., Li, G.H., Sun, B.L.: Unsupervised kernel learning for abnormal events detection. Vis. Comput. **31**, 245–255 (2015)
23. Simoncelli, E.P., Adelson, E.H.: Noise removal via Bayesian wavelet coring. In: Proceedings of 3rd IEEE International Conference on Image Processing, vol. 1, pp. 379–382 (1996)
24. Elad, M., Aharon, M.: Image denoising via sparse and redundant representations over learned dictionaries. IEEE Trans. Image Process. **15**(12), 3736–3745 (2006)
25. Engan, K., Aase, S.O., Husy, J.H.: Multi-frame compression: theory and design. Sig. Process. **80**(10), 2121–2140 (2000)
26. Jiang, Z., Lin, Z., Davis, L.S.: Learning a discriminative dictionary for sparse coding via label consistent K-SVD. In: 2011 IEEE Conference on Computer Vision and Pattern Recognition, pp. 1697–1704 (2011)
27. Li, N., Wu, X., Xu, D., et al.: Spatio-temporal context analysis within video volumes for anomalous-event detection and localization. Neurocomputing **155**, 309–319 (2015)

28. Lee, H., Battle, A., Raina, R., et al.: Efficient sparse coding algorithms. In: Advances in Neural Information Processing Systems, pp. 801–808 (2007)

29. Horn, B., Schunck, B.: Determining optical flow. In: Technical Symposium East, pp. 319–331 (1981)

30. Uijlings, J., Duta, I.C., Sangineto, E., et al.: Video classification with densely extracted HOG/HOF/MBH features: an evaluation of the accuracy/computational efficiency trade-off. Int. J. Multimed. Inf. Retr. 4(1), 33–44 (2015)

31. Nguyen, T.V., Phung, D., Gupta, S., et al.: Interactive browsing system for anomaly video surveillance. In: IEEE Eighth International Conference on Intelligent Sensors, pp. 384–389 (2013)

32. Cong, Y., Yuan, J.S., Tang, Y.D.: Video anomaly search in crowded scenes via spatio-temporal motion context. IEEE Trans. Inf. Forensics Secur. 8(10), 1590–1599 (2013)

33. Alvar, M., Torsello, A., Sanchez-Miralles, A., et al.: Abnormal behavior detection using dominant sets. Mach. Vis. Appl. 25(5), 1351–1368 (2014)

34. Jin, D., Zhu, S., Wu, S., et al.: Sparse representation and weighted clustering based abnormal behavior detection. In: 2018 24th International Conference on Pattern Recognition, pp. 1574–1579 (2018)

35. Cong, Y., Yuan, J., Liu, J.: Abnormal event detection in crowded scenes using sparse representation. Pattern Recogn. 46(7), 1851–1864 (2013)

36. Xu, D., Wu, X., Song, D., et al.: Hierarchical activity discovery within spatio-temporal context for video anomaly detection. In: 2013 IEEE International Conference on Image Processing, pp. 3597–3601 (2013)

37. Cong, Y., Yuan, J., Liu, J.: Sparse reconstruction cost for abnormal event detection. In: 2011 IEEE Conference on Computer Vision and Pattern Recognition, pp. 3449–3456 (2009)

38. Yuan, Y., Feng, Y., Lu, X.: Statistical hypothesis detector for abnormal event detection in crowded scenes. IEEE Trans. Cybern. 99, 1–12 (2017)

39. Fan, Y., Wen, G., Qiu, S., et al.: Detecting anomalies in crowded scenes via locality-constrained affine subspace coding. J. Electron. Imaging 26(4), 1–9 (2017)

A Distributed Scheduling Framework of Service Based ETL Process

DongJu Yang[1,2(✉)] and ChenYang Xu[3]

[1] Beijing Key Laboratory on Integration and Analysis of Large-Scale
Stream Data, Beijing, China
yangdongju@ncut.edu.cn
[2] Data Engineering Institute, North China University of Technology,
Beijing, China
[3] Institute of Scientific and Technical Information of China, Beijing, China
xucy@istic.ac.cn

Abstract. The use of service oriented computing paradigm and ETL (Extract-Transform-Load) technology has recently received significant attention to enable data warehouse construction and data integration. Aiming at improving scheduling and execution efficiency of service based ETL process, this paper proposes a distributed scheduling and execution framework for ETL process and a corresponding method. Firstly, add different weights to the ETL process to ensure the loading efficiency of core business data. Secondly, the scheduler selects the executors according to the performance and load, then allocates the ETL process execution request based on the greedy balance (GB) algorithm to make the load of the executor balancing. Thirdly, the executors parses ETL process to ETL services, then selects one or more executors to deploy and execute the ETL service according to the locality-aware strategy, that is, the amount of data involved and the distance of the node network which service involved, which can reduce the network overhead and improve execution efficiency. Finally, the effectiveness of the proposed method is verified by experimental comparison.

Keywords: Data integration · Service based ETL process · Scheduling · ETL services · Locality-aware

1 Introduction

The large amount of data, complex business, and variable demand are the difficult problems in building a data warehouse under. The traditional ETL technology is used to extract, transform and load distributed and heterogeneous data into a target database. Due to the tight coupling and indivisibility of ETL components, it is difficult to meet the requirements of scalability and flexibility in data integration. It is a better solution to use service-oriented architecture to segment ETL tasks and encapsulate them into fine-grained service components, and realize data integration through loosely coupled integration and interoperability of service components. Simplify ETL task construction,

This work is supported by the Key projects of the National Natural Science Foundation of China: Research on the theory and method of big service in big data environment (No. 61832004).

improve scalability and flexibility through reuse of service components [1]. How to efficiently schedule and execute these service based ETL processes in a big data environment is a key issue to be solved in building a data warehouse.

In the previous work, Yang et al. [2] divided the ETL task into three stages: data extraction, data transformation and data loading. Encapsulate data processing components such as different data extraction rules, data cleaning algorithms, data transformation processing methods, and data loading methods into a standard RESTful service. When constructing the ETL task, the corresponding data processing services of each stage are combined to construct the ETL process. Data extraction, transformation and loading are realized through the parsing and execution of ETL process.

For example, during the integration of the technology resource management data, there are various data about project, topic, unit, personnel, etc. which are distributed in different phases, such as pre-declaration, declaration, and project approval. Aiming at integrating them into the target database, we construct the service based ETL process described in a script file, which details which ETL services are used during the extraction, transformation and loading phases, and how the interfaces interoperate. The file is stored as metadata in the ETL process repository. In view of the large amount of data, high execution frequency, and complex business of the integrated service, the cluster has been adopted for ETL job scheduling. However, there are problems as follows: 1. For different business data, the requirements of integration time are different. Without considering the importance and hierarchy of ETL task-related data, the ETL task execution waiting time related to the core business is too long. 2. When the task execution request is assigned to the executor, the polling algorithm is used. The task is scheduled in the order of the executor. Without considering the different execution time for different task, the load of executors is unbalanced. 3. Accessing a large amount of data during the execution of an ETL service is a huge challenge to network overhead, bandwidth overhead, and execution efficiency. How to reduce the network overhead in service execution and improve the efficiency of service execution is a key problem to be solved.

Aiming at the above problems, this paper proposes a distributed scheduling and execution framework of service based ETL process based on cluster scheduling. The framework is divided into two stages: ETL process scheduling, ETL service deployment and execution.

2   Related Work

Scheduling has experienced single-machine scheduling, distributed scheduling, etc., and there are multiple algorithms used in scheduling, like polling scheduling algorithm, Min-Min algorithm, particle swarm algorithm, etc. However, the efficiency of these algorithms applied to distributed environments is not high. Frequently, the single node load is too high, and some nodes are idle. Based on the distributed scheduling environment [3], Kokilavani et al. [4] proposed an improved algorithm based on Min-Min algorithm, and improved the scheduling service quality of distributed platform through algorithm optimization. The research focused on how to distribute the process to the executor effectively and reasonably, and there is no much consideration for the performance difference of the node. In addition, Wang et al. [5] proposed a load fuzzy classification and local rescheduling algorithm based on Min-Min limit compression algorithm, and introduced the idea of fuzzy classification to improve the load balancing problem of the algorithm. Song et al. [6] proposed the idea of load balancing data layout method in data-intensive computing, Zhang et al. [7] studied the related scheduling algorithm based on node load balancing, but the above research did not consider the problem of process level and process execution stage. Chen et al. [8] divide the performance of the node into the positive domain, the middle domain, and the negative domain by defined the rough set, and distribute the process at different node levels. However, the above allocation strategy is based on the load balancing indicator of the number of process connections. Because the amount of data contained in the process is a key factor that affects the execution time of the process. So in a distributed environment, subject to the difference in node performance, how to dynamically allocate the process and use effective algorithms to optimize the scheduling strategy is a difficult point of distributed scheduling.

In the big data environment, the principle of locality has been used in scheduling increasingly. Jin et al. [9] proposed a novel data location-aware scheduling algorithm. The scheduling algorithm combines data placement and graphics topology to reduce data transmission costs. Yekkehkhany [10] proposed a near-data scheduling method under data center with multi-level data locality, which improves the service rate and execution efficiency of the process. Shang et al. [11] improved the delay-priority scheduling strategy based on data locality, and made significant progress in resource utilization. Tao et al. [12] proposed a resource scheduling scheme based on load feedback, and a data locality scheme based on dynamic migration. The two schemes can balance distributed resource performance and meet data locality. Combined with the above research, how to effectively use the locality principle to optimize the efficiency of ETL service execution in the implementation stage is one of the key issues.

3 Distributed Scheduling and Execution Framework of ETL Process

3.1 Description of ETL Process

An ETL process includes a data extraction service E-Service, a data transformation service T-Service, and a data loading service L-Service, and each service includes a plurality of data that can be independently deployed and executed by corresponding data processing algorithms. Processing services include A11, A12, A21, A22, A31, A32, etc. An ETL process can be expressed as:

Start
E-Service(A11, A12,...) //extraction service, include multiple ETL services, like A11 and A12
T-Service(A21, A22,...) //transformation service, include multiple ETL services, like A21 and A22
L-Service(A31, A32,...) //loading service, include multiple ETL services, like A31 and A32
End

The structure of ETL process is described in the script file, detailing which ETL services are used during the collection, conversion, and loading phases, and how the interfaces are interoperable. E-Service includes some services related to data extraction, including data field selection, data extraction, etc. T-Service contains some service related to data transformation, including field format conversion, semantic conversion, mapping, etc. L-Service contains some service related to data loading, mainly about loading, mapping, etc. The ETL process script is as follows:

```
<ETL Process Info>
    <Stage Extract>
        <SourceDataBase name="warehouse_zdyf">  <SourceTable name="red_project">
        <Service name=E-Service>
            <E-Service:extract>  <E-Service:clean>
        </Service name=E-Service>
    </Stage Extract>
    <Stage Tranform>
        <Service name=T-Service>
            <T-Service:format transformation >  <T-Service:mapping>
            <T-Service:semantic transformation>  <T-Service:data correction>
        </Service name=T-Service>
    </Stage Tranform>
    <Stage Load>
        <TargetDataBase name="rws_zdyf">  <TargetTable name="project">
        <Service name=L-Service>
            <L-Service:load>
        </Service name=L-Service>
    </Stage Load>
</ETL Process Info>
```

3.2   Distributed Scheduling and Execution Framework

The distributed scheduling framework of the ETL process proposed in this paper is shown in Fig. 1. The framework includes two parts: the scheduler and the executor. The schedule include two stages: ETL process scheduling and deployment. The scheduler is responsible for scheduling the ETL process to the appropriate executor, and the executor is responsible for parsing ETL process to ETL services and deploying ETL services. The scheduler and the executor are separated from each other, and the executors can cooperate with each other to distribute and receive deployment execution requests of the ETL services. The scheduler reads a batch of ETL processes periodically from the ETL process repository.

In the phase of ETL process scheduling, the weight model of the ETL process is designed, and the weight of ETL process is set according to the weight of the tables for data to load, and the ETL process is inserted into the scheduling queue according to the sorting result of the weight. So ETL tasks related to core business data can be scheduled first. Next, the scheduler obtains real-time performance information to determine the current load of each executor, such as CPU and memory usage of each executor. This determines which executor can be allocated the next task. In this way, a dynamic task allocation strategy for load balancing between nodes is achieved. After determining the executor, the ETL process execution request is delivered according to the Greedy Balance (GB) algorithm.

During the ETL service deployment and execution phase, the executor parses the ETL process, obtains the ETL services that need to be used in each stage of extraction, conversion, and loading, and selects an appropriate executor to execute. If the amount of data accessed by the service is large, data transmission will bring huge network overhead, which greatly affects the efficiency of service execution. Therefore, when performing the ETL services, according to the locality-aware strategy, the data size, node distance and other factors are considered to select one executor to deploy and execute the ETL services.

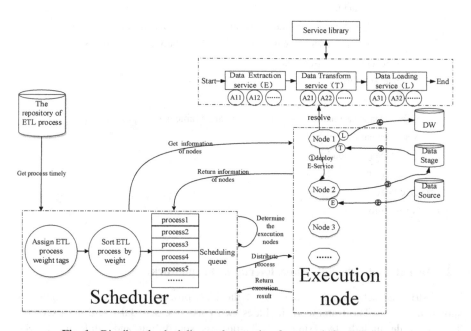

Fig. 1. Distributed scheduling and execution framework for ETL process

4   The Method of ETL Process Scheduling and Execution

Based on the distributed scheduling and execution framework of ETL process, this paper proposes an ETL process scheduling and execution method, including ETL process scheduling and allocation and ETL service deployment and execution. The method workflow is shown in Fig. 2.

The method is described as follows:

(1) The scheduler obtains a batch of ETL processes from the ETL process repository which need to be scheduled and executed;

(2) Determine the ETL process weight level, and insert the ETL process into the to-be-scheduled queue according to the weighted order;

(3) The scheduler obtains performance information of all executors at the current time, and determine the load level of the node. There are three load levels, which is low, medium and high;

(4) If there are Low-Load node, the request are sequentially allocated to the Low-Load executor according to the GB algorithm; if there are no Low-Load node, the request are allocated to the medium-load executor according to the GB algorithm; If there are no medium-load node, the request are allocated to the high-load executor according to the GB algorithm;

(5) The executor parses the ETL process, and obtains services of data processing algorithm that need to be executed at each stage of data extraction, transformation, and loading;

(6) The executor calculates the amount of data that each ETL service needs to process. If the amount of data is large, the nearest executor to the data is selected to deploy and execute services according to the node distance;

(7) After the execution of the entire ETL process is completed, the executor returns status information to the scheduler whether the execution is successful.

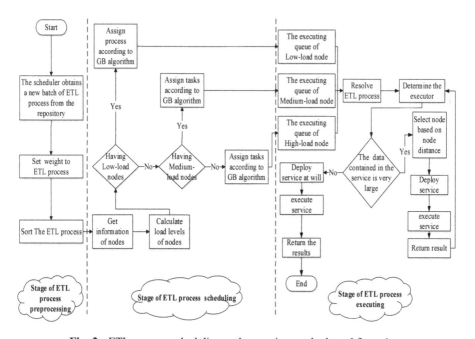

Fig. 2. ETL process scheduling and execution method workflow chart

4.1   ETL Process Weight Model

The data integration contains many ETL processes under the data warehouse, including the extraction, transformation and loading of heterogeneous data and log information. Under the big data environment, many enterprises have very high efficiency requirements for loading core business data. So a scheduling priority is required for the ETL process at the beginning of the process scheduling. Therefore, this paper designs a weight model to support the division of weight levels of business entities. Then define the weight of the target table to be loaded by the business data, and set the weight to the ETL process related to the target table. So The ETL process is scheduled by weight.

$$\text{ETL process}: \begin{cases} \textit{the extraction of heterogeneous data} \\ \textit{the transformation of heterogeneous data} \\ \textit{the integration of heterogeneous data} \\ \textit{extraction and integration of } \log \textit{ information} \\ \dots\dots\dots\dots\dots\dots\dots\dots\dots\dots\dots\dots\dots\dots \end{cases}$$

This paper takes the technology resource management data integration as an example to describe the weight definition. The weighted data model architecture of target database is shown in Fig. 3. Given Project, Topic, Unit, and Person are four entities. There is a one-to-many relationship between the Project and the Topic. $wl1$ is the association weight between the entity and the subsidiary table. $wl2$ is the association weight between the entity and the dimension table. $wl3$ is the association weight between one entity and another. The weight of the business entity(W) is formalized as:

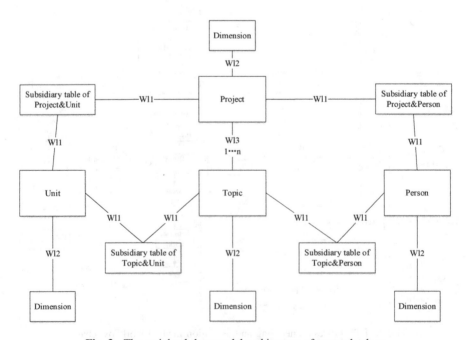

Fig. 3. The weighted data model architecture of target database

$$\text{Weight(W)} = \frac{n1 * Wl1 + n2 * Wl2 + n3 * Wl3}{n1 + n2 + n3} \, (ni \text{ is the number of } Wli) \quad (1)$$

The weight is set to the ETL process according to the weight of the target table, then set the weight to the batch ETL process. This paper uses a stable fast sorting algorithm to sort the ETL processes, and get an optimal expected scheduled sequence.

4.2 ETL Process Scheduling and Allocation

The scheduling framework of this paper is based on the distributed scheduling framework in the cluster environment. The scheduler and the executor are separate, and the executors are distributed. In the distributed environment, the number of processes and the amount of data contained in processes are different for different executors. So the performance and load of each executor at the same time is different. Therefore, the scheduler needs to allocate ETL processes according to the performance of the executor to achieve distributed environment balanced.

The scheduler obtains the execution request of an ETL process from the scheduling queue, and sends an access request to each executor. If the network is connected, the executor returns the information about usage of CPU and memory (RAM). After receive the performance information of each executor, the scheduler calculates the load level according to performance. We define the Activity of each executor to determine whether the node is able to receive the process execution requirements.

Activity of each executor is:

$$\text{Activity (A)} = \frac{(1-C) + (1-R)}{2} \quad (0 < C \le 1; 0 < R \le 1) \quad (2)$$

A is an executor; C is the CPU usage; R is the memory usage.

The current load level is determined by the Activity A of each executor. Low Activity means that the current load of the executor is high, and high Activity means low load. There are three load levels of the executor, which are High-Load, Medium-Load and Low-Load.

$$\text{Activity(A)} = \begin{cases} 0 \sim 35\% & High - Load \\ 36\% \sim 75\% & Medium - Load \\ 76\% \sim 100\% & Low - Load \end{cases} \quad (3)$$

If the Activity is between 0 and 35%, the executor is High-Load. If the Activity is between 36% and 75%, the executor is Medium-Load. If the Activity is between 76% and 100%, the executor is Low-Load. So there are three groups according to the load level, and there are zero to multiple executors in each group. Executors in Low-Load groups have the lowest load and the ability to accept process execution request is the strongest. Therefore, the process scheduling request should be assigned to the Low-Load executor. If there are no Low-Load executor, the process is assigned to the Medium-Load executor, and so on. If there are no Low-Load and Medium-Load executors, that is, the current load of the entire executor environment is high, the efficiency of current process scheduling and execution will be very low. Once the entire executor environment is in High-Load level for a long time, the performance of the distributed environment needs to be increased or the number of executors increased, to improve the performance of the entire distributed environment.

The Greedy algorithm means that the overall optimal solution to the problem can be achieved through a series of locally optimal choices, namely Greedy choices [13]. This is the first basic element of a Greedy algorithm. The Greedy choice is to make successive choices from top to bottom iteratively. Each time a Greedy choice is made, the problem is reduced to a smaller sub-problem. That is, from the local optimum to the overall optimal [14]. Based on the Greedy algorithm [15], this paper introduces the amount of process data, adds the optimal expected distribution process data volume, and proposes the Greedy Balance algorithm (GB algorithm, for short). The GB algorithm allocation process is further described by the mathematical model. First assume that the initial processing capabilities of each node are the same in a distributed environment, and each node can work independently. That is, no assistance from other nodes is required, and the influence of the external environment on the server node (such as temperature, etc.) is shielded. The relevant definitions are as follows:

Assume $E = \{e1, e2, e3, \ldots, en\}$ represents a new batch of ETL process to be schedule, ei is the ith process; $D = \{d1, d2, d3 \ldots dn\}$ represents a collection of data volumes contained in n ETL processes, di is the amount of data contained in the ith process; $N = \{n1, n2, n3 \ldots nj\}$ represents a collection of distributed cluster executors, the number of nodes is j, ni is the executor i. dni_{pre} represents the amount of data contained in the ETL process on the executor i, When the process is assigned, dni_{aft} represents the amount of data contained in the ETL process on the executor i, the total amount of data contained in the ETL process on the executor participating in the execution is $Data = \sum_{i=0}^{j} dni_{aft}$; Optimal amount of task data to be expected for each executor Opt_i is:

$$Opt_i = \frac{Data}{j} - dni_{pre} = \frac{\sum_{i=0}^{j} dni_{aft}}{j} - dni_{pre} \tag{4}$$

The executor data load index μ is represented by the variance of the data amount, Then the data load index μi of each executor ni can be expressed as

$$\mu_i = \left(dni_{aft} - dni_{pre} - Opt_i\right)^2 \tag{5}$$

The data load index μ of the executor population can be expressed as:

$$\mu = \frac{\sum_{i=0}^{j} \left(dni_{aft} - dni_{pre} - Opt_i\right)^2}{j} \tag{6}$$

During the task distribution process, it is necessary to ensure load balanced about the cluster resources, the μ is relatively small. Define the maximum value of μ during task distribution by defining threshold δ. Calculate the value of μ in real time during task distribution, if $\mu \succ \delta$, then it need to adjust the task distribution strategy to ensure the cluster resource load balancing.

The specific steps of the GB algorithm to allocate ETL process are as follows:

(1) Initialize the ETL process collection: $E = \{e1, e2, e3, \ldots, en\}$ first, A collection of data contained in an ETL process: $D = \{d1, d2, d3 \ldots dn\}$, Set of executors to be assigned: $N = \{n1, n2, n3 \ldots nj\}$;

(2) Sort batch ETL tasks in descending order of the amount of data, and store them in the queue Q. $Q = \{q1, q2, q3, q4, \ldots qn\}$, and $q1$ is *(e1, d1)*, $q2$ is *(e2, d2)*, ...qn is *(en, dn)*, $d1 \geq d2 \geq dn$;

(3) Calculate the data load index $\mu1, \mu2, \mu3, \ldots \mu j$, about the executors of number j in real-time, which are in the executor set N. Adjust the order between nodes based on the data load index. For example, if $\mu1 \prec \mu2 \prec \mu3 \prec \ldots \prec \mu j$, adjust the node order to $n1, n2, n3, \ldots, nj$;

(4) Define the variable K as the number of nodes, which can allocated at this time. Assign the number of nodes j to K, that is, $K = j$;

(5) For the tasks of number n, which are in the Q, if $n > K$, then take tasks of number j out, and distribute them to nodes of number j. If $0 < n \leq K$, then all tasks are fetched, and distributed them to nodes of number j, for example, $e1$ distributed to $n1$, $e2$ distributed to $n2$; if $n = 0$, the algorithm ends;

(6) In the task distribution process, calculate the values of μ_i and μ in real time, if $\mu \geq \delta$, then execute (7); if $\mu \prec \delta$, then execute (4);

(7) Select the executors that performs node data load index in $\mu_i \prec \delta$, and count the number of nodes, then execute (4), if there are no nodes, and node load index is less than δ, which indicates that the distributed environment load is too high at this time. It is necessary to stop distribute the task temporarily or add new executors.

4.3   Deployment and Execution of ETL Services

After the ETL process is distributed to the executor (Executor A), it needs to go through two stages, process analysis and service execution.

In the ETL process analysis phase, get the specific ETL service in the extraction, transformation and loading phase, recorded as E-Service, T-Service and L-Service. Each ETL service contains a set of executable data processing algorithm services.

During the ETL service execution phase,Firstly, the E-Service is called to complete the data extraction. The data corresponding to the service is the source data, includes data extraction service and data cleaning service. After the E-Service is executed, the extracted data is stored in a temporary database (Data Stage). At this point, the T-Service is called to complete the data transformation. T-Service is more complicated due to involve many conversions, and the execution time is slightly longer. It includes data format conversion service, data semantic conversion service, etc. After the T-Service service is executed, the L-Service is called to load the converted data into the target database. In order to reduce the network bandwidth overhead, considering the factors such as the amount of data and the distance between nodes, the locality-aware strategy is used to optimize the scheduling and execution.

The executor first judges the amount of data accessed by each service. If the amount of data is large, the executor (executor B) with the smallest node network distance is selected. Then obtain the corresponding service entity from the service repository and deploy for execution. Taking the data extraction service as an example, it is responsible for extracting corresponding data from the data source, and performing operations such as cleaning, completion, and drying, and then handing it over to the next stage. If the amount of data contained in the extracted data source is large, it will compute the network distance between nodes, select the executor (executor B) which have the smallest distance from the data node, and send the service deployment execution request to executor B. After receiving the service deployment execution request, the executor B obtains the corresponding one or more service entities from the service repository first, deploys for execution, and returns the result to the executor A after execution. Figure 4 shows the specific ETL service deployment and execution flow.

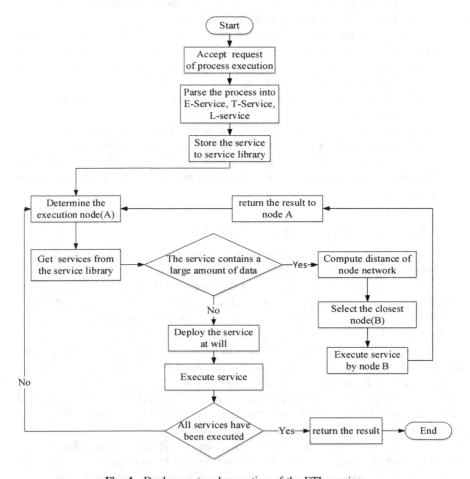

Fig. 4. Deployment and execution of the ETL services

For the calculation of the network distance between nodes, we refer to the calculation method of the network topology distance defined in Hadoop (https://hadoop.apache.org). The node network distance is used to indicate the distance between two nodes in the network topology. In the network topology tree, the distance from the child node to the parent node is set to 1, and the network distance between any two nodes is the sum of their distances to the nearest common ancestor node. Assume that the two data centers are represented as d1, d2, the rack is represented as r1, r2, r3, and the server node is represented as n1, n2, n3, n4. Then the network distance between two nodes is calculated as follows:

1. distance(d1/r1/n1, d1/r1/n1) = 0 (Same node)
2. distance(d1/r1/n1, d1/r1/n2) = 2 (Different nodes in the same rack)
3. distance(d1/r1/n1, d1/r2/n3) = 4 (Different racks in the same data center)
4. distance(d1/r1/n1, d2/r3/n4) = 6 (Different data centers)

For the calculation of the position and distance between the node and the data, mark the location of the source data first, then the temporary data, and then the target data. After determining the executor of the ETL process, the executor of the service is determined according to the method described above and the location of the related data. In this way, the services move with the data due of the dynamic deployment of services.

Algorithm description: Input:An ETL process to be executed; Output:The execution result

Start

Var Node A; Var Node B; Var process P;

Receive request of process execution;

Parse process to E-Service;T-Service;L-Service;

Store the service to service repository;

Confirm executor A;

While(!All services have been executed){

  Node.A.do{

    Get services from the service repository;

    While(the_data_size in the process is big){

      calculate distance about node and data;

      determine the Node B; then deploy the Service;

      Node B.do{

        execute Service;

      }return;

    };

    deploy Service at will; then execute service;

    }return;

  }return;

  return results;

End

5 Experiment Analysis

5.1 Experimental Environment

In order to verify the performance of the scheduling and execution method, a distributed environment is constructed, include one master node (scheduler n1) and three slave nodes (executors n2, n3, n4). The data sources accessed by ETL are located in n3 and n4. The specific configuration is shown in Table 1. In order to verify the performance of the scheduling and execution method in this paper, under the same conditions, compare the scheduling and execution result of the ETL process. One scheme is use the Min-Min algorithm to schedule and execute with the FCFS algorithm. Another solution is use the polling algorithm to schedule and execute with the FCFS algorithm.

Table 1. Configuration information of each node in the distributed environment

Node type	CPU	RAM	Operating system	Network speed	Network location
(Scheduler n1)	8 cores Intel 3.40 GHZ	8G	Linux	500 M/s	d1, r1
Slave node 1 (executors n2)	8 cores Intel 3.40 GHZ	8G	Linux	100 M/s	d1, r1
Slave node 2 (executors n3)	8 cores Intel 3.40 GHZ	8G	Linux	100 M/s	d1, r2
Slave node 3 (executors n4)	8 cores Intel 3.40 GHZ	8G	Linux	100 M/s	d1, r2

5.2 Analysis of Experimental Results

This paper analyzes the results of three scheduling and execution methods from the ETL process scheduling and execution time, and makes a verification analysis on the two aspects of the average utilization of distributed system resources and the load balance between the executors. In this experiment, we define the execution time of the ETL process scheduling as: in the case of multiple processes, the scheduler obtains the entire duration of the process to be scheduled from the ETL process repository to the completion of all processes. The comparison of ETL process scheduling and execution time results is shown in Fig. 5. In the same test environment, for the number of different processes [10, 50, 100, 300], the three scheduling and execution methods are tested separately. The result data is shown in Table 2.

Table 2. Data about ETL process scheduling and execution time result

Number \ Methods / Time	Poll+FCFS	Min-Min+FCFS	Scheduling and execution method in this paper
10	90s	70s	55s
50	140s	110s	80s
100	245s	185s	155s
300	400s	320s	255s

It can be concluded from Table 2 and Fig. 5. The scheduling and execution method of this paper consumes less time than the polling + FCFS algorithm scheduling and execution method and the Min-Min + FCFS algorithm scheduling and execution method in the case of different ETL processes. As the number of processes increases, the time spent on scheduling and execution methods in this paper grows more slowly.

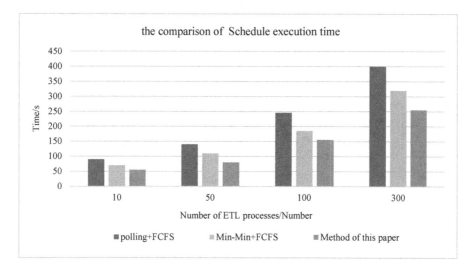

Fig. 5. The chart ETL process scheduling and execution time comparison

When the number of ETL processes is 100, the time taken by the polling + FCFS algorithm and the Min-Min + FCFS algorithm and the ETL process scheduling and execution method are [245 s, 185 s, 155 s]. The resource utilization of the three scheduling and execution methods at time nodes [45 s, 85 s, 110 s, 150 s] is calculated, and the average utilization of distributed system resources is calculated. The average CPU and RAM utilization rate of the polling +FCFS scheduling and execution scheme is [59%, 71%], and the average CPU and RAM utilization of the Min-Min + FCFS scheduling and execution scheme is [49%, 59%]. The average CPU and RAM utilization of the solution is [41%, 52%], and the comparative analysis chart is shown in Fig. 6.

From the data results and Fig. 6, we can conclude that the scheduling and execution method of this paper is smaller than the other two scheduling and execution methods in the CPU and memory utilization of each node, which means that the scheduling and execution method can make full use of the node resources.

In terms of load balancing, when the number of ETL processes is 300, the ETL process scheduling and execution method takes 255 s which is defined in this paper, and the CPU and memory usage of three nodes at different time nodes [1 min, 2 min, 3 min, 4 min] are counted. As can be seen from Figs. 7 and 8, the CPU utilization of the three executors fluctuates between [51%–56%] in CPU usage, and the memory utilization rate is between [62%–68%]. Fluctuation, the overall fluctuation difference is around 6%. It can be concluded that the load between the executors is relatively balanced, which effectively verifies the advantages of the scheduling and execution method in this paper.

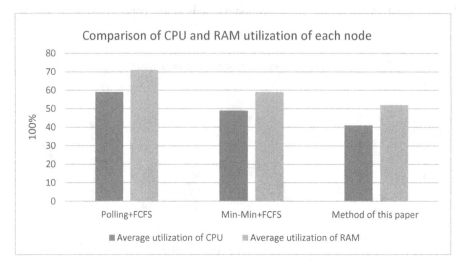

Fig. 6. Comparison of CPU and memory utilization of each node

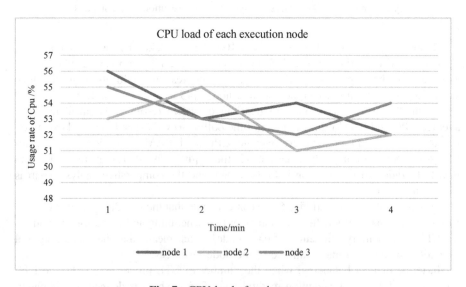

Fig. 7. CPU load of each executor

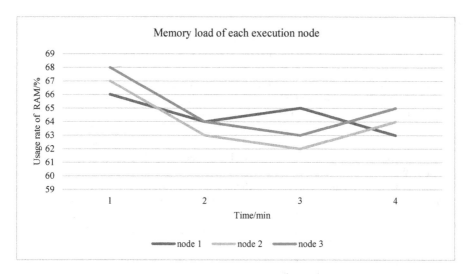

Fig. 8. Memory load of each execution nod

6   Conclusion

This paper focuses on the efficient scheduling and execution optimization requirements of the ETL process constructed with the service-oriented architecture in the big data environment. In order to solve the data integration efficiency slowly under the core business, waste of task scheduling node resources in current distributed environment, bandwidth consumption is too large, execution efficiency is not high, and other issues. This paper propose a distributed scheduling and execution framework and an ETL process scheduling and execution method. By dividing the ETL process into weights, the efficiency of core business loading is guaranteed first.

Secondly, in the distributed scheduling environment, the scheduler obtains the current performance information of each node first. Under the condition of fully considering the difference in node performance, the GB algorithm and the load balancing index are used to allocate the ETL process to the executor to perform scheduling dynamically and reasonably in order to achieve load balancing between distributed nodes.

Thirdly, for the scheduling and execution of the ETL service parsed after the ETL process, adopted the locality-aware strategy, Considering the amount of data involved in the service and the network distance of the node, select the appropriate executor to complete the deployment of the service, reduce network consumption and improve the service execution efficiency.

Finally, the effectiveness of the research scheme is verified by experiments, and it also provides a reference for the ETL process scheduling scheme. Considering the characteristics of different ETL services, performing personalized and precise scheduling will be the focus of the next step.

References

1. Yi, X.: Design and implementation of an SCA-based ETL architecture. Donghua University (2014)
2. Yang, D., Xu, C.: Research on data quality assurance technology based on meta model control in big data environment. Comput. Eng. Sci. **41**(2), 197–206 (2019)
3. Mirzayi, S., Rafe, V.: A hybrid heuristic workflow scheduling algorithm for cloud computing environments. J. Exp. Theor. Artif. Intell. **27**(6), 721–735 (2015)
4. Kokilavani, T., George Amalarethinam, D.I.: Load balanced MinMin algorithm for static metatask scheduling in grid computing. Int. J. Comput. Appl. **20**(2), 43–49 (2011)
5. Wang, W., Yan, Y., Zhou, J.: Optimization of Min-Min task scheduling algorithm based on load balancing. J. Nanjing Univ. Sci. Technol. (Nat. Sci.) **39**(4), 398–404 (2015)
6. Song, J., Li, T., Yan, Z., et al.: Data layout method for load balancing in data-intensive computing. J. Beijing Univ. Posts Telecommun. **36**(4), 76–80 (2013)
7. Zhang, L., Liu, S., Han, Yu.: Task scheduling algorithm based on load balancing. J. Jilin Univ. (Sci. Ed.) **52**(4), 769–772 (2014)
8. Chen, L., Wang, J., et al.: Research on load balancing algorithm based on rough set. Comput. Eng. Sci. **32**(1), 101–104 (2010)
9. Jin, J., Luo, J., Du, M., et al.: A data-locality-aware task scheduler for distributed social graph queries. Future Gener. Comput. Syst. **93**(68), 1010–1022 (2019)
10. Yekkehkhany, A.: Near data scheduling for data centers with multi levels of data locality (2017)
11. Shang, F., Chen, X., Yan, C.: A strategy for scheduling reduce task based on intermediate data locality of the MapReduce. Clust. Comput. **20**(4), 2821–2831 (2017)
12. Tao, D., Lin, Z., Wang, B.: Load feedback-based resource scheduling and dynamic migration-based data locality for virtual hadoop clusters in openstack-based clouds. Tsinghua Sci. Technol. **22**(2), 149–159 (2017)
13. Song, Y., Shi, W., Hu, B., et al.: Gain maximization team construction algorithm based on multi-objective greedy strategy. High-Tech Commun. **28**(4), 279–290 (2018)
14. Wang, S., Chen, K., et al.: Research on task scheduling method based on greedy algorithm in ETL. Microelectron. Comput. **26**(7), 130–133 (2009)
15. Ying, K.C., Lin, S.W., Cheng, C.Y., He, C.D.: Iterated reference greedy algorithm for solving distributed no-idle permutation flowshop scheduling problems. Comput. Ind. Eng. **110**(36), 413–423 (2017)

A Probabilistic Soft Logic Reasoning Model with Automatic Rule Learning

Jia Zhang[1], Hui Zhang[2(✉)], Bo Li[1,3], Chunming Yang[1],
and Xujian Zhao[1]

[1] School of Computer Science and Technology,
Southwest University of Science and Technology, Mianyang 621010, China
[2] School of Science, Southwest University of Science and Technology,
Mianyang 621010, China
zhanghui@swust.edu.cn
[3] School of Computer and Technology, University of Science and Technology
of China, Hefei 230027, China

Abstract. Probabilistic Soft Logic (PSL), as a declarative rule-based probability model, has strong extensibility and multi-domain adaptability and has
been applied in many domains. In practice, a main difficult is that a lot of
common sense and domain knowledge need to be set manually as preconditions
for rule establishment, and the acquisition of these knowledge is often very
expensive. To alleviate this dilemma, this paper has worked on two aspects:
First, a rule automatic learning method was proposed, which combined AMIE+
algorithm and PSL to form a new reasoning model. Second, a multi-level
method was used to improve the reasoning efficiency of the model. The
experimental results showed that the proposed methods are feasible.

Keywords: Probabilistic soft logic · Rules automatically extracted ·
Multi-level approach · Machine learning

1 Introduction

In 2013, Kimmig proposed probabilistic soft logic (PSL) [1]. Similar to Markov Logic
Network (MLN) and other statistical relational learning methods, PSL also uses
weighted first-order logic rules model the dependencies in the problem. However,
unlike MLN, the logic represented by PSL is to use soft truth values in the interval
[0, 1] instead of boolean value 0 or 1 to represent the atoms in the domain with a
probabilistic manner, which allows PSL's reasoning become a continuous optimization
problem [1]. In addition, as a probability model based on declarative rules, PSL can
flexibly add useful priori domain knowledge as rule input in solving new domain
problems, and its declarative rules are understandable to both machine and human. The
model is easier to man-machine after it is built.

However, one of the major challenges that PSL is facing today is that the required
declarative rules are completely artificially generated. Such rules are often constructed
in a very expensive way. Moreover, human-acquired knowledge is often distorted due
to perceived deviations and the variability of problem itself. The knowledge inevitably

© Springer Nature Singapore Pte Ltd. 2019
H. Jin et al. (Eds.): BigData 2019, CCIS 1120, pp. 33–45, 2019.
https://doi.org/10.1007/978-981-15-1899-7_3

contains incorrect information, which may increase the uncertainty of the reasoning model. This paper is trying to introduce the rules of automatic extraction method to meet the above challenges, to fill in the blank of probabilistic soft logic rules automatic learning.

In this paper, we proposed an automatic learning method of rules for probabilistic soft logic reasoning model, which uses the AMIE+ algorithm [2] to extract rules from RDF, and transforms these rules into adaptive probabilistic soft logic models. By evaluating the method on two real data set, the results showed that the method was feasible, and the model with multi-level method had better effect than the non-hierarchical model.

The main contributions of this paper are as follows: (1) We proposed an automatic rule-building approach for PSL, which can greatly reduce manual work. (2) The multi-level reasoning method was modeled, so that the inference result of one predicate can be used as the input condition of another predicate, and thus improved the efficiency of model inference.

2   Related Work

In recent years, PSL has been widely applied to emotion classification, entity recognition, knowledge graph, link prediction, image processing and many other issues [1].

Tomkins et al. defined a variety of attributes manually based on time-series data [7]. PSL was applied to the decomposition of energy consumption of household appliances and obtained good results. It has found a feasible starting point for reducing energy consumption and wasting resources. This method can easily incorporate a variety of information, while its accuracy requires a lot of manual integration of knowledge as a support, which makes it difficult to transplant the method to other problems. Huang et al. used PSL to establish a model of social trust [6], which modeled the spread of social influence by defining a large number of rules, validating that people trust their family more than their trusted colleagues in the life issue, and trust their colleagues more in their work about the intuition of career advice, which is modeled by unsupervised and clustering methods, but not all the methods are PSL-rule inputs. In image restoration, based on the work of Poon and Domingos et al., the LINQS team performed pixel-level restoration of images by PSL, which is faster than SPN (Sum-Product Networks) [5]. However, the model they built the number of PSL rules up to tens of thousands, need to spend a lot of labor costs. Shobeir Fakhraei et al. used PSL to construct models by using drug similarity features to predict drug interactions [6]. However, the practicality of the rules used in the method is difficult to verify. Jay Pujara constructed a general model of entity recognition using PSL relational features in a knowledge graph [9], whereas the rules governing relational dominance are difficult to fully define for large-scale data.

In a word, PSL has strong adaptability, related research and application has spanned many areas, but so far, almost all work related to the definition of each rule based on the manual and the workload will be incalculable as the complexity of the problem increases. Our method is trying to solve the above problems. As the same as the previous works, we still use PSL as the modeling foundation to keep the reasoning flexibility. But the difference between our work and the previous is that all of our rules are generated by AMIE+ algorithm.

3   Probabilistic Soft Logic

3.1   PSL Theoretical Basis

In PSL, the continuous soft truth value with closed atomic probability [0, 1] is denoted as $I(a)$, and the probability that the logical rule r holds is denoted as $I(r)$, usually using Lukasiewicz logic to compute $I(r)$, the Lukasiewicz logic can be expressed as Eqs. (1), (2) and (3)

$$I(l_1 \wedge l2) = max\{I(l1) + I(l2) - 1, 0\} \tag{1}$$

$$I(l1 \vee l2) = min\{I(l1) + I(l2), 1\} \tag{2}$$

$$I(\neg l2) = 1 - I(l1) \tag{3}$$

A rule r in PSL can be described as $rbody \rightarrow rhead$, which is satisfied when $I(rbody) \leq I(rhead)$, i.e. $I(r) = 1$. Otherwise, measure the extent to which the logic rules are satisfied by calculating distance satisfaction $d(r)$, $d(r)$ as shown in Eq. (4)

$$d(r) = max\{0, I(rbody) - I(rhead)\} \tag{4}$$

For example, we can calculate the logical rules $friends\,(a, b) \wedge like\_eat\,(a, b) \rightarrow$ $0.3, like\_eat\,(a, c)a, c) \rightarrow like\_eat\,(b, c)$.   Satisfaction: $I(friends\,(a, b) \wedge like\_eat\,(a, c)) = max\{0, 1 + 0.9 - 1\} = 0.9$, $d(r) = max\{0, 0.9 - 0.3\} = 0.6$. By d(r), the PSL defines the probability of interpretation of all closed atoms for the probability distribution:

$$p(I) = \frac{1}{Z} exp\left\{-1 * \sum_{r \in R} \lambda_r(d(r))^p\right\} \tag{5}$$

Where Z is a normalized constant, λr is the weight of the rule r, R is the set of all rules, p defines the loss function, and the PSL will seek an explanation with the minimum satisfaction distance $d(r)$ as much as all possible rules.

3.2   PSL Rule Definition

The rules in the PSL are composed as follows:

$$P1(X, Y) \wedge P2(Y, Z) > > P2(X, Z) : weight \tag{6}$$

P1 and *P2* are called predicate, which define the relationship between random variables *X*, *Y* and *Z*. *weight* stands for weight and represents the importance of each rule in reasoning [3]. In previous studies, such rules were almost always generated manually.

4 Methods: Probabilistic Soft Logic Reasoning Model with Automatic Rule Learning

Figures 1 and 2 show the architecture of the reasoning model, in the model system, the data partitioning is based on each target relationship R. The training set and the test set are divided. The rule learning model is used to mine rules from the RDF data, and the output is a series including head coverage, confidence, and for the rule of PCA confidence [2], the rule optimization model can assign weights to the rules, and then filter high-quality rules. The function of the reasoning module is to apply the rules to the PSL for reasoning. In the paper, in order to distinguish the inference results of the rules corresponding to one relationship as the multiple effects of the condition of another relationship on the reasoning of other relationships, we introduce a multi-level method into the model to obtain a multi-level probabilistic soft logic inference model for automatic learning of rules, as shown in Fig. 2.

There is the reasoning module (dashed line) in Figs. 1 and 2, the reasoning of each relationship corresponding to the inference part of the non-hierarchical method is independent of each other and does not affect each other; in the multi-level method, the inference module first runs the relationship R_1. And then the rules input the result of the inference as part of the data from R_2 to R_n until the last relationship is completed.

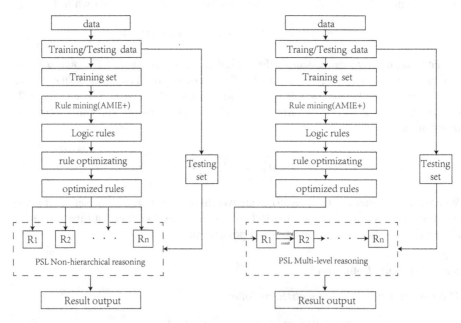

Fig. 1. Non-hierarchical reasoning model for automatic learning of rules

Fig. 2. Multi-level reasoning model for automatic learning of rules

4.1   Rule Mining

AMIE+ is the rule mining algorithm of this model. It can iteratively output rules from RDF. If the rule becomes a closed rule, AMIE+ will detect the rule according to the initial threshold of the model and add the qualified rule to the rule queue. The iterative process will continue until no new rules are generated.

In addition to the previous steps, AMIE+ also provides pruning strategies and approximation strategies that enable the algorithm to run more efficiently. The AMIE+ algorithm takes the data K, the head coverage threshold minHC, the maximum rule length threshold maxLen and the minimum confidence threshold minConf as inputs. The AMIE+ rule queue initially only contains the rule head (the rule body is empty), if the rule meets the output requirements, it will be selected into the output sequence, if it does not meet the requirements, it will be redefined by more running processes, and repeating the process until the queue is empty.

Algorithm.1 AMIE+ Rule mining algorithm

```
1:    function AMIE(KB K, minHC, maxLen, minConf)
2:         q = [r1(x, y), r2(x, y) ... rm(x, y)]
3:         out = <>
4:         while ¬q.isEmpty() do
5:              r = q.dequeue()
6:              if AcceptedForOutput(r, out, minConf) then
7:                   out.add(r)
8:              end if
9:              if length(r) < maxLen then
10:                  R(r) = Refine(r)
11:                  for all rules r_c ∈ R(r) do
12:                       if hc(r_c) ≥ minHC & r_c ∉ q then
13:                            q.enqueue(r_c)
14:                       end if
15:                  end for
16:              end if
17:         end while
18:         return out
19:    end function
```

An example of the AMIE+ mining rules is given below. If you enter the triples shown in Table 1, AMIE+ can learn the rules from these triples.

Table 1. Training data example

Subject	Relation	Object
Arthur	Son	Penelope
Arthur	Son	Christopher
Christopher	Father	Arthur
Christopher	Husband	Penelope
Arthur	Husband	Margaret
Arthur	Uncle	Charlotte
...	...	...

If we want to mine the rules of relationship Son, then AMIE+ will first set the rule head (query predicate) to *Son*, and then add atoms (evidence predicates) in the body of the rule. There are many predicates to choose from, such as *Son, Father, Husband,* etc. each option can be followed by a new choice. If the rule becomes a closed rule and meets the initial set threshold, it will be output. According to Table 1, AMIE+ can get the following rules:

$$Son\,(C,A)\ \&\&\ Husband\,(A,B)\ >\ >\ Son\,(C,B)$$
$$Father\,(A,B)\ >\ >\ Son\,(B,A)$$

Where *A, B,* and *C* are variables that can represent any instantiated constants, so if there are the following triples in the test data:

$$(Smith,\ Son,\ Johnson)$$
$$(Johnson,\ Husband,\ Taylor)$$

Then we will easily get the prediction "*(Smith, Son, Taylor)*" by entering the rules into the PSL.

4.2   Rule Optimization

The quality of the rules obtained by rule mining module cannot be completely guaranteed. It is easy to exclude some valuable rules by simply specifying the threshold. Starting from the literal meaning of the computer program, each rule obtained by the rule mining model has specific training data corresponding to it, but not all rules are applicable to the real situation corresponding to the problem. For example, the rules we got in the test like the following three rules:

$$0.8 : Wife\,(A,B)\ >\ >\ Husband\,(B,A)$$
$$0.25 : Mother\,(E,A)\ \&\&\ Wife\,(E,B)\ >\ >\ Son\,(A,B)$$
$$0.67 : Daughter\,(B,A)\ >\ >\ Mother\,(A,B)$$

Combined with the reality we understand, it is obvious that the first rule is completely established, and the second rule are correct in most cases, but the third rule is not completely correct. However, the weight of the third rule is higher than that of the second rule. If the weight obtained by the average of the head coverage and the PCA confidence is only determined by setting the threshold to determine whether the rule is retained, then the loss may be large. Some of the rules with the same weight value but the real applicability is the same as the second rule above. Therefore, this paper will firstly update the weight of each rule by Maximum Likelihood Estimation (MLE). Then, according to the range of rule weights corresponding to the same relationship, multiple thresholds are divided for testing, and finally the best performance is selected as the threshold of rule optimization. In this paper, the number of thresholds was set to 5, and the number of rules is less than 5, and the rules were tested by setting the number of rules from 4 to 1.

4.3   Reasoning Model

Non-hierarchical Reasoning. Non-hierarchical reasoning means that all the rules of the target relationship $R_1, R_2, \ldots, R_n$ are applied to the reasoning when their belongings are juxtaposed with each other and do not affect each other. Simply put, only one output will be output per run. Therefore, without considering the iterative optimization, to obtain the inference results of n kinds of relationships, the inference model needs to run n times. This approach can protect the current relationship from the uncertain reasoning data brought about by the reasoning of other rules, and at the same time facilitate us to make a quality assessment of the reasoning results. However, as a cost, non-hierarchical reasoning loses the opportunity to predict multiple relationships. For example, we dig into the following two rules from different relationships:

$$Father\,(B,A) \,>\, >\, Child\,(A,B)$$
$$Mother\,(E,A)\;\&\&\;Wife\,(E,B)\,>\,>\,Father\,(B,A)$$

We could have predicted that the child of A is B in the test data according to the above two rules. However, for non-hierarchical reasoning, this step cannot be performed.

Multi-level Reasoning. The multi-level reasoning model aims to overcome the shortcomings of the non-hierarchical reasoning model, so that the inference model can deal with multiple relationships while minimizing the uncertainty of the inference results. Figure 3 shows in detail the inference part of the model. The inference model first divides the input rules into n sub-models according to the target relationship, and each sub-model uses a subset of rules corresponding to its relationship. In the reasoning, let the R_1 sub-model run first, get the inference result, then pass the result of R_1 to the R_2 sub-model as its additional known condition, then R_2 infers the result, and the result is passed along with the inference result of R_1 to R_3. Repeat this process until all sub-models have finished running.

Logic rules & Testing set

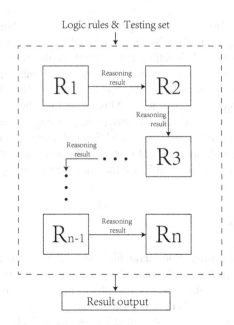

Fig. 3. PSL Multi-level reasoning module

5 Experimental Results

5.1 Datasets

We use two publicly available preprocessed versions of the dataset. One of them is the Kinship dataset from UCI [16], which is a small data set that can be used to help us understand the reasoning results. The other is YAGO [17], a semantic knowledge base derived from Wikipedia and GeoNames, which has been updated to the YAGO 3.1 version. The data used in this paper is a high-quality subset of YAGO2, which contains 36 pairs of 948358 triples.

5.2 Preparatory Work

Before mining the rules, the parameters of AMIE+ are set as shown in Table 2. *minHC* is the minimum head coverage; *maxLen* is the regular length threshold, which indicates the maximum number of atoms that can be accommodated by the rule (the sum of the number of rules and rule heads); *minConf* is the minimum *PCA* confidence that the rule should satisfy; *minInitialSup* is the support threshold of the head, which represents the minimum number of facts that satisfy the rule header [2].

Table 2. Rule mining module parameter setting

minHC	maxLen	minConf	minInitialSup
0.01	3	0.1	3

This paper only mines rules with a rule length of less than or equal to 3, because the time spent by AMIE+ increases exponentially as the length of the rule increases. As shown in Table 3, when the maximum rule length threshold is set to 4, it takes nearly 20 s on the kinship data, and the rule mining model has been difficult when the maximum rule length threshold is set to 5. Getting results in a short time, this is only the time spent on the rule mining model, adding the time of the optimization model and the inference model, the total time spent by the model system will become immeasurable. In addition, in general, a long rule can be similarly effected by multiple inferences by a rule of length 3. Therefore, in terms of integrated time cost, only rules with a rule length of 3 or less will be considered.

Table 3. The time it takes to mine rules under different rule length settings

maxLen	Time (ms)
2	78
3	660
4	17570
5	3500530 (Not finished)

For each relationship, this paper divides the corresponding facts into 70% training data and 30% test data.

5.3   Results

This paper uses the number of rules, accuracy, F1, area under the receiver operating characteristic curve (AuROC), and the average precision to evaluate the model. It is worth mentioning that, in addition to the non-hierarchical and multi-level models constructed in this paper, we have not set up additional comparison models. The reason is that this paper is the first article to apply multi-level methods and automatic rule mining techniques to PSL reasoning.

The rule mining module has tapped a total of 88 rules from YAGO2, that all can be understood by us, as shown below:

$$isMarriedTo\,(B,A) \,>\,> \,isMarriedTo\,(A,B)$$
$$isMarriedTo\,(A,F)\,\&\&\,livesIn\,(F,B)\,>\,>\,livesIn\,(A,B)$$
$$hasAcademicAdvisor\,(A,F)\,\&\&\,worksAt\,(F,B)\,>\,>\,worksAt\,(A,B)$$

In addition, the PCA confidence roughly represents the precision of each rule [2]. Table 4 is the average precision of the top n rules. It can be seen that as n increases, the precision decreases. Therefore, rule optimization is necessary.

Table 4. The average precision of the top *n* rules

Top *n*	Average precision
10	0.943
25	0.812
50	0.600
88	0.426

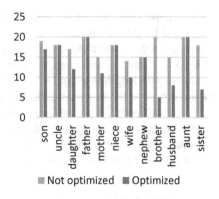

Fig. 4. Number of rules

In order to explain the problem more intuitively, we choose the relationship-intensive dataset Kinship to illustrate. Figure 4 is the comparison of the number of rules before and after the rule optimization in the case of ensuring the maximum F1 value. The target of rule optimization is to achieve the same excellent results with fewer rules.

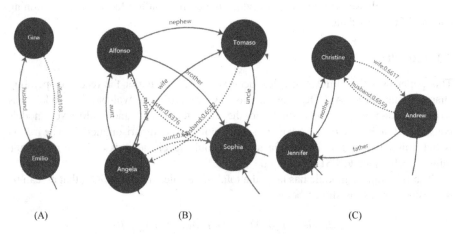

Fig. 5. Correct prediction example (Color figure online)

In the visual relationship diagram of the inference results, many graphs like A, B, and C sub-graph in Fig. 5 predict very accurate results (the red dotted line in the figure is the prediction result). At the same time, there are interesting examples like the one shown in Fig. 6, but the authenticity can't be fully verified. Just as the Sophia's grandfather is the father of her uncle, this is the prediction result of the multiple reasoning relationship under the action of multi-level reasoning.

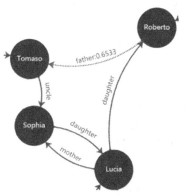

Fig. 6. Interesting example

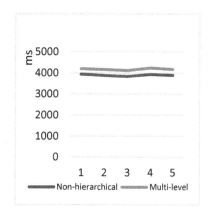

Fig. 7. Time spent

Through experiments on Kinship data, we found that the entire process from rule mining, rule optimization to execution reasoning took only a short time, which was unmatched by manual definition rules. At the same time, non-hierarchical method and multi-level method took almost the same amount of time (Fig. 7), but they achieved very different results (Fig. 8).

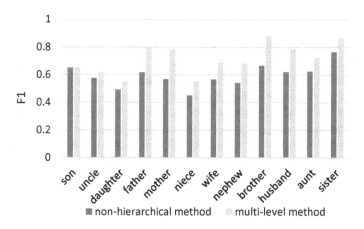

Fig. 8. Average F1 for each relation tested on the Kinship.

Table 5. Average F1 and AUROC for each relation tested on the YAGO2

Relation	F1		AUROC	
	Non	Multi	Non	Multi
worksAt	0.000	0.000	/	/
wasBornIn	0.191	0.191	0.491	0.491
hasCapital	0.000	0.000	/	/
isKnownFor	0.000	0.000	/	/
isLeaderOf	0.017	0.017	0.508	0.508
actedIn	0.010	**0.120**	0.745	**0.790**
exports	0.080	0.080	0.652	0.652
graduatedFrom	0.333	0.333	0.500	0.500
directed	0.054	0.054	0.572	0.572
isMarriedTo	0.824	**0.864**	0.571	**0.619**
diedIn	0.072	**0.399**	0.493	**0.832**
dealsWith	0.233	0.233	0.656	0.656
isPoliticianOf	0.034	0.034	0.393	0.393
isCitizenOf	0.133	**0.293**	0.231	0.504
created	0.351	**0.939**	0.427	**0.464**
livesIn	0.038	**0.222**	0.178	**0.594**
produced	0.054	**0.091**	0.157	**0.484**
imports	0.269	**0.440**	0.606	**0.823**

By constructing the PSL Reasoning Model with automatic rule Learning, we obtained the experimental results of Fig. 8 and Table 5, which are from the Kinship dataset and the YAGO2 dataset, respectively. The experimental results on the two datasets are very different, because Kinship is a relation-intensive data set. However, YAGO2 is an open relational sparse data set. The reasoning results of the inference model are difficult to verify all in the test data, thus resulting in a lower F1 value. In addition, the main purpose of measuring with F1 and AUROC is to compare the performance of non-hierarchical methods and multi-level methods. Based on this, we can still get the conclusion that multi-level methods are better than non-hierarchical methods.

6   Conclusion

We have offered two main contributions to the task of automatic rule Learning for PSL. First, we have presented a new method for combining AMIE + Algorithm and PSL into an automatic reasoning model that is much more effective than manual method in prior work. This allowed us apply methods introduced previously to much wider field, running inference on two datasets combined with a few millions of relations. Second, we have introduced a multi-level reasoning method for the PSL model with automatic rule Learning, by enhancing the dependency of the inference relationship before and after, achieving the purpose of improving the accuracy of reasoning. The experimental results have shown that the proposed method is feasible, and the multi-level methods are better than non-hierarchical methods.

References

1. Kimmig, A., Bach, S., Broecheler, M., et al.: A short introduction to probabilistic soft logic. In: Proceedings of the NIPS Workshop on Probabilistic Programming: Foundations and Applications, pp. 1–4 (2012)
2. Galárraga, L., Teflioudi, C., Hose, K., et al.: Fast rule mining in ontological knowledge bases with AMIE+. VLDB J.—Int. J. Very Large Data Bases 24(6), 707–730 (2015)
3. Bach, S.H., Broecheler, M., Huang, B., et al.: Hinge-loss markov random fields and probabilistic soft logic. J. Mach. Learn. Res. 18(109), 1–67 (2017)
4. Pellissier Tanon, T., Stepanova, D., Razniewski, S., Mirza, P., Weikum, G.: Completeness-aware rule learning from knowledge graphs. In: d'Amato, C., et al. (eds.) ISWC 2017. LNCS, vol. 10587, pp. 507–525. Springer, Cham (2017). https://doi.org/10.1007/978-3-319-68288-4_30
5. Poon, H., Domingos, P.: Sum-product networks: a new deep architecture. In: 2011 IEEE International Conference on Computer Vision Workshops (ICCV Workshops), pp. 689–690. IEEE (2011)
6. Huang, B., Kimmig, A., Getoor, L., et al.: A flexible framework for probabilistic models of social trust. In: Greenberg, A.M., Kennedy, W.G., Bos, N.D. (eds.) SBP 2013. LNCS, vol. 7812, pp. 265–273. Springer, Heidelberg (2013). https://doi.org/10.1007/978-3-642-37210-0_29
7. Tomkins, S., Pujara, J., Getoor, L.: Disambiguating energy disaggregation: a collective probabilistic approach. In: Proceedings of the 26th International Joint Conference on Artificial Intelligence, pp. 2857–2863. AAAI Press (2017)
8. Sridhar, D., Fakhraei, S., Getoor, L.: A probabilistic approach for collective similarity-based drug–drug interaction prediction. Bioinformatics 32(20), 3175–3182 (2016)
9. Pujara, J.: Probabilistic models for scalable knowledge graph construction. Doctoral dissertation, University of Maryland, College Park (2016)
10. Kouki, P., Pujara, J., Marcum, C., et al.: Collective entity resolution in multi-relational familial networks. Knowl. Inf. Syst., 1–35 (2018)
11. Farnadi, G., Babaki, B., Getoor, L.: Fairness-aware relational learning and inference. In: Workshops at the Thirty-Second AAAI Conference on Artificial Intelligence (2018)
12. Embar, V.R., Farnadi, G., Pujara, J., et al.: Aligning product categories using anchor products. In: First Workshop on Knowledge Base Construction, Reasoning and Mining (2018)
13. Srinivasan, S., Babaki, B., Farnadi, G., et al.: Lifted Hinge-Loss Markov Random Fields (2019)
14. Pujara, J., Augustine, E., Getoor, L.: Sparsity and noise: where knowledge graph embeddings fall short. In: Proceedings of the 2017 Conference on Empirical Methods in Natural Language Processing, pp. 1751–1756 (2017)
15. Kouki, P., Pujara, J., Marcum, C., et al.: Collective entity resolution in familial networks. Under Review (2017)
16. Kinship dataset. https://archive.ics.uci.edu/ml/datasets/kinship. Accessed 27 Feb 2019
17. YAGO2 dataset. http://resources.mpi-inf.mpg.de/yago-naga/amie/data/yago2/yago2core_facts.clean.notypes.tsv.7z. Accessed 27 Feb 2019

Inferring How Novice Students Learn to Code: Integrating Automated Program Repair with Cognitive Model

Yu Liang[1,2](✉) [iD], Wenjun Wu[1](✉), Lisha Wu[1](✉), and Meng Wang[1](✉)

[1] School of Computer Science and Engineering, Beihang University, Beijing, China
{liangyu,wwj,wulisha,wangmeng}@nlsde.buaa.edu.cn
[2] Shen Yuan Honors College, Beihang University, Beijing, China

Abstract. Learning to code on Massive Open Online Courses (MOOCs) has become more and more popular among novice students while inferring how the students learn programming on MOOCs is a challenging task. To solve this challenge, we build a novel Intelligent Programming Tutor (IPT) which integrates the Automated Program Repair (APR) and student cognitive model. We improve an efficient APR engine, which can not only obtain repair results but also identify the types of programming errors. Based on APR, we extend the Conjunctive Factor Model (CFM) by using programming error classification as cognitive skill representation to support the student cognitive model on learning programming. We validate our IPT with the real dataset collected from a Python programming course. The results show that compared with the original CFM, our model can represent programming learning outcomes of students and predict their future performance more reliably. We also compare our student cognitive model with the state-of-the-art Deep Knowledge Tracing (DKT) model. Our model requires less training data and is higher interpretable than the DKT model.

Keywords: Intelligent tutoring system · Automated Program Repair · Student cognitive model

1 Introduction

Massive Open Online Courses (MOOCs) have become a popular way of programming education. Because of the disparity between the number of instructors and students, several challenges for MOOCs that aim to teach programming have arisen. A common strategy used to solve those challenges is to study on intelligent tutoring systems [2,8]. The systems created for programming education are always called Intelligent Programming Tutors (IPTs) [2].

There exist two major tasks for IPT. One is Automated Programming Repair (APR), which can correct student codes without the intervention of teachers. Because of the high student-teacher ratio, it is infeasible for the instructors to correct each student's programs. Thus, APR is crucial for novice programmers on

© Springer Nature Singapore Pte Ltd. 2019
H. Jin et al. (Eds.): BigData 2019, CCIS 1120, pp. 46–56, 2019.
https://doi.org/10.1007/978-981-15-1899-7_4

MOOCs. The other task is student cognitive model, which can accurately infer novice students' learning state and predict their future performance by observing their current performance in solving programming exercises. Such a student cognitive model on learning programming enables instructors to understand students' learning process and assess the different challenges for the students during programming. Based on the information of the student model, teachers can give more personalized instructional suggestions to the students. Ideally, both tasks of IPT should work synergistically to check students' codes and assess their learning states. The learning process of novice students in programming courses often involves the development of latent cognitive skills that correspond to the major concepts in programming languages such as *basic types* and *if-else conditional structures*. The outputs of the APR task generate relevant features, including student program structures and programming errors, which can be potentially used to evaluate the levels of student cognitive skills.

However, to the best of our knowledge, most IPT systems design their APR and student model in a separate way. APR implementations [4] concentrate on fault localization and code fixing, and do not examine in detail error types in students' source code and associate them with programming concepts. Towards the student model on learning programming, there are two major approaches to modeling latent programming skills: learning factor analysis such as the conjunctive factor model (CFM) [1], and deep knowledge tracing (DKT) model [12]. CFM needs to define a Q-Matrix to associate skills with every program exercise. Then, CFM models the students' cognitive learning state by multiplication of skill parameters from the Q-Matrix. The formation of Q-Matrix depends upon human input and expert knowledge. In contrast to the CFM, DKT model does not specify the probabilistic relationship between latent skills and student code. Instead, it builds programming embeddings and uses RNN to track the latent student state by the support of massive learning data. Neither the CFM nor DKT fully exploits the rich features from the output of the APR and has its limitations in modeling cognitive development of novice student in programming courses.

In this paper, we build an IPT that adequately takes into consideration both the automated approach for program repair and student cognitive model on learning programming, and studies the integration of them. We improve an efficient program repair engine, which can not only obtain repair results but also identify types of programming errors. These classifications of errors in the multiple versions of student code throughout their learning attempts can be related to the students' understanding of key programming concepts, which is essentially the formation of the Q-Matrix. Based on the outputs of the APR, we extend CFM by incorporate the error classifications as skill indicators into the formation of personalized Q-Matrix (see Fig. 1 for an overview).

From an introductory Python programming course, we collect a real dataset of 6885 attempts, including nine assignments and 1037 novice students. We design a comparative experiment and validate our student model with this dataset. The results show that our model can represent students' learning

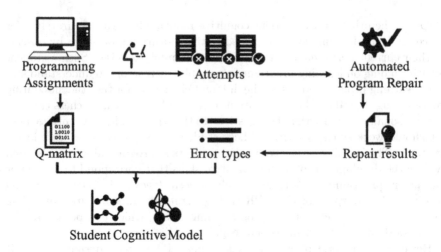

Fig. 1. Overview of our IPT

outcomes and predict their future performance accurately. Comparing the state-of-the-art Deep Knowledge Tracing (DKT) model, our model requires less training data and is higher interpretable. Furthermore, our IPT makes sense to not only the novice students but also the instructors. It can help the students in learning programming and improve teaching quality on MOOCs.

The rest of the paper is organized as follows. We present an overview of the related work in Sect. 2. In Sect. 3, we describe the details of our model. In Sect. 4, we discuss our implementation and experiments. We give conclusions and future work of our paper in Sect. 5.

2   Related Work

Many different IPT techniques have been applied to tutor students in programming. Crow et al. [2] list a series of existing IPTs and report the critical information about them. The authors also discuss the prevalence and difference within them. Nesbit et al. [8] publish a survey of the effectiveness of different IPTs for programming education field. Based on those works, there are two typical tasks of IPT. One is how to repair students' incorrect programs automatically, while another one is how to describe the students' cognitive state using some student model.

Automated Program Repair. Prophet [6] and Angelix [7] are two state-of-the-art APR tools, and they both work well in large, real-world code repair applications. They can automatically find patches to repair defects. However, the approaches are not suitable for the educational field because the programs written by beginners always have very significant mistakes. Qlose [3] is an approach to repair students' program attempts in education field automatically. Its main technique is based on different program distances. AutoGrader [11] is a tool that aims to find a series of minimal corrections for incorrect programs based on program synthesis analysis. The tool needs the course teachers to provide some base material, such as a list of potential corrections based on some known expression rewrite rules and a series of existing solutions.

Gulwani et al. [4] propose a novel APR technique for introductory programming assignments. The authors use some correct student solutions, which are already existed to fix the newcome incorrect attempts. A limitation with the work is that it may not provide more educational feedback to the students and the instructors. Yi et al. [13] study the feasibility of combining the intelligent programming tutoring with some real-world APR tools (e.g. Prophet [6] and Angelix [7]) in the education field. However, the system can be used to generate partial repairs rather than complete repairs for the programming beginner in most cases.

Unfortunately, those above tools are limited to fixing the wrong code or getting the right repair results. None of them examine in detail the types of mistakes students make or try to integrate the outputs with the student model. In fact, these error types contain lots of useful information, which is very useful in intelligent tutoring field.

Student Cognitive Model. The learning process of novice students in programming courses involves the development of multiple latent cognitive skills. Yudelson et al. [14] report a list of automated models on learning programming courses and modeling student knowledge state when solving programming exercises. Rivers et al. [10] analysis the programs in detail by using learning curve analysis. The authors aim to find out which programming skills the programmers struggle with the most while having a Python course. Bayesian knowledge tracing (BKT) model is a typical one to modeling students' cognitive states. However, the conventional BKT model [5] does not support multi-dimensional skill model and must work with other algorithms to perform Q-Matrix discovery.

Cen et al. [1] propose a novel cognitive model named conjunctive factor model (CFM) based on machine learning methods. The authors intend to establish a better cognitive relationship based on student learning data. The core concept of CFM is that the item (problem) which contains various skills is commonly more complicated than that requires only one sample skill. The CFM considers both student and skill factor and performs better when handling conjunctive skills. However, the boolean Q-Matrix is the pre-required matrix for CFM to describe the relationship between items and skills, but it does not treat multiple skills in one item differently, which might lead to inaccurate skill assessment.

Piech et al. [9] propose a graphical model which can show how learners in programming courses make progress through they make attempts for assignments. The model is also based on a set of machine learning methods. Wang et al. [12] use a recurrent neural network (RNN) and focus on students' sequences of submissions within a single programming exercise to predict future performance. The DKT approach has two major limitations: first, it needs a mass of student programming data to train the neural network, which is often infeasible to collect on an ordinary MOOC course with only hundreds of active enrollment in each session; second, the DKT model has poor interpretability because of the black box nature of the deep neural network. It does not specify the probabilistic relationship between latent skills and student code in the form of Q-Matrix, which makes it hard for instructors to understand the analysis results of the DKT model.

3 Model Details

3.1 Automated Program Repair

As mentioned previously, the student-teacher ratio can be very high on MOOCs. A MOOC instructor may not have enough time to correct each student's programs manually. Thus, it is crucial to introduce automated program repair for novice programmers in IPT for online programming courses.

In our IPT, we use the CLARA engine [4] to localize errors in student code submissions and repair them. The CLARA engine mainly includes two steps: (1) For a given programming assignment, it can automatically cluster abstract syntax trees (AST) of correct student codes using dynamic program analysis. From each cluster, we select one of the programs as a specification. (2) Given a wrong student attempt, it start to run a repair procedure on the student submission against all source code specifications and automatically generates minimal repair patches based on one of them.

We extend CLARA in three ways. Figure 2 shows the whole process of our APR approach. First, we visualize the clustering result of a programming assignment and present it as a three-level tree diagram (seeing from the right part of Fig. 2), which is convenient for students and instructors to understand. Second, when repairing an incorrect student program, we give preference to the specifications from the clusters which contain the most of correct attempts. These specifications are usually the most common solutions and highly possible to generate valid repairs for the wrong program. In Fig. 2, the specification $c\_5.c$ contains the most three correct attempts, so we choose it as the first template to repair a wrong attempt. In this way, we improve the efficiency of the program repair process. Third, we accurately categorize the types of program errors and relate them with the concepts of a programming language. As a result, these error types can represent the student's misunderstanding of certain programming concepts or low cognitive skills to construct program components using these concepts. Examining the error types of students programming code in detail can model their learning process. The grouping concepts and error types are the bases of the student cognitive model on learning to code, which will be introduced in the next sections.

3.2 Student Cognitive Model

A student cognitive model in an IPT gives a quantitative assessment of how students solve program problems and predict their future performance in new

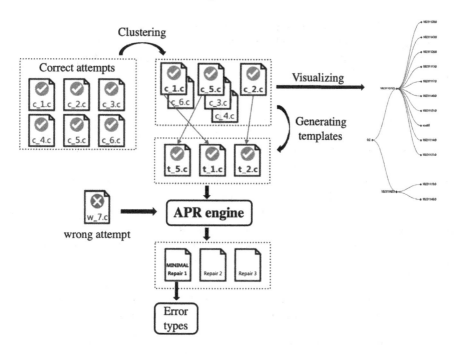

Fig. 2. Process of our APR approach

exercises. Cognitive skills for programming depend upon students' understanding of knowledge concepts and their capability of program construction using these concepts. Researchers in the field of educational data mining often refer to these skill requirements as knowledge components (KC) [10]. One program assignment always involves a conjunction of multiple KCs, which are associated with programming. The Conjunctive Factor Model (CFM) is a useful modeling framework for our IPT, which can describe a multiplication of cognitive skills in a programming assignment [1]. CFM uses a fixed Q-matrix to describe the association between program exercises and cognitive skills, which often needs experts to label every exercise manually.

In our IPT framework depicted in Fig. 1, we propose a new approach to generate the Q-matrix for the CFM model. Our improved APR can not only produce the repair results of a student's wrong attempts but also identify the exact error types in his code. It indicates that a programming exercise may pose different challenges on the individual student and the relevant concepts (skills) defined in the Q-matrix of the exercise should have different influence on the success probability of different students. Specifically, if a student correctly finishes the programming assignment, it means the student already grasps all the concepts required in this assignment. Otherwise, he may encounter learning challenges on the specific concepts suggested by the error types in his code. Based on the assumption, we extend CFM by using programming error classification as an indicator of cognitive skill representation to express whether the student masters the skills of the Q-Matrix. Therefore, we introduce a new parameter to distinguish the different levels of skill mastery (correct, wrong, or unknown during predicting). We call the new extended model PFM (short for personalized factor model) as shown in Eq. 1.

$$p_{sp} = Prob\left(R_{sp} = 1|\theta_s, \boldsymbol{\beta}, \boldsymbol{\gamma}\right) = \prod_{c=1}^{C} \left(\frac{e^{\theta_s + \beta_c + \gamma_c T_{sc}}}{1 + e^{\theta_s + \beta_c + \gamma_c T_{sc}}}\right)^{q_{pc} u_{sc}} \tag{1}$$

Where the parameters of the original CFM model: R_{sp} indicates the response of student s on the programming assignment p; θ_s indicates the student s's mastery level (a.k.a. proficiency) of programming; β_c is the coefficient for the difficulty of concept (skill) c; γ_c is the coefficient for the learning rate of concept c, reflecting how easy the concept is to grasp; T_{sc} is the number of previous times (a.k.a. practice opportunities) of student s to use the concept c; q_{pc} is 1 if assignment p includes concept c otherwise it is 0; C is the total number of concepts in the Q-matrix. The parameter we extend:

$$u_{sc} = \begin{cases} \varepsilon, \varepsilon \in (0,1) & \text{if student } \boldsymbol{s} \text{ correctly uses concept (skill) } c \\ 1 & \text{incorrectly or unknown during predicting} \end{cases}$$

The meaning of PFM is that the probability of whether the student s could get the programming assignment p correct is related to the number of required concepts (skills) which the student grasps, and the difficulty level of the concepts, and the student's general capability, and the sum of learning results of every practice opportunities. Notably, for incorrect attempts, we weaken the impact of correct concepts while strengthening the impact of wrong concepts (because they lead to incorrect attempts), treating multiple skills in an exercise differently.

4   Implementation and Experiments

4.1   Dataset

We design the following experiments to perform our methods. We collect a dataset which includes 6885 programs completed by 1037 novice students through nine assignments. The dataset is acquired from a Python Programming Introductory course collected from EduCoder[1]. The course is composed of a list of programming assignments (called challenges). Multiple attempts towards one challenge are allowed, so the collected programs include not only the final submission but also the intermediate versions. The students' submissions sequence on the challenges can be recorded in the course environment. The challenges are designed around 11 Python concepts, which are *constants(CO)*, *variables(VA)*, *operators(OP)*, *strings(ST)*, *expressions(EX)*, *lists(LI)*, *tuples(TU)*, *dictionaries(DI)*, *conditionals(CD)*, *loops(LO)* and *print statements(PR)*. The last challenge(C-469) is a comprehensive problem which includes nine main concepts. The overview of our dataset is described in Table 1.

---

[1] https://www.educoder.net.

Table 1. Dataset overview

Challenge	Topic	# Students	# Programs	# Correct	# Wrong	Concepts
C-1	String concatenation	513	807	501	306	VA, OP, ST, EX, PR
C-2	Changing list elements	289	1002	288	714	CO, VA, OP, EX, LI, PR
C-3	Sorting tuple elements	318	391	317	74	VA, EX, TU, PR
C-4	Changing dictionary elements	315	379	315	64	CO, VA, ST, EX, DI, PR
C-5	Calculating quantities	392	602	381	221	CO, VA, OP, EX, PR
C-6	Data classification	118	207	118	89	CO, VA, ST, EX, CD, PR
C-7	Computing factorials	200	520	193	327	CO, VA, OP, ST, EX, CD, PR
C-8	Data traversal	666	1293	653	640	CO, VA, OP, ST, EX, CD, LO, PR
C-9	Comprehensive problem	645	1684	634	1050	CO, VA, OP, ST, EX, LI, CD, LO, PR

4.2 Experimental Setup and Results

The first step of our ITP is using the improved repair tools to fix students' wrong programs and generate the error types of each buggy attempt. The evaluation summary is in Table 2. Because the selected challenges are for beginners, each challenge is simple and only has a few correct templates. As a result, our ITP can easily fix the most of buggy attempts (88.75%) and get their error types (2.37 errors per program on average) in a short time (3.2 s on average).

Table 2. Program repair results

Challenge	# Fixed	Rate	# Error types (avg.)	Time
C-1	306	100.00%	1.31	1.2 s
C-2	695	97.34%	2.85	1.6 s
C-3	70	94.59%	0.50	1.8 s
C-4	60	93.75%	0.64	1.7 s
C-5	199	90.05%	1.22	2.1 s
C-6	85	95.51%	1.71	1.6 s
C-7	289	88.38%	2.25	3.3 s
C-8	545	85.16%	2.24	4.0 s
C-9	844	80.38%	3.72	5.5 s
Total	3093	88.75%	2.37	3.2 s

Then we apply our PFM model to the collected programs with error types. In this experiment, we design the following training task: *based on a student's sequence of submission attempts on a list of programming challenges, predict whether the student will complete or fail the next programming challenge within the same course.* Specifically, we choose the first eight challenges (C-1 to C-8) as the training dataset to get the parameters and predict the student performances on the last comprehensive problem (C-9). Empirically, we set $\varepsilon = 0.5$. For comparison, we also implement the original CMF and DKT on the same dataset.

Table 3 shows the predicted results of three student cognitive models. We can find that our PFM performs the best on the accuracy, precision, recall, F1-score, and area under the curve (AUC).

Table 3. Results of three student cognitive models

	Accuracy	Precision	Recall	F1-score	AUC
PFM	0.790	0.837	0.549	0.663	0.833
DKT	0.728	0.667	0.423	0.518	0.672
CFM	0.625	0.528	0.030	0.057	0.616

Note that all the evaluation measures of DKT are lower than them represented in [12]. One main reason is that DKT needs a mass of training data to achieve good prediction effect. In [12], the dataset contains over one million code submissions, which is very hard to collect on ordinary MOOC courses. The recall of CFM is obviously low, which means lots of correct attempts are predicted to be wrong. The reason may be that CFM does not take into account the error types. For example, during CFM training, when a student makes mistakes in a program, the coefficient for difficulty (β) of all the concepts in this program will increase. However, only the β of "wrong" concepts (a.k.a. error types) should increase while the β of "correct" concepts which the student already grasp and correctly use should stay relatively stable. Meanwhile, it has an additional effect on the coefficient for the learning rate (γ) of those "correct" concepts and the coefficient for proficiency (θ) of the student. As a result, in the stage of prediction, the possibility that student could make a correct attempt would decrease. In contrast, PFM does better on recall, which proofs our extension for CFM works well.

5   Conclusions and Future Work

In this paper, we build an IPT that effectively integrates the APR and student cognitive model. Our improved APR engine can fix the most of programming errors and identify the types of them. Based on the outputs of the APR, we build the PFM which incorporating the error classifications as skill indicators into the formation of personalized Q-Matrix. The experience results show that compared with the original CFM and DKT model, our improved PFM has better prediction effect. Meanwhile, PFM requires less training data and has higher interpretability than the DKT model, which makes PFM suitable for the novice learners on ordinary MOOC programming courses.

In the future, we want to pursue several directions. (1) Improve the APR tool further so that it can support more languages (i.e., C and Java) and compare it with some current state-of-the-art techniques. (2) Extent more parameters of student model on programming by modifying the conventional knowledge tracing models. (3) Use the results of program repair and student model to generate personalized instructional feedback for the novice learner, which could help them understand the errors and improve their learning outcomes.

Acknowledgments. This work is supported in part by the National Key Research and Development Program of China (Funding No. 2018YFB1004502), the National Natural Science Foundation of China (Grant No. 61532004) and the State Key Laboratory of Software Development Environment (Funding No. SKLSDE-2017ZX-03).

References

1. Cen, H., Koedinger, K., Junker, B.: Comparing two IRT models for conjunctive skills. In: Woolf, B.P., Aïmeur, E., Nkambou, R., Lajoie, S. (eds.) ITS 2008. LNCS, vol. 5091, pp. 796–798. Springer, Heidelberg (2008). https://doi.org/10.1007/978-3-540-69132-7_111

2. Crow, T., Luxton-Reilly, A., Wuensche, B.: Intelligent tutoring systems for programming education: a systematic review. In: Proceedings of the 20th Australasian Computing Education Conference, pp. 53–62. ACM Press, Brisbane (2018). https://doi.org/10.1145/3160489.3160492

3. D'Antoni, L., Samanta, R., Singh, R.: QLOSE: program repair with quantitative objectives. In: Chaudhuri, S., Farzan, A. (eds.) CAV 2016. LNCS, vol. 9780, pp. 383–401. Springer, Cham (2016). https://doi.org/10.1007/978-3-319-41540-6_21

4. Gulwani, S., Radicek, I., Zuleger, F.: Automated clustering and program repair for introductory programming assignments. In: Proceedings of the 39th ACM SIGPLAN Conference on Programming Language Design and Implementation, pp. 465–480. ACM Press, Philadelphia (2018). https://doi.org/10.1145/3192366.3192387

5. Kasurinen, J., Nikula, U.: Estimating programming knowledge with Bayesian knowledge tracing. In: Proceedings of the 14th Annual SIGCSE Conference on Innovation and Technology in Computer Science Education, pp. 313–317. ACM Press, Paris (2009). https://doi.org/10.1145/1562877.1562972

6. Long, F., Rinard, M.: Automatic patch generation by learning correct code. In: Proceedings of the 43rd Annual ACM SIGPLAN-SIGACT Symposium on Principles of Programming Languages, pp. 298–312. ACM Press, St. Petersburg (2016). https://doi.org/10.1145/2837614.2837617

7. Mechtaev, S., Yi, J., Roychoudhury, A.: Angelix: scalable multiline program patch synthesis via symbolic analysis. In: Proceedings of the 38th International Conference on Software Engineering, pp. 691–701. ACM Press, Austin (2016). https://doi.org/10.1145/2884781.2884807

8. Nesbit, J.C., Adesope, O.O., Liu, Q., Ma, W.: How effective are intelligent tutoring systems in computer science education? In: Proceedings of the 14th IEEE International Conference on Advanced Learning Technologies, pp. 99–103. IEEE Computer Society, Athens (2014). https://doi.org/10.1109/ICALT.2014.38

9. Piech, C., Sahami, M., Koller, D., Cooper, S., Blikstein, P.: Modeling how students learn to program. In: Proceedings of the 43rd ACM Technical Symposium on Computer Science Education, pp. 153–160. ACM Press, Raleigh (2012). https://doi.org/10.1145/2157136.2157182

10. Rivers, K., Harpstead, E., Koedinger, K.R.: Learning curve analysis for programming: which concepts do students struggle with? In: Proceedings of the 2016 ACM Conference on International Computing Education Research, pp. 143–151. ACM Press, Melbourne (2016). https://doi.org/10.1145/2960310.2960333

11. Singh, R., Gulwani, S., Solar-Lezama, A.: Automated feedback generation for introductory programming assignments. In: Proceedings of the ACM SIGPLAN Conference on Programming Language Design and Implementation, pp. 15–26. ACM Press, Seattle (2013). https://doi.org/10.1145/2491956.2462195

12. Wang, L., Sy, A., Liu, L., Piech, C.: Deep knowledge tracing on programming exercises. In: Proceedings of the Fourth ACM Conference on Learning @ Scale, pp. 201–204. ACM Press, Cambridge (2017). https://doi.org/10.1145/3051457.3053985

13. Yi, J., Ahmed, U.Z., Karkare, A., Tan, S.H., Roychoudhury, A.: A feasibility study of using automated program repair for introductory programming assignments. In: Proceedings of the 2017 11th Joint Meeting on Foundations of Software Engineering, pp. 740–751. ACM Press, Paderborn (2017). https://doi.org/10.1145/3106237.3106262

14. Yudelson, M., Hosseini, R., Vihavainen, A., Brusilovsky, P.: Investigating automated student modeling in a Java MOOC. In: Proceedings of the 7th International Conference on Educational Data Mining, pp. 261–264. International Educational Data Mining Society (IEDMS), London (2014)

Predicting Friendship Using a Unified Probability Model

Zhijuan Kou, Hua Wang, Pingpeng Yuan$^{(\boxtimes)}$, Hai Jin, and Xia Xie

National Engineering Research Center for Big Data Technology and System,
Service Computing Technology and System Lab,
Cluster and Grid Computing Lab,
School of Computer Science and Technology,
Huazhong University of Science and Technology, Wuhan 430074, China
{zhijuankou,whua,ppyuan,hjin,shelicy}@hust.edu.cn

Abstract. Now, it is popular for people to share their feelings, activities tagged with geography and temporal information in Online Social Networks (OSNs). The spatial and temporal interactions occurred in OSNs contain a wealth of information to indicate friendship between persons. Existing researches generally focused on single dimension: spatial or temporal dimension. The simplified model only works in limited scenarios. Here, we aim to understand the probability of friendship and the place and time of interactions. First, spatial similarity of interactions is defined as a vector of places where persons checked in. Second, we employ exponential functions to characterize the change of strength of interactions as time goes on. Finally, a unified probability model to predict friendship between two persons is given. The model contains two sub-models based on spatial similarity and temporal similarity respectively. The experimental results on four data sets including spatial data sets (Gowalla and Weeplaces) and temporal data sets (Higgs Twitter Data set, High school Call Data set) show that our model works as expected.

Keywords: Spatial similarity · Temporal similarity · Models

1 Introduction

Now, it is popular for people to share their feelings, activities tagged with geography and temporal information in online social networks (OSNs), such as Twitter or Weibo. As a result, OSNs generated huge amount of both location-related data and time-related data. The data contain rich information which indicates the relationship between friendship and persons' activities. For example, people tend to visit the locations where their friends used to travel [4]. In addition, the spatial and temporal data are potential in the field of data mining and available in some applications, such as recommended system [20].

The issue how to measure relationship between people in physical world through analyzing spatial and temporal data causes widespread concern. Some

© Springer Nature Singapore Pte Ltd. 2019
H. Jin et al. (Eds.): BigData 2019, CCIS 1120, pp. 57–72, 2019.
https://doi.org/10.1007/978-981-15-1899-7_5

research has been employed in studying trajectory similarity [21]. Generally, location histories can be acquired from GPS logs explicitly [8,18], or from check-ins or contents labeled with geographic information from location-related social media implicitly. Xiao et al. [18] captured semantic location sequences (e.g., mall→cafe→museum) from GPS trajectories and defined maximal subsequences. The user similarity was related to the length of maximal subsequences and granularity levels. However, in actual, there are few same visited routes between people. The visited histories obtained from social media are more fragmented when compared to the visited histories obtained from GPS tracks. This paper attempts to propose a new method to store the location histories and the number of visits. Moreover, travel distance is related to human activities. For example, when people plan a long trip, they tend to visit the place where their friends live [4]. Thus, we consider travel distance when measuring the similarity of visited histories. In addition, measuring the relationship between people by analyzing interactions has been used [17]. Social closeness was proposed by considering interaction duration and frequency [12]. Besides that, it found that relationships between two people would change over time. Furthermore, interaction-based social graph in which the weight of edges represented relationship between individuals and their interacting objects was proposed [14]. The weight was measured by the frequency, recency, and direction of interaction. However, few works take the duration of interaction, the number of interactions, and attenuation of interaction influence over time into account. We design integral function to describe the duration of interaction and the change of interaction weight over time. Besides, previous researches generally focused on only spatial dimension or only temporal dimension. We propose a mixed model to predict friendship on two dimensions.

Generally, people tend to travel and interact with their friends. In other words, when and where activities occur can indicate the friendship between people. First, trajectory vector is utilized to record visited locations and the times of check-ins. According to the previous description, we propose travel distance to describe distance between home of individual and the check-in location. Thus, we can get trajectory similarity by considering the location history, times of check-ins, and travel distance. Then we evaluate spatial similarity based on trajectory similarity. Second, we weight relationship between people in social networks by taking **When** (the time interactions occurred) and **How** long (time the interaction lasts) into account. Finally, we combine spatial similarity and temporal similarity to predict the friendship of two persons.

In summary, the main contributions of the paper include:

- We study spatial and temporal factors which play a crucial role in friendship prediction. According to this, we introduce two similarity definitions based on space and time.
- We propose a novel unified model based on two sub-models which establish connections between the probability of being friends and spatial similarity or temporal similarity.

– We conduct extensive experiments to verify the regular patterns and the correctness of models, and the results show that our model work well both in spatial data sets and temporal data sets.

The rest of the paper is organized as follows. Section 2 describes a summary of prior works. Section 3 defines spatial similarity and temporal similarity, and proposes the probability models based on spatial and temporal similarity. Section 4 shows the results of experiments. Section 5 is the conclusion of this paper.

2   Related Work

Driven by advanced positioning technology, analysis and mining of spatial and temporal data have received much attention. Some works has studied the relationship between friendship and human behavior [3,4]. Previous studies have researched relationship based on location histories [8,18], human interactions [14], meeting events [5] and etc.

The similarity of location histories has been used to measure the similarity between people. Persons with greater trajectory similarity usually have similar behaviors and hobbies, as well as greater strength of relationship. Li et al. [8] modeled location histories as hierarchical graph, in which the granularity of top geographic regions is coarser than that of bottom geographic regions. The length of similar location sequences shared by two people and hierarchy were taken into account to get the user similarity. Xiao et al. [18] attached semantic factors to visited histories and identified GPS trajectories as semantic location sequences. The popularity of location, sequences similarity, and granularity were considered.

Interaction between people are related to the strength of relationship in social networks. Interaction duration and frequency were taken into account to measure the closeness between two people in [12]. The frequency of interaction between individuals, the time difference from current time, and the direction of interaction were used to reflect the relationship between individuals in [14].

Another common mining method is to analyze meeting events or co-occurrences. Crandall et al. [5] proposed co-occurrences (i.e., two individuals appeared in the adjacent areas at roughly same time) and discussed the influence of spatial span and temporal span. Then probabilistic models were presented to measure social relationship. However, the inequality of time and space in meeting events are not considered. Cranshaw et al. [6] introduced factors based on location, such as location entropy which indicated the location popularity. It found that invalid random occurrence of meeting events (e.g., one met strangers rather than friends) were more likely in the locations with high entropy. In addition, friends are more likely to encounter with each other in various of places instead of single place. Thus, diversity of the co-occurrence itself needs to be considered. Pham et al. [11] studied the diversity of co-occurrences and defined weight frequency based on location entropy to answer the question whether the location is a popular place. Moreover, other factors related to co-occurrences were also taken into account. Wang et al. [15] studied personal factors which explained the possibility of an individual going to a location and global factors which indicated the degree of crowding in the area. Then, a unified model was proposed to measure the relationship based on personal, global, and temporal factors. Other than this, people who have more interactions and same interests with each other tend to form a cluster or community [13]. Yin et al. [19] attempted to take advantage of encounters and semantic similarity to identify community.

3 Probability Model to Predict Friendship

Most of existing methods build a weighted social network, and compute similarity between two persons based on the weight of relationship with his neighbors [2,9]. Here, we explore spatial similarity and temporal similarity.

3.1 Spatial Similarity

Generally, people appear at the same locations many times due to some reasons. For example, persons tend to share the messages or photos when they travel to a place. The behaviors actually produce a set of check-in events, each of which refers to the event that person appeared at the location at a certain moment.

Definition 1 (Check-in event). *A check-in event is a triple* $[u, loc, t]$ *where* t *is the time stamp when person* u *checked in at location* loc. *Each location can be identified by its longitude and latitude.*

The distance a person travels indicates the travel is for casual or friendship. For example, a person may go to a market near his home for shopping. He may also visit a friend living in another city.

Definition 2 (Travel Distance). *Assume the home of person* u *is* h_u. *When person* u *checks in location* loc, *the travel distance is the distance from current location* loc *to his home. So, the travel distance of* u *in the location* loc *is the Euclidean distance between* h_u *and* loc: $D_u(loc) = dis(loc, h_u)$.

Definition 3 (Total Travel Distance). *Assume the homes of person* u, v *are* h_u, h_v *respectively. Person* u *and* v *travel to location* loc. *The total travel distance of* u, v *in* loc *is*

$$\lambda(loc) = D_u(loc) + D_v(loc)$$

Definition 4 (Trajectory). *A trajectory of a person is the sequence of locations where the person checked in. Person* u_i's $t_{u_i} = (c_0, c_1, \ldots, c_k, \ldots, c_{n-1})$ *in which* c_k $(0 \leq k < n)$ *is the times* u_i *checks in at* loc_k. *If* $c_k = 0$, *so person* u_i *did not check in at* loc_k.

The trajectories of two persons is shown in Fig. 1. In Fig. 1, the red line and blue line indicate the trajectory of two people respectively, and each cylinder represents a location with different latitude and longitude, the height of the cylinder represents the times that user checked in at the location. The trajectory of each user consists of many different cylinders. For example, the trajectory vector $t_{u_1} = (1, 2, 1, 0, 0, 0)$ shows user u_1 visited the first three locations (loc_1, loc_2, loc_3) and did not go to the last three places. u_1 visited the first and the third location once and the second place twice.

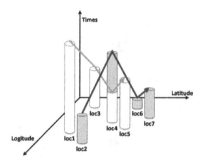

Fig. 1. The trajectory of persons (Color figure online)

We evaluate spatial similarity between two persons by computing cosine similarity instead of dot product of two trajectories. The reason is that the dot products of any two trajectories varies much.

$$SS_{u_i, u_j} = \frac{t_{u_i} \bullet t_{u_j}}{||t_{u_i}|| \times ||t_{u_j}||}$$

The greater SS_{u_i, u_j} is, the closer two persons are. From previous work and our experiments (In section VI) we know that the probability that two persons are friends (P) increases as SS_{u_i, u_j} increases. In the experiments (Section VI), we will set a baseline distance d, and filter out person pairs who once checked in at the location loc that $\lambda(loc) \geq d$. Based on the filtered-out person pairs, we explore the laws between P and SS_{u_i, u_j}.

3.2 Temporal Similarity

Users in online social networks interact with others by replying, mentioning, retweeting other persons' messages. Each interaction is also tagged with temporal information. In the following, we will explore friendship of persons by evaluating interactions between them.

Interactions Between Persons. An interaction beginning at time t_s and ending at time t_e between person u_s and u_o is a tuple $[u_s, u_o, t_s, t_e]$. For those instant interactions, such as replying, mentioning, retweeting, which do not last for some time, let $t_e = t_s + \delta$, δ is a small positive number.

The interaction between persons can be modeled as interaction graph like Fig. 2. In Fig. 2, vertices correspond to persons and each of the interaction pairs

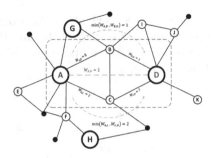

Fig. 2. The interactions between persons

of persons corresponds to an edge (solid line). The times, intervals, and duration of interactions between two persons can have important implications for the prediction of friendship. Different interactions may show different intimating degrees. So, a weight is assigned as a label to each edge of interaction graph.

Interaction Weight. Similar as news, the weight of an interaction decays as time goes by. Motivated by [16], we employ exponential functions to characterize the change of weight. Equation 1 is a decay function. $f(t)$ represents the interaction weight density at the time t, and t_τ (based on different application scenarios) is the current time, or the latest time in the data set. In a time interval $[b, e]$, there are many interactions. In order to compute the weight of an interaction in $[b, e]$, we evaluate the integrals of the Eq. 1 from time b to e (Eq. 2).

$$f(t) = e^{\alpha(t-t_\tau)}, (\alpha > 0) \tag{1}$$

$$W_m = \int_b^e f(t)\, dt \tag{2}$$

If two persons interacted with each other, they are connected directly in the interaction graph. It is reasonable to sum all the interaction weight between two persons as the connected inter-person weight between them. In Eq. 3, there are n interactions between u_i and u_j, and the inter-person weight between them is defined as $W_{i,j}$.

$$W_{i,j} = \sum_{m=1}^{n} W_m \tag{3}$$

If there is no interaction between two persons, they are unconnected in interaction graph. In social networks such as Twitter, phone call network, if two persons are unconnected, the closeness between them decrease rapidly with hop count. It is reasonable that we only calculate temporal similarity between person pairs within two hops.

We can calculate weight between connected persons directly using Eq. 3, as for those who are unconnected, we propose another method to calculate weight between them. We can see the part in the blue dotted box in Fig. 2, we want to calculate weight between vertex A and other vertices within two hops. $W_{X,Y}$

indicates the connected inter-person weight between vertex X and vertex Y, $W_{A,B} = 8$, $W_{A,C} = 2$, $W_{B,D} = 1$, $W_{C,D} = 7$. There are two paths from vertex A to vertex D, $A - B - D$ and $A - C - D$. We choose $\min(W_{A,B}, W_{B,D})$ as the unconnected inter-person weight between vertex A and vertex D on the path $A - B - D$, and choose $\min(W_{A,C}, W_{C,D})$ as the unconnected inter-person weight between vertex A and vertex D on the path $A - C - D$. In the end, the definition of the unconnected inter-person weight between vertex A and vertex D is calculated in Eq. 4.

$$W_{A,D} = \max(\min(W_{A,B}, W_{B,D}), \min(W_{A,C}, W_{C,D})) \tag{4}$$

Estimating Parameters. Considering the function image of Eq. 1, α affects the curve of the function image, which is also the convergence speed of the function. In different data sets, the value where Eq. 1 will converge would be different. In this paper, we use $\hat{w}_i$ to estimate convergence position, and $\hat{w}_i$ is the average of the intervals from t_s to t_τ in all interactions (Eq. 5). In Eq. 5, n indicates the number of all interactions in data set, t_{s_i} represents the start time of interaction i. In most spatial and temporal data sets, there are always existing extremely large values, which cannot reflect the general laws of most data. So, if we regard the maximum in the dataset as the convergence position, most values will be much smaller than the maximum, and the convergence effect of the function will be very poor. Because of that we take $k \times \hat{w}_i$ as the convergence position in which $k > 0$. The experimental results show that if k is $6 \sim 8$, the function has the best convergence effect.

$$\hat{w}_i = \frac{\sum_{i=1}^{n}(t_\tau - t_{s_i})}{n} \tag{5}$$

$$\alpha = \frac{1}{k \times \hat{w}_i}, (k = 6, 7, 8) \tag{6}$$

According to the analysis above, we can calculate inter-person weights with a suitable α.

Temporal Similarity of Two Persons. In fact, the weight of an edge represents the temporal similarity between two persons. But the interaction weight is $\in [0, \frac{1}{\alpha}]$, and inter-person weight is $\in [0, +\infty)$. We hope to define the temporal similarity to $[0, 1]$. In this paper, we define a variable β (Eq. 7), the $\hat{w}_u$ in Eq. 7 indicates the average of inter-person weight in various person pairs. In order to reduce the impact of extreme values, we regard $k \times \hat{w}_u$ as the maximum of inter-person weight in the dataset, and we define temporal similarity as Eq. 8. In the experiments below we will calculate temporal similarity following Eq. 8. By the way, we have a tip on the temporal similarity between two-hop person pair. Because the closeness between persons would decrease sharply if they are unconnected, we divide the temporal similarity between unconnected persons by a constant integer to distinguish connected persons and the unconnected.

$$\beta = \frac{1}{k \times \hat{w}_u}, (k = 6, 7, 8) \tag{7}$$

$$TS_{u_i,u_j} = \begin{cases} \beta \times W_{i,j} & , W_{i,j} < \frac{1}{\beta} \\ 1 & , W_{i,j} \geq \frac{1}{\beta} \end{cases} \qquad (8)$$

3.3 The Probability Model Based on Spatial and Temporal Similarity

We explore the models that reflect the relationship between similarity and the probability of being friends between two users P. For the cases, we evaluate the conditional probability of two persons being friends given spatial similarity and temporal similarity.

There must be some connections between spatial similarity and P, but there also are shortcomings if we consider the spatial similarity only. So, we consider the probability under the given suitable spatial similarity S and given distance d and take into account the times that two persons checked in at the same locations. Assuming the spatial similarity of person u_i and u_j, $SS_{u_i,u_j} \geq S$ and $\lambda(loc) \geq d$, we first filter out person pairs whose $SS_{u_i,u_j} \geq S$ and they checked in at the same locations where $\lambda(loc) \geq d$ for many times. On the filtered person pairs, we use Fourier function in the style of Eq. 9 to model the increasing trend of $P_s(x)$ with the number of times(x) that they checked in at the same locations where $\lambda(loc) \geq d$.

$$P_s(x) = P_{u_i,u_j}(x)|_{SS_{u_i,u_j} \geq S,\ \lambda(loc) \geq d} = a + b \times \cos(\omega x) + c \times \sin(\omega x) \qquad (9)$$

We also try to figure out the specific laws between temporal similarity and $P_t(x)$. Assuming the temporal similarity of person u_i and u_j, $TS_{u_i,u_j} \geq T$ (T is the given suitable temporal similarity), we filter out person pairs whose $TS_{u_i,u_j} \geq T$, and on the filtered person pairs, we explore the temporal models. In this part, based on the features of temporal data, a rational function as Eq. 10 is used to model the increasing trend of $P_t(x)$ with the temporal similarity(x).

$$P_t(x) = \frac{ax + b}{x + c} \qquad (10)$$

As for the data sets that have both spatial and temporal dimensions, we can combine these two models and choose the maximum as a unified probability model (Eq. 11).

$$P(x) = \max(P_s(x), P_t(x)) \qquad (11)$$

4 Evaluation

There are four different spatial and temporal data sets for experiments. All the experiments were run on a CentOS machine with 28 Intel Xeon E5-2680 v4 CPUs with 2.40 GHz and 256GB RAM. All the programs were implemented in Spark 2.3.0, Hadoop 2.7.5, Scala 2.10, and Python 3.6.4. Because no data sets contain both spatial temporal information and friendships, we verify spatial model and temporal model respectively. The probability is the maximal of two probabilities, it does not affect the correctness of our method.

4.1 Experiments on Spatial Data

We first use Gowalla and Weeplaces from LBSNs to evaluate the spatial model.

Gowalla was a location-based social network application. Because most of Gowalla users are Americans, we exclude the records from the non-USA users. The number of users from USA is 49,981 and the number of their check-ins is 3,545,251. Gowalla also indicates whether two persons are friends.

Weeplaces [1] is another location-based social network. There are 15,799 users and 1,098,204 locations in Weeplaces data set. Weeplaces also provides a friendship network.

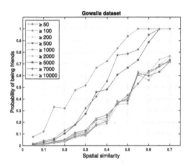

Fig. 3. Probability of being friends increases as the spatial similarity increases when set a condition that $\lambda \geq d$

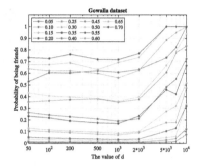

Fig. 4. Probability of being friends increases as d increases when spatial similarity $\geq S$

First, we compute the home of every user. The center of the area where user appears most frequently is always regarded as his home. Concretely, the USA area is divided into 50 by 50 km cells. For user u, the cell with the most check-in records will be chosen. Then we cluster the locations using K-means clustering algorithm, and choose the center of the cluster as home of the user u.

The Impact of Travel Distance on Probability of Being Friends. We first consider the impact of the total travel distance on probability. We define a variable d, and assign d from 50 to 10000. In the condition that $\lambda \geq d$, we filter out the person pairs (u, v) who checked in at any location loc where $\lambda(loc) = D_u(loc) + D_v(loc) \geq d$. We would like to explore how P changes as the spatial similarity increases on different filtered-out user pairs. From Fig. 3, first we can discover that as the spatial similarity increases, P will increase to at least 0.71, which means if spatial similarity is large enough, we have a great possibility to judge the friendship between two persons. Other than this, we assign a variable S from 0.05 to 0.7 by 0.05. We want to explore the change of P as d increases under the condition that $SS(u, v) \geq S$. We classify the person pairs by their spatial similarity. We can see from Fig. 4, P will increase as d, especially when d is greater than or equal to 2000KM, we can see that P has a larger increase. It shows if a pair of persons checked in farther away from their home, they are more likely to be friends. If two persons checked in at a location where λ is greater than or equal to 5000KM, we even have the chance to confirm that they must be friends. Moreover, the small d has little effect on P. It is easy to explain that if people appear at the same place very far from their home, they have a great possibility to travel together or have the same target like for conference or something, they have a high probability of being friends. So even if users have a small spatial similarity, we still have a chance to predict that they are friends if they checked in at a location very far from their home.

We did the same experiments on the Weeplaces, and the results are shown in Fig. 5. It shows the same regular pattern with that in Gowalla.

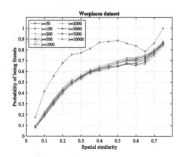

Fig. 5. Probability of being friends increases as spatial similarity increases in Weeplaces

The Impact of Times on Probability of Being Friends. When we set a condition that $\lambda \geq d$, the probability of predicting user friendship obviously increases. So, when the spatial similarity is small, we can also predict user friendship by λ. Although a large d is useful to improve the probability of being friends, we could discover that only when d is great enough, can P have a huge leap. When the spatial similarity is less than 0.2, P would have a huge leap when λ is greater than 5000KM even 7000KM. But the driving distance from New York USA to Vancouver Canada is about 5000KM, so it is very difficult to meet the condition that two persons checked in at a location where $\lambda \geq 5000$KM. In most situations, λ would be less than 500KM. How to predict friendship between persons within a relatively small distance is another problem to discuss.

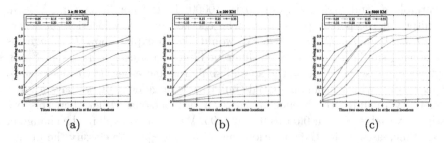

Fig. 6. (a), (b) respectively show when d is small, the growth trend of P with times that users checked in at the same location under the condition that S is from 0.05 to 0.35; (c) shows when d is large ($\lambda \geq 5000$KM), P increases rapidly with times that users checked in at the same location.

If two persons checked in at the same location far from their homes once, it could be a coincidence. But if they checked in at the same locations far from their homes many times, it must be planned in advance. We take the times that two persons checked in at the same location where $\lambda \geq d$ into account to decrease the probability of coincidence. The above experiments are the cases in which times is equal to 1. We can guess if two persons checked in at the same locations for more times, they indeed have closer contact. A variable ct is defined as the times two persons checked in at the same locations that satisfies the condition that $\lambda \geq d$. ct is an integer assigned from 1 to 10 by 1. We filter out the person pairs who checked in at several same locations under the conditions that $\lambda \geq d$ and times $\geq ct$. It means the filtered-out person pairs checked in at several same locations, at least ct of which satisfies $\lambda \geq d$. We explore how P changes as ct increases on the filtered-out person pairs. From Fig. 6 it is obvious that with the number of times that checked in at the same locations increases, the probability of being friends increases fast, especially for the case that spatial similarity is great. We can see from the experiments result, when we have a scenario that persons move in the range of a relative short distance (Fig. 6(a) (b)), like $\lambda \geq$ 200KM, in most cases, the spatial similarity between persons is small, and it is hard to predict friendship between persons. But when we take the times into account, it could have a higher probability to predict friendship. If the range of person movement is relatively small, people will have a greater chance to check in at the same location, so it is reasonable to take the times into account. When the spatial similarity is greater than 0.25, we even could have a more than 80% chance to predict the friendship. When λ is very large like 5000KM (Fig. 6(c)), we could see a huge leap in P with the times increase. The same situation also appears in the results of Weeplaces's experiments in Fig. 7.

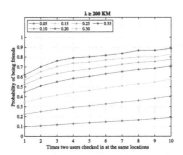

Fig. 7. Espectively shows when d is small, the growth trend of P with times that users checked in at the same location under the condition that S is from 0.05 to 0.35 in Weeplaces

So P is inseparable from the check-in location and the times of checking at the same locations. When λ and the times are both considered, we can have a great probability to predict whether two persons are friends, even if when $SS_{u,v}$ is small, we can also have a great chance to predict friendship between them.

Spatial Model Evaluation. The experiment results above prove that if λ and times are taken into account, we can predict the person friendship in the scenario with limited range of activities. If the spatial model we proposed works, we can predict the person friendship in other similar data sets.

We mainly want to explore that under the conditions that $SS(u,v) \geq S$ and $\lambda \geq d$, how P changes as the times that two persons checked in at the same locations increases. So we set $SS(u,v) \geq 0.2$ and $\lambda \geq 200$. We execute function fitting with the Fourier function in the style of Eq. 9. The function fitting result in Gowalla is shown in Fig. 8(a), the RMSE of the fitting function is 0.0051, and the R-square is 0.9997. The result in Weeplaces is shown in Fig. 8(b), the RMSE of the fitting function is 0.0094, and R-square is 0.9916. So from the results, the fitting effect of the Fourier function on the experiment results are excellent. Because of such excellent function fitting effects, we have a conclusion that in other spatial data sets, the probability of being friends between two persons and the times that two persons checked in at the same location also follow the growth trend of the Fourier function like Eq. 9, it also means that the spatial model we proposed works.

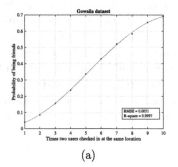
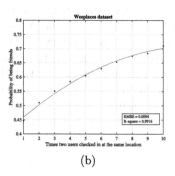

(a)   (b)

Fig. 8. Figure (a) shows P increases as the spatial similarity increases in Gowalla; figure (b) shows P increases as the spatial similarity increases in Weeplaces.

4.2   Experiments on Temporal Data

Higgs Twitter Data set [7] and High school Call Dataset [10] were used in the experiments. **Higgs Twitter Data set** contains 304,691 users and 563,069 records. The records are events occurring in Twitter from July 4–7, 2012 after the announcement of the discovery of the elusive Higgs boson. Record $[u_s, u_o, t_s, t_e]$ shows an interaction (re-tweeting, replying, or mentioning) beginning at time t_s and ending at t_e between two persons u_s and u_o. Two persons are considered as friends if they follow each other in Twitter. **High school Call Data set** shows the phone calls between students from a high school of Marseilles, France in December 2013. These two datasets also provide friendship networks.

Temporal Similarity. On Twitter, each user can re-tweet, or mention/reply others he does not know in reality. Therefore, it is more difficult to predict the friendship by using data set from Twitter.

In our experiments, we compute temporal similarity between all the users and their two-hop users in data set. We assign a variable ts from 0 to 1 by 0.025, we classify user pairs by their temporal similarity under the condition that $TS(u, v) \geq ts$, and P is calculated in different cases. We can see from Fig. 9, with the increase of temporal similarity, P has a rapid rise in the early stage, and the upward trend is slower in the later. It is in line with the physical connection between users, the growth trend would not always remain rapid, because the closeness between users should have a limit. With the increase of temporal similarity, the difference of P between people with great closeness will be smaller and smaller. As for the result of High school Call Dataset in Fig. 9, the increase trend is a little different from Higgs Twitter Dataset, we believe that because the data size of High school Call Dataset is small, the experiment results are affected by extreme values and noise more easily. But the growth trends of the two datasets are still roughly the same.

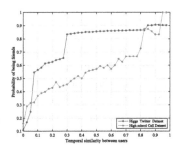

Fig. 9. Probability of being friends increases as temporal similarity

Temporal Model Evaluation. Here we want to evaluate the correctness of our temporal model. We also execute function fitting using the model of Eq. 10. We did the same operation on these two datasets. We can see from Fig. 10(a), in Higgs Twitter Dataset the coefficients that are calculated automatically using nonlinear least squares method, and the RMSE of the fitting function is 0.0610, the R-square is 0.9267. In Fig. 10(b), in High School Call Dataset, the RMSE is 0.0543 and R-square is 0.9207. The fitting effect of these sample points is also very good, which also verifies our temporal model. So we think in temporal data set, P and the temporal similarity between users follow rational function in a style of Eq. 10.

In the experimental section, we introduced several data sets on which we did our experiments. In spatial part, through experiments we found the effect of λ and times on P. Based on this, we evaluated our spatial model proposed above

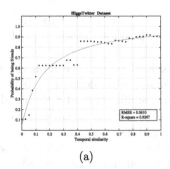
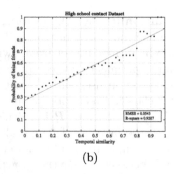

(a)   (b)

Fig. 10. (a) shows function fitting results for Higgs Twitter Dataset; (b) shows function fitting results for High School Call Dataset.

and found it works well. In temporal part, we also found a close relationship between P and temporal similarity. The correctness of temporal model was also proven. In the end, based on the probability theory, we can prove the correctness of our unified probability model.

5 Conclusions

In this paper, we propose a spatial and temporal model to evaluate the friendship probability. In spatial model, we use a location vector to indicate trajectory, and evaluate the spatial similarity of two users based on location vectors. In temporal model, we consider the time when an interaction happens, the duration of interaction, and the numbers of interactions between users. We combine spatial model and temporal model to deliver a unified probability model. The experimental results verify our model.

Acknowledgment. The research is supported by The National Key Research & Development Program of China (No. 2018YFB1004002), NSFC (No. 61672255), Science and Technology Planning Project of Guangdong Province, China (No. 2016B030306003 and 2016B030305002), and the Fundamental Research Funds for the Central Universities, HUST.

References

1. Baral, R., Wang, D., Li, T., Chen, S.C.: GeoTeCS: exploiting geographical, temporal, categorical and social aspects for personalized poi recommendation. In: Proceedings of the 17th International Conference on Information Reuse and Integration, pp. 94–101. IEEE (2016)
2. Chen, H.H., Gou, L., Zhang, X., Giles, C.L.: Capturing missing edges in social networks using vertex similarity. In: Proceedings of the 6th International Conference on Knowledge Capture, pp. 195–196. ACM (2011)

3. Cheng, R., Pang, J., Zhang, Y.: Inferring friendship from check-in data of location-based social networks. In: Proceedings of the 7th IEEE/ACM International Conference on Advances in Social Networks Analysis and Mining, pp. 1284–1291. ACM (2015)
4. Cho, E., Myers, S.A., Leskovec, J.: Friendship and mobility: user movement in location-based social networks. In: Proceedings of the 17th ACM SIGKDD International Conference on Knowledge Discovery and Data Mining, pp. 1082–1090. ACM (2011)
5. Crandall, D.J., Backstrom, L., Cosley, D., Suri, S., Huttenlocher, D., Kleinberg, J.: Inferring social ties from geographic coincidences. Proc. Nat. Acad. Sci. **107**(52), 22436–22441 (2010)
6. Cranshaw, J., Toch, E., Hong, J., Kittur, A., Sadeh, N.: Bridging the gap between physical location and online social networks. In: Proceedings of the 12th ACM International Conference on Ubiquitous Computing, pp. 119–128. ACM (2010)
7. De Domenico, M., Lima, A., Mougel, P., Musolesi, M.: The anatomy of a scientific rumor. Sci. Rep. **3**, 2980 (2013)
8. Li, Q., Zheng, Y., Xie, X., Chen, Y., Liu, W., Ma, W.Y.: Mining user similarity based on location history. In: Proceedings of the 16th ACM SIGSPATIAL International Symposium on Advances in Geographic Information Systems, p. 34. ACM (2008)
9. Li, Y., Luo, P., Wu, C.: A new network node similarity measure method and its applications. arXiv preprint arXiv:1403.4303 (2014)
10. Mastrandrea, R., Fournet, J., Barrat, A.: Contact patterns in a high school: a comparison between data collected using wearable sensors, contact diaries and friendship surveys. PLoS ONE **10**(9), e0136497 (2015)
11. Pham, H., Shahabi, C., Liu, Y.: EBM: an entropy-based model to infer social strength from spatiotemporal data. In: Proceedings of the 40th ACM SIGMOD International Conference on Management of Data, pp. 265–276. ACM (2013)
12. Phithakkitnukoon, S., Dantu, R.: Mobile social closeness and communication patterns. In: Proceedings of the 7th IEEE Consumer Communications and Networking Conference, pp. 1–5. IEEE (2010)
13. Qi, G.J., Aggarwal, C.C., Huang, T.: Community detection with edge content in social media networks. In: Proceedings of the 28th International Conference on Data Engineering, pp. 534–545. IEEE (2012)
14. Roth, M., et al.: Suggesting friends using the implicit social graph. In: Proceedings of the 16th ACM SIGKDD International Conference on Knowledge Discovery and Data Mining, pp. 233–242. ACM (2010)
15. Wang, H., Li, Z., Lee, W.C.: PGT: Measuring mobility relationship using personal, global and temporal factors. In: Proceedings of the 14th IEEE International Conference on Data Mining, pp. 570–579. IEEE (2014)
16. Wu, H., Zhao, Y., Cheng, J., Yan, D.: Efficient processing of growing temporal graphs. In: Candan, S., Chen, L., Pedersen, T.B., Chang, L., Hua, W. (eds.) DASFAA 2017. LNCS, vol. 10178, pp. 387–403. Springer, Cham (2017). https://doi.org/10.1007/978-3-319-55699-4_24
17. Xiang, R., Neville, J., Rogati, M.: Modeling relationship strength in online social networks. In: Proceedings of the 19th International Conference on World Wide Web, pp. 981–990. ACM (2010)
18. Xiao, X., Zheng, Y., Luo, Q., Xie, X.: Finding similar users using category-based location history. In: Proceedings of the 18th SIGSPATIAL International Conference on Advances in Geographic Information Systems, pp. 442–445. ACM (2010)

19. Yin, H., et al.: Discovering interpretable geo-social communities for user behavior prediction. In: Proceedings of the 32nd IEEE International Conference on Data Engineering, pp. 942–953. IEEE (2016)
20. Zheng, V.W., Zheng, Y., Xie, X., Yang, Q.: Collaborative location and activity recommendations with GPS history data. In: Proceedings of the 19th International Conference on World Wide Web, pp. 1029–1038. ACM (2010)
21. Zheng, Y., Zhou, X.: Computing with Spatial Trajectories. Springer, New York (2011). https://doi.org/10.1007/978-1-4614-1629-6

Product Feature Extraction via Topic Model and Synonym Recognition Approach

Jun Feng[1], Wen Yang[1], Cheng Gong[1], Xiaodong Li[1],
and Rongrong Bo[2(✉)]

[1] School of Computer and Information, Hohai University,
Nanjing 211100, China
[2] School of Foreign Languages, Nanjing Medical University,
Nanjing 211166, China
rrbo@njmu.edu.cn

Abstract. As e-commerce is becoming more and more popular, sentiment analysis of online reviews has become one of the most important studies in text mining. The main task of sentiment analysis is to analyze the user's attitude towards different product features. Product feature extraction refers to extracting the product features of user evaluation from reviews, which is the first step to achieve further sentiment analysis tasks. The existing product feature extraction methods do not address flexibility and randomness of online reviews. Moreover, these methods have defects such as low accuracy and recall rate. In this study, we propose a product feature extraction method based on topic model and synonym recognition. Firstly, we set a threshold that TF-IDF value of a product feature noun must reach to filter meaningless words in reviews, and select the threshold by grid search. Secondly, considering the occurrence rule of different product features in reviews, we propose a novel product similarity calculation, which also performs weighted fusion based on information entropy with a variety of general similarity calculation methods. Finally, compared with traditional methods, the experimental results show that the product feature extraction method proposed in this paper can effectively improve $F1$ and recall score of product feature extraction.

Keywords: Product feature extraction · LDA · Synonym recognition · Shopping reviews

1 Introduction

With the rapid advance of e-commerce technology, online shopping has gradually become the preferred way of daily consumption. At the same time, a large number of reviews on various products and services have been shared over the Internet. It is important to analyze the emotional information expressed by reviews, which can not only help manufacturers find defects in their products but also help consumers make purchase decisions. For online product reviews, sentiment analysis includes sentiment extraction, sentiment classification, sentiment retrieval and summarization. As an essential first step towards achieving sentiment classification and deeper retrieval and

© Springer Nature Singapore Pte Ltd. 2019
H. Jin et al. (Eds.): BigData 2019, CCIS 1120, pp. 73–88, 2019.
https://doi.org/10.1007/978-981-15-1899-7_6

induction, product feature extraction refers to extracting the product features of user evaluation from reviews.

With the urgent need of fine-grained sentiment analysis in practical applications, product feature extraction has gradually become a research hotspot. Product features can be divided into implicit features and explicit features, and the current research mainly focuses on the latter. And these studies fall into two categories: by supervised method and by unsupervised method. If the annotation data is sufficient and accurate, the supervised explicit feature extraction method can achieve better results. Moreover, the common unsupervised extraction method determines product features by mining nouns and noun phrases that occur frequently [1]. In recent years, more and more scholars used the topic model [2] and its improvement to extract product features. However, the topic model can only extract coarser-grained global features. The existing product feature extraction methods do not address characteristics of flexibility and randomness of online reviews, and there are still defects such as low accuracy and recall rate.

In this paper, we propose a product feature extraction method based on topic model (LDA, Latent Dirichlet Allocation) and synonym recognition, which makes up for the defect that the topic model can only extract coarse-grained global product features. According to the product features in reviews, we define the product feature similarity rules and propose a product similarity calculation method. We also perform weighted fusion based on information entropy with a variety of general similarity calculation methods. The experimental results show that our algorithm has better performance of product feature extraction compared with the traditional method.

The rest of the paper is structured as follows. Section 2 discusses the related work. Section 3 presents our method for extracting the product features from shopping reviews. Section 4 describes our experimental setup and results. And the conclusions are presented in Sect. 5.

2   Related Work

The most representative study began with a summary of digital product features by Hu and Liu [3] in 2004, and they divided product features into implicit features and explicit features. The present research focuses on the latter, which refers to the evaluation object or features expressed by users in reviews with specific words [4]. At present, the research on product feature extraction mainly focuses on extracting explicit features. And the research methods can be categorized as supervised methods and unsupervised methods [5].

The supervised methods treat product feature extraction as a sequence labeling or classification problem. And many supervised algorithms can be applied to product feature extraction. Li et al. [6] proposed Skip-CRF and Tree-CRF based on Conditional Random Field (CRF), using Skip-CRF to solve the long-distance dependence problem between words and using Tree-CRF to learn the grammatical relationships in reviews. As for the unsupervised methods, which can be further divided into three classes: methods on statistical features, methods on relationship between emotional words [7], and methods on topic model [8]. With the development of the topic model, some scholars use it to extract product feature. Mei et al. [9] proposed a joint model based on PLSA topic model, while most others' research was based on the extension of the LDA topic model [10]. However, most of the research on topic model is concerned with the appearance of high-frequency global product features and emotional words, which cannot extract local features with finer granularity.

In addition to methods based on topic model, some scholars use synonym lexicons to identify synonyms of product features. Due to the different expression habits of people, the same product features can be described by different words or phrases the different expression habits of people. Only by identifying synonyms of these product features, can we better extract product features and summarize viewpoints. Semantic lexicons such as "Synonym Lin", "HowNet" and "WordNet" [11–13] are often used to identify synonyms for product features. Moreover, some scholars proposed a series of methods for calculating the similarity of product features, such as TF-IDF [14], Sim-Rank [15] etc. The TF-IDF considered the context information of the product features in reviews, took the words around the product features as features of their vector representation, and used the TF-IDF values of words as the weight of their vector. The SimRank constructed the connected relationship of product features in online reviews, and used the graph structure to calculate the similarity of product features. Besides, word2vec [16] was used to mine text semantics. Luo et al. [17] used word2vec to calculate the similarity between words in the domain text, and realized the clustering of domain words.

However, these lexicons or similarity calculation methods described above do not consider the expression characteristics of the shopping reviews. In this paper, we analyze the expression characteristics of the shopping reviews and propose a novel method for calculating the similarity of product features. And we compare similarity calculation methods based on TF-IDF and word2vec in test.

3   Model

Considering the word frequency in parameter inference process of topic model, the product features extracted by LDA in the model are mostly nouns (global features) with high occurrences and specific interpretations and descriptions (local features) are always ignored. To solve the problem mentioned above, we propose a product feature extraction method based on LDA and synonym recognition. Figure 1 shows the framework of our method. Firstly, we use TF-IDF to filter nouns that do not represent any product features. Secondly, we use the perplexity to determine the parameters of LDA and extract the global features. Finally, we use synonym lexicons and similarity algorithm to supplement local product features and then get all the product features.

Fig. 1. Framework of method used for extracting product features.

3.1   Extract Global Product Features by LDA

Data Preprocessing. Since we cannot directly analyze the unstructured online customer reviews, it is necessary to preprocess the data according to word segmentation and Part-of-Speech (POS) tagging. According to the observation, product features generally appear in the form of nouns and noun phrases in shopping reviews, while emotional words used to express opinions are usually adjectives and verbs. In this paper, we use the jieba [18] toolkit for Chinese word segmentation and word form tagging, and use NLTK [19] for English stem segmentation and lemmatization. No matter in English or in Chinese, there are a large number of words that do not contain any meaning (e.g. "是", "的" in Chinese and "is", "the" in English). In order to reduce the negative impact of these meaningless high frequency words in our text analysis, we need to filter them out by using the stop words processing operation. So we use the words in the stop words list as the Chinese stop words, and the words in the NLTK stop words module as the English stop words.

Nouns Filtering. The product features are often presented in nouns or noun phrases. Because of the colloquial and non-standard expressions in reviews, not all the nouns in shopping reviews can be used as candidate product features, and these nouns are usually not included in any existing stop words list. To overcome this problem, it is necessary to filter the nouns and noun phrases obtained after the word segmentation (remove many nouns that do not represent product features, such as "things", "time", etc.). Moreover, due to the irregularity and colloquial-ism of shopping reviews, such term occupies a high percentage in reviews. When using LDA for product feature extraction, we need to filter these terms.

The object of evaluation should be a term that frequently appears in reviews of a class of goods and rarely appears in others. Therefore, we use TF-IDF to select the object of evaluation. The calculation formula of TF-IDF is as follows:

$$TFIDF(word) = TF(word) \times IDF(word) \tag{1}$$

where $TF(word)$ is the frequency of words appearing in the document and $IDF(word)$ is the inverse document frequency. It is worth noting that the same word has different TF-IDF values in different documents. So even if there are many reviews for each type of product, we only need to combine the reviews of similar products into one document.

According to Eq. (1), we can get the TF-IDF values of all words in different documents. At the same time, the nouns with higher TF-IDF values are related to features of the evaluated products. For example, we perform a TF-IDF calculation on a large number of reviews about the hotels. It can be found that the term such as "restaurant", "bathroom" and "air conditioner" are closely related to the hotel and the TF-IDF value is higher. When commenting on a hotel, we inevitably refer to nouns such as "restaurant" and "bathroom", which are rarely used to review the other products. We set a threshold ε as the TF-IDF value that the evaluation object must reach in the experiment.

Model Parameter Setting. By using the LDA model, the probability distribution for each word under each topic can be obtained as shown:

$$P(w|z = j) = \{p_{wj1}, p_{wj2}, \ldots, p_{wjv}\} \tag{2}$$

where $P(w|z = j)$ is the probability distribution of each word under topic j; p_{wji} is the probability of word w_i under topic j. After getting all the words in each topic of probability distributions, according to Eq. (3), we get the global feature set S_1:

$$S_1 = \{w_{ij}|p_{ij} \gg \sigma\} \tag{3}$$

Parameters of the LDA model need to be input: (1) Hyperparameters α, β of the LDA model; (2) the total number of topics K; (3) TF-IDF threshold ε for initial noun filtering; (4) probability threshold σ of global feature candidate words under each topic. Next, the settings of the above four types of parameters will be described respectively.

First, α is $K/50$, and β is 0.01, both of which are the artificially specified, using empirical values [20].

Then, using the perplexity to determine the value of the K. In information theory, the perplexity is an indicator, which measures the quality of the probability distribution or probability model to predict samples. The definition of perplexity can be expressed by the following formula:

$$perplexity = e^{\frac{-\sum \log(p(w))}{N}} \tag{4}$$

where $p(w)$ represents the probability of occurrence of each word in the test set, and is specifically calculated into the LDA model as follows:

$$p(w) = \sum_{z,d} p(z|d) \times p(w|z) \tag{5}$$

where z represents the topic that has been trained and d represents each document in the test set. The denominator N in the formula (4) represents the number of all words contained in the test set (the total length). For LDA topic model, the lower the perplexity is, the better model performance is. In this paper, we use the log-perplexity function in the genism module in Python to calculate the perplexity.

For the TF-IDF threshold ε of the initial noun filtering and the probability threshold σ parameter of the global feature candidate under each topic, we draw a method in [21], determined by the following way.

Since the threshold ε whose range is (0,1) determines the TF-IDF values of words in reviews, the larger the ε is, the more nouns are filtered out. Similarly, since the threshold σ whose range is (0,1) limits the probability of occurrence of words under each topic, the larger σ is, the fewer nouns are as global features. Therefore, for a given number of topics K, set $\varepsilon = m\beta$ and $\sigma = n\beta$. In this paper, we mainly test and compare three indicators: accuracy, recall and $F1$ index. $F1$ can be considered as a function of m and n as the following formula:

$$F1 = f(m, n) \tag{6}$$

In the process of model training, we can use the grid search to adjust the parameters, the final result is the best performing parameter after trying each possibility in all candidate parameters by loop traversal. We set the parameters of the model, input the review text into the LDA topic model, and extract the top 5 most probable nouns or noun phrases for each topic. Similarly, take the hotel comment as an example, and the results are shown in the following table.

The nouns shown in Table 1 are mostly global features and are frequently used in reviews on hotels. However, some specific local features such as "temperature" or "noise" have not been success-fully mined. Therefore, the next part of this paper will study local feature extraction based on synonym discovery.

Table 1. The topic model extracts the global features of hotel reviews

Topic 1	Topic 2	Topic 3
Service	Location	Restaurant
Hotel	Hotel	Facility
Attitude	Air conditioning	Environment
Air conditioning	Distance	Breakfast
Customer service	Subway	Health

3.2   Extract Local Product Features by Synonym Recognition

Due to different expression habits, different words or phrases are often used in reviews to describe the same product feature. For example, "shape" and "appearance" in the clothing reviews indicate the same product feature. The reviews on a class of products often contain hundreds or thousands of product features. It is time-consuming to manually label synonyms of product features, so we need to find an automatic method to identify synonyms of product features.

Synonym Lexicon and Product Feature Similarity. Synonymous supplementation of product features, based on synonym lexicons, is one of the most popular and convenient methods in this research field. Considering the validity and convenience of lexicons, we use a public synonym lexicon to supplement product features that are mined by the LDA.

Select Cilin as the Chinese synonym lexicon and WordNet as the English synonym dictionary. Although the two lexicons are not organized in the same way, both of them can be used to search for a set of synonyms of a word. The synonym expansion algorithm is as follows:

Algorithm 1. Mine product features by synonym.

Input:	LDA extracts global feature set S_1, *Synonym*(synonym lexicon),
	D corpus (sets of reviews)
Output:	The extended global feature set S_2

1	**While** S_1 is not change
2	**for** each word w_i in S_1
3	**If** w_i has near-synset t in *Synonym*
4	put t in S_1 if t appear in D & t not in S_1
5	**end for**
6	**end while**, $S_2 = |S_1|$
7	**return** S_2

First, we traverse each word w_i in the global feature S_1. Second, we search for a synonym set t of the word according to the synonym lexicon, add words set t that have appeared in corpus to set S_1, and continue to traverse S_1 until it no longer changes. Finally, we remove the words that are repeated, and get a set of feature words that are augmented by the synonym lexicon.

Although we use the synonym lexicon to supplement product features in part, there are still some limitations. Some words or phrases that are used to describe the same feature in shopping reviews are not synonyms in the lexicon. For example, "appearance" and "styling" represent the same feature in digital product reviews, but they cannot be classified as synonyms in a synonym lexicon. The main reason is that synonym lexicon is not designed for specific product reviews, but for common sense. Therefore, we focus on product review corpus, mining the similarities between features, and further expanding the product features.

In shopping reviews, feature synonyms that describe the same product feature tend to have similar contexts. However, the similarity based on the semantic dictionary does not consider the context, that is, it does not take advantage of context in-formation in product review features. By analyzing the product features in corpus, we find that users always like to start from global evaluation and then fall on the details, such as "The hotel bathroom is not good, and the toilet is broken." Therefore, for product features in shopping reviews, we propose the similarity principles:

- Rule 1: If product features m_i and m_j appear in the same review, they are considered to have potential similarities, such as "restaurant food".

- Rule 2: If product features m_i and m_j have a side-by-side relationship or affiliation, they are considered to have strong similarities, such as "sheets and pillows", "toilet in the bathroom", etc.
- Rule 3: If product features m_i and m_j have a turning point, they are considered to have no similarities, such as "although the location is far, the price is low".

For Rule 2 and Rule 3, it is easy to use the syntactic dependency analysis and adversative relation [22] to judge the two product features (noun /noun phrase) in reviews. In this paper, we use Stanford's syntactic dependency analysis tree Stanford parser [23] to analyze comment corpus. The similarity algorithm (MRBPF) proposed in this paper is as follows:

Algorithm 2. Mine relationship between product features (MRBPF)

Input	$M = \{m_1, m_2, ..., m_p\}, M' = \{m'_1, m'_2, ..., m'_t\}, D = \{d_1, d_2, ..., d_n\}$
Output:	Matrix $C[t][p]$
1	$i = 1, C[t][p] = zero\ matrix$
2	**while** $i \leq t$ **do**
3	$j = 1$
4	**while** $j \leq p$ **do**
5	**for** each $d \in D$ **do**
6	**if** (m'_i, m_j) in d && $m'_i \neq m_j$ **then**
7	$C[i][j] = C[i][j] + 1$
8	**if** (m'_i, m_j) hasrelationA in d && **then**
9	$C[i][j] = C[i][j] + 1$
10	**if** (m'_i, m_j) hasrelationB in d && **then**
11	$C[i][j] = C[i][j] - 1$
12	$j = j + 1$
13	$i = i + 1$
14	**for** each row $row \in C[t][p]$ **do**
15	row = normal(row)
16	**return** C_1

$M' = \{m'_1, m'_2, ..., m'_t\}$ is the product feature set obtained, $M = \{m_1, m_2, ..., m_p\}$ is the noun set of all the nouns in the corpus (excluding the noun below the threshold) and $D = \{d_1, d_2, ..., d_n\}$ is the shopping reviews set. *hasrelationA* and *hasrelationB* means Rule 2 and Rule 3 have been met. *normal* is a normalization function which is used to ensure that each dimension in the vector is within the interval $[0, 1]$. Finally, we use the matrix C to store the similarity between global feature nouns and the remaining nouns.

Similarity Fusion Based on Information Entropy. Considering the diversity of similarity calculation methods, the same two words will get different similarities under different models and calculation methods. Therefore, in order to mine product features as comprehensively and accurately as possible, we consider two other common similarity calculation methods, based on the word2vec method and the TF-IDF method, and store the results in the matrices C_1, C_2 and C_3. Similarly, we normalize each row of the matrix, and mark the similarity of the same words as 0. Therefore, through the matrices C_1, C_2 and C_3, we can get three different similarities of MRBPF, word2vec and TF-IDF proposed in this paper, which are respectively recorded as sim_m, sim_w and sim_t.

Define w_1, w_2 and w_3 respectively to represent the weight of three kinds of similarity. As there is little knowledge about the difference between different similarity degrees, we adopt a popular method to determine the weight of three different similarity degrees. A basic idea of the entropy weight method is to determine the objective weight according to the variability of the indicator. In general, if the information entropy of an indicator is smaller, the index is more variable and more information is provided. The greater the role that can be played in the comprehensive evaluation is, the greater the weight will be. We use the entropy weight method to calculate the weights of the above three similarities, and the specific steps are as follows:

- The similarity obtained by the three different methods is normalized. Since we need to calculate the entropy of each similarity, it is necessary to standardize the values of all similarities. Among them, C_1, C_2 and C_3 store the similarity of any two product features in the two sets M' and M, and each matrix size is $t * p$. For ease to calculate, the three kinds of similarity matrix are transformed into three arrays X_1, X_2, X_3. Where $X_i = \{x_1, x_{2,...}, x_{t*p}\}$, $x_1 \sim x_n$ represents all $t * p$ similarities obtained by each similarity algorithm. It is assumed that the standardized value is Y_1, Y_2, Y_3. The formula is as follows:

$$Y_{ij} = \frac{X_{ij} - min(X_i)}{max(X_i) - min(X_i)} \tag{7}$$

- Calculate the information entropy of each index. According to the definition of information entropy in information theory, the formula for calculating the entropy value e_k of the kth similarity is as follows:

$$e_k = \frac{1}{\ln t * p} \sum_{i=1}^{t*p} p_{ik} \ln \frac{1}{p_{ik}} \tag{8}$$

If $p_{ik} = 0$, then $\lim_{p_{ij} \to 0} p_{ij} \log(p_{ij}) = 0$.

- Determine the weight of each indicator. According to the calculation formula of information entropy, the information entropies of the three similarities can be calculated as e_1, e_2 and e_3, respectively. The weight of each similarity is calculated by information entropy, and the formula is as follows:

$$w_i = \frac{1 - e_i}{k - \sum e_i} (i = 1, 2, 3) \tag{9}$$

Therefore, by calculating the information entropy contained in the similarity of product features under different similarity algorithms, the weights of the three similarities can be obtained. Finally, the similarity between the two product features can be calculated by the following formula which is called MMRBPF:

$$similarity\left(m_i, m_j'\right) = w_1 sim\_m\left(m_i, m_j'\right) + w_2 sim\_w\left(m_i, m_j'\right) + w_3 sim\_t\left(m_i, m_j'\right) \quad (10)$$

In addition, it should be noted that the MMRBPF can obtain similarity because it is weighted and summed based on three different similarity methods. For the sake of comparison, it is also normalized, and the final result stored in a matrix with same structure. Finally, we continue to extract product features by using similarity between nouns about global features and nouns about detailed features.

Table 2. Hu and Liu's collection of product reviews

Product name	Number of sentences	Number of product features
Digital camera (Canon)	597	237
Digital camera (Nikon)	346	174
Cell phone (Nokia)	546	302
MP3 player (Creative)	1716	674
DVD player (Apex)	740	296

4 Experiments

4.1 Dataset

This paper mainly tests the product feature extraction method in the field of English product reviews. We use datasets collected by Hu and Liu [3], which are widely used by many researchers, and reviews contained in corpus are shown in the table below (Table 2).

4.2 Evaluation Metric

In this paper, we use the accuracy rate P, recall rate R and $F1$ values as metrics. The evaluation formula is:

$$P = \frac{FP}{F} * 100\% \quad (11)$$

$$R = \frac{FP}{FE} * 100\% \quad (12)$$

$$F1 = \frac{2 \times P \times R}{P + R} \quad (13)$$

In the above formula, *FR* represents the number of correct product features extracted, *F* represents the total number of product features extracted, and *FE* represents the number of real product features contained in actual corpus.

4.3   Experimental Setup and Results Analysis

The experiments in this paper are divided into two parts: In the first part, the test using TF-IDF threshold ε to filter nouns. In the second part, we compare the product feature similarity algorithm proposed in this paper with the traditional similarity algorithm to verify the effectiveness of this method in product feature extraction.

Test 1: Test the effectiveness of the TF-IDF threshold ε

In order to test the influence of different threshold ε on product feature extraction, it is concluded that the ε selected by the final model is mostly around 0.3. Therefore, the value of ε starts from 0, and is gradually increased from 0.05 to 0.3, and calculate the P, R, *F*1 separately. Moreover, the other parameters of the model are fixed at this time. α and β are determined by manual experience, and K is determined by the degree of perplexity. As for the probability threshold σ of global feature candidate for each topic, determined by performing the grid search optimal parameter on the training set. Next, for the subsequent synonym expansion, we adopt the MMRBPF similarity algorithm proposed in this paper.

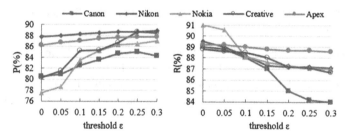

Fig. 2.   Accuracy rate P and recall rate R with threshold ε change line graph

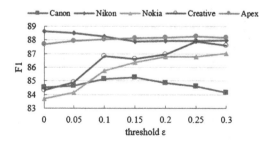

Fig. 3.   *F*1 line graph with threshold value ε

The experimental results are shown in Figs. 2 and 3. With the increase of ε, the accuracy P of five types of products have been improved, but the recall rate of the products has been reduced to extent. For $F1$, within a certain threshold range (e.g. $0 < ε \leq 0.25$), four of the five categories of products have been improved (relative to the threshold ε of 0), and only $F1$ of Nikon's products has declined slightly.

Thus, using the TF-IDF to set threshold ε to filter the nouns in corpus can significantly improve the accuracy of product feature extraction. Second, as the threshold ε increases, some of the correct product features are filtered because of the low frequency of occurrence. Therefore, the recall of product feature extraction will continue to decrease. It also shows that the improvement of the correct rate of product feature extraction is due to the decrease in the number of candidate product features extracted, that is F in Eq. (11). In addition, compared with the threshold ε of 0, as the increase of the threshold ε, Creative and Nokia's $F1$ is greatly improved, Cannon's $F1$ is slightly increased within a certain range (e.g. $0 < ε \leq 0.25$), while Nikon's $F1$ is slightly decreased. This is because the total number of sentences and words contained in Creative and Nokia are relatively rich, but Nikon contains relatively few sentences and words. As the threshold ε increases, the impact of the Nikon product feature extraction recall rate is greater than the accuracy rate. In summary, we adopt TF-IDF to set the threshold ε to filter is effective for the product feature extraction method, which can improve the accuracy and favor the $F1$ promotion of certain commodity categories, and more suitable for the number of reviews and vocabulary richer corpus. From above, the optimal threshold ε for different categories of goods are different, and there is no rule. Therefore, it is necessary to select the parameter of the grid search on the training set.

Test 2: Test the effectiveness of the MRBPF and MMRBPF methods for product feature extraction.

We compare MRBPF and MMRBPF with two general word similarity algorithms are used: based on word2vec and based on TF-IDF. Similarly, we compare three metrics P, R, and $F1$ of product feature extraction results. For these four different similarity calculation methods, we set the remaining parameters of the model to be consistent.

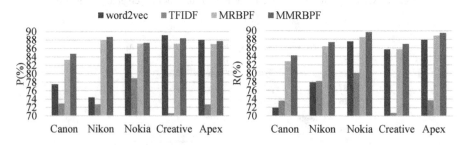

Fig. 4. Accuracy rate P and recall rate R under different similarity algorithms

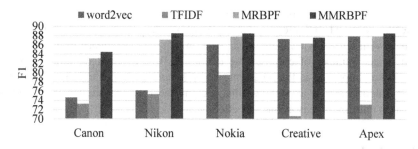

Fig. 5. Product feature extraction $F1$ under different similarity algorithms

The experimental results are shown in Figs. 4 and 5. For the accuracy rate P, the semantic similarity based on TF-IDF is the worst in five categories of products. In addition, the word2vec method performs better in Creative and Apex because of the richer reviews, but poor performance in Nikon and Cannon due to fewer reviews. While the similarity algorithm MRBPF and MMRBPF algorithm proposed in this paper are relatively stable, and show good performance in product feature extraction accuracy of five types of products. For the recall rate R, the performance of TF-IDF is still the worst, word2vec is not good in the Nikon and Cannon products with less reviews. The MRBPF and MMRBPF methods proposed in this paper are relatively stable, and MMRBPF achieved the best results in the recall rate R of the five types of product feature extraction. Similarly, for product feature extraction $F1$, the performance MRBPF and MMRBPF method are relatively stable, MMRBPF also have achieved the best results.

The results reveal that use the TF-IDF similarity to extract product features is not ideal because only the contextual TF-IDF values are considered. Although it works well in some corpus by using the word2vec to calculate the similarity extraction feature, it dependents on the number of corpora. If the corpus is not rich enough, the performance of word2vec will drop dramatically. While the MRBPF method has achieved good and stable effects in different types of products based on the specific characteristics of product reviews. Secondly, the MMRBPF method proposed in this paper combines the above three similarity algorithms. Although the accuracy rate is not always optimal, the recall rate is improved, and the final F1 results are also optimal. The average of the weights (30 times) obtained by the three kinds of similarities through the information entropy is shown (Table 3):

Table 3. Average weight of three different similarities

Product name	word2vec	TF-IFF	MRBPF
Digital camera (Canon)	0.301	0.015	0.684
Digital camera (Nikon)	0.183	0.065	0.752
Cell phone (Nokia)	0.341	0.135	0.524
MP3 player (Creative)	0.379	0.154	0.467
DVD player (Apex)	0.314	0.113	0.573

From above, the weights of the similarities obtained by the five types of commodities based on TF-IDF are very small. Word2vec has a smaller weight in Cannon and Nikon with fewer reviews, and it can explain that the dependence of the word2vec method on the number of corpus. Due to the MMRBPF based on three similarity fusions, it has improved in the recall rate and $F1$, compared to the three independent similarity algorithms. It also shows that weight the different similarities based on information entropy is reasonable and effective.

5 Conclusions

Product feature extraction is an essential part of sentiment analysis of online product reviews. In this paper, we propose a product feature extraction method based on LDA and synonym recognition. Firstly, we consider the TF-IDF value of a product feature noun must reach as threshold to filter out meaningless words, and use grid search to determine the threshold. Secondly, considering the occurrence rule of different product features in the reviews, we propose a novel product feature similarity calculation algorithm MRBPF. Moreover, we propose another similarity calculation algorithm MMRBPF by weighting fusion of MRBPF with two popular similarity calculation methods TF-IDF and word2vec. Finally, we conduct experiments on English product shopping reviews for product feature extraction.

From the experimental results, we find that the $F1$ and accuracy of product extraction will improve by setting TF-IDF threshold because that can filter nouns that do not represent any product features. Compared with TF-IDF and word2vec, the MRBPF improves the accuracy though the corpus is not sufficient. In addition, we find that the MMRBPF has the best recall and $F1$, thus proving the effectiveness of the similarity fusion based on information entropy.

References

1. Popescu, A.M., Etzioni, O.: Extracting product features and opinions from reviews. In: Kao, A., Poteet, S.R. (eds.) Natural Language Processing and Text Mining, pp. 9–28. Springer, London (2007). https://doi.org/10.1007/978-1-84628-754-1_2
2. Blei, D., Carin, L., Dunson, D.: Probabilistic topic models. IEEE Signal Process. Mag. 27(6), 55–65 (2010)
3. Hu, M., Liu, B.: Mining and summarizing customer reviews. In: KDD-2004 - Proceedings of the Tenth ACM SIGKDD International Conference on Knowledge Discovery and Data Mining, Seattle, WA, United States, pp. 168–177. Association for Computing Machinery (2004)
4. Rana, T.A., Cheah, Y.-N.: Aspect extraction in sentiment analysis: comparative analysis and survey. Artif. Intell. Rev. 46(4), 459–483 (2016)

5. Qian, L.: Research on approaches to opinion target extraction in opinion minging. Southeast University (2016)
6. Li, F., Han, C., Huang, M.: Structure-aware review mining and summarization. In: Proceedings of the 23rd International Conference on Computational Linguistics, vol. 2, pp. 653–661, Beijing, China. Tsinghua University Press (2010)
7. Qiu, G.A., Liu, B., Bu, J.J.: Opinion word expansion and target extraction through double propagation. Comput. Linguist. **37**(1), 9–27 (2011)
8. Schouten, K., Frasincar, F.: Survey on aspect-level sentiment analysis. IEEE Trans. Knowl. Data Eng. J. **28**(3), 813–830 (2016)
9. Mei, Q., Ling, X., Wondra, M.: Topic sentiment mixture: modeling facets and opinions in weblogs. In: 16th International World Wide Web Conference, WWW 2007, Banff, AB, Canada, pp. 171–180. Association for Computing Machinery (2007)
10. Brody, S., Elhadad, N.: An unsupervised aspect-sentiment model for online reviews. In: NAACL HLT 2010 - Human Language Technologies: The 2010 Annual Conference of the North American Chapter of the Association for Computational Linguistics, Proceedings of the Main Conference, Los Angeles, CA, United States, pp. 804–812. Association for Computational Linguistics (ACL) (2010)
11. Tian, J., Zhao, W.: Words similarity algorithm based on Tongyici Cilin in semantic web adaptive learning system. J. Jilin Univ. (Inf. Sci. Ed.) **28**(06), 602–608 (2010)
12. Xu, L., Lin, H., Pan, Y.: Constructing the affective lexicon ontology. **2**(2008), 602–608 (2010)
13. Lopez-Arevalo, I., Sosa-Sosa, V.J., Rojas-Lopez, F.: Improving selection of synsets from WordNet for domain-specific word sense disambiguation. Comput. Speech Lang. **41**, 128–145 (2017)
14. Xi, Y.: Recognizing the feature synonyms in product review. **30**(4), 150–158 (2016)
15. Jeh, G., Widom, J.: SimRank: a measure of structural-context similarity. In: Proceedings of the ACM SIGKDD International Conference on Knowledge Discovery and Data Mining, Edmonton, Alta, Canada, pp. 538–543. Association for Computing Machinery (2002)
16. Mikolov, T., Chen, K., Corrado, G.: Efficient estimation of word representations in vector space. Comput. Sci. 1301–3781 (2013)
17. Luo, J., Wang, Q., Li, Y.: Word clustering based on word2vec and semantic similarity. In: Proceedings of the 33rd Chinese Control Conference, CCC 2014, Nanjing, China, pp. 804–812. IEEE Computer Society (2014)
18. Python Software Foundation. jieba0.39. https://pypi.org/project/jieba/. Accessed 28 Aug 2017
19. Natural Language Toolkit. NLTK3.4 documentation. http://www.nltk.org/. Accessed 7 Nov 2018
20. Turney, P.D., Littman, M.L.: Measuring praise and criticism: inference of semantic orientation from association. ACM Trans. Inf. Syst. **21**(4), 315–346 (2003)

21. Ma, B., Yan, Z.: Product features extraction of online reviews based on LDA model. **20**(1), 98–103 (2014)
22. Wu, S.: Sentiment polarity unit extraction of web-based financial information based on dependency parsing. Jiangxi University of Finance & Economics (2015)
23. De Marneffe, M.C., Manning, C.D.: The Stanford typed dependencies representation. In: Coling 2008: Proceedings of the Workshop on Cross-Framework and Cross-Domain Parser Evaluation, pp. 1–8 (2008)

Vietnamese Noun Phrase Chunking Based on BiLSTM-CRF Model and Constraint Rules

Hua Lai[1,2], Chen Zhao[1,2], Zhengtao Yu[1,2(✉)], Shengxiang Gao[1,2],
and Yu Xu[1,2]

[1] Faculty of Information Engineering and Automation,
Kunming University of Science and Technology, Kunming 650500, China
ztyu@hotmail.com
[2] Yunnan Key Laboratory of Artificial Intelligence,
Kunming University of Science and Technology, Kunming 650500, China

Abstract. In natural language processing, the use of chunk analysis instead of parsing can greatly reduce the complexity of parsing. Noun phrase chunks, as one of the chunks, exist in a large number of sentences and play important syntactic roles such as subject and object. Therefore, it is very important to achieve high-performance recognition of noun phrase chunks for syntactic analysis. This paper presents a Vietnamese noun phrase block recognition method based on BiLSTM-CRF model and constraint rules. This method first carries out part-of-speech tagging, and integrates the marked part-of-speech features into the input vector of the model in the form of splicing. Secondly, the constraints rules are obtained by analyzing the Vietnamese noun phrase blocks. Finally, the constraints rules are integrated into the output layer of the model to realize the further optimization of the model. The experimental results show that the accuracy, recall and F-value of the method are 88.08%, 88.73% and 88.40% respectively.

Keywords: Vietnamese · Noun phrase · BiLSTM-CRF · Constraint rules

1 Introduction

Syntactic parsing is one of the key underlying technologies in natural language processing, and it is also the link between the upper application and the lower processing. It provides direct and indirect assistance for achieving the ultimate goal, such as machine translation, information retrieval, question and answer system. [1] However, due to the large amount of syntactic information to be determined by complete syntactic analysis, the implementation performance is unsatisfactory, the shallow syntactic analysis came into being. The chunk was proposed by Abney in 1991. [2] He believes that the replacement of complete syntactic analysis by chunk analysis can greatly reduce the complexity of complete syntactic analysis and reduce the difficulty of syntactic analysis. Noun phrase chunking is one of the chunk parsing. The noun phrase is abundant in the sentence, and it mainly bears the important syntactic role of the subject and object. So implementing a high-performance noun phrase chunking method is a very significant issue.

© Springer Nature Singapore Pte Ltd. 2019
H. Jin et al. (Eds.): BigData 2019, CCIS 1120, pp. 89–104, 2019.
https://doi.org/10.1007/978-981-15-1899-7_7

At present, in English and Chinese, many scholars have proposed various methods in the noun phrase chunking task, and have achieved good results. For example, rule-based methods [3–5]; statistical-based methods [6–8]; hybrid-based methods [9, 10]; neural network-based methods [11, 12]. In Vietnamese, syntactic analysis is often used as a feature in the Chinese-Vietnamese Machine Translation model and always plays a certain role. However, in small languages like Vietnamese, there are some problems, such as the scarcity of corpus resources in the network, the small size of corpus and the lack of experimental training corpus. Therefore, data sparseness often occurs in the process of model training, which makes the performance of the experiment poor.

In Vietnamese noun phrase chunking, Thao [13] and others proposed a method based on conditional random field model to solve the problem of Vietnamese noun phrase chunking. The precision and recall of the method were 82.62% and 82.72%, respectively. Yanchao [14] and others proposed a method based on conditional random fields and error-driven learning transformation for Vietnamese chunk parsing. The precision, recall and F-value of the method is 89.7%, 82.48% and 86.25%. Nguyen [15] and others integrated part-of-speech features into CRF, SVM and other models respectively, Experiment in the same corpus, the results show that the performance of CRF model is better than other models.

Aiming at the problem of noun phrase chunking in Vietnamese, this paper proposes a method combining language features and deep learning. By analyzing the existing neural network model, the BiLSTM-CRF model was selected. Among them, the BiLSTM model retains the advantage that the LSTM model can overcome the gradient to solve long-distance dependence, and can effectively use the context information to deal with the good effect of the noun phrase chunking chunking. And it can also effectively use contextual information for noun phrase chunking and achieve good results. However, since the output of BiLSTM is a separate tag, it is neglected that there is a certain relationship between the tags in the case of BIO tagging. For example, I-NP or O will appear after B-NP, and B-NP will not appear. Therefore, this paper integrates the CRF layer in the decoding process of BiLSTM, so that the output will not be independent of each other, but the optimal tag sequence, which is more suitable for the noun phrase chunking problem. In the BiLSTM-CRF model used in this paper, the input vector of the model is composed of word vector and part of speech vector. And the constraint rules are integrated into the output layer of the model to further optimize the model.

2   Construction of Vietnamese Noun Phrase Constraint Rule

Noun phrases refer to phrases whose grammatical functions are equivalent to nouns. In Vietnamese, there are two kinds of noun phrase: One of them is composed of a noun header and a modifier, which is generally divided into pre-modifier and post-modifier. In this case, the phrase is relatively fixed; The other is composed of several non-noun words to express noun semantics. In the first case, according to the Vietnamese grammar knowledge in literature [2] and the statistics of the corpus in this paper, Nine Vietnamese noun phrase constraint rules are constructed as shown in Table 1. Among $TAG = (tag_i | i = 1, 2, \ldots, m)$.

Table 1. Establishes the constraint rules.

Constraint condition	Constraint rules	Example
NP = N + N	$tag_T = N$ AND $tag_{T+1} = N$	Hố/n bom/n
NP = N + A	$tag_T = N$ AND $tag_{T+1} = A$	Báo/N nổi tiếng/A
NP = L + N_C + N	$tag_T = L$ AND $tag_{T+1} = N_C$ AND $tag_{T+1} = N$	một/M con/N_C bò/N
NP = L + N	$tag_T = L$ AND $tag_{T+1} = N_C$	Hai/M đứa/N
NP = N + P	$tag_T = N$ AND $tag_{T+1} = P$	Đường/N này/P
NP = N + C + N	$tag_T = N$ AND $tag_{T+1} = C$ AND $tag_{T+1} = N$	Món/N và/C cơm/N
NP = A + N + A	$tag_T = A$ AND $tag_{T+1} = N$ AND $tag_{T+1} = A$	Nhiều/A hoàn_cảnh/N khác/A
NP = P + C + N	$tag_T = P$ AND $tag_{T+1} = C$ AND $tag_{T+1} = N$	Mình/P và/Cđồng đội/N
NP = L + N + A	$tag_T = L$ AND $tag_{T+1} = N$ AND $tag_{T+1} = A$	một hoa đẹp

After obtaining the constraint rules, this paper combines the constraint rules with the part-of-speech tags sequences in the input corpus to get the noun phrase tag sequence. Transform it into vector and integrate it into the output layer of BiLSTM-CRF model, Constraining the output of the model. The matching process of constraint rules is shown in Fig. 1. The input example sentence is "Mảnh đất của đạn bom không còn người nghèo." the part-of-speech tags equences "Nc N E N N R V N A CH", The output noun phrase tag sequence is "B-NP I-NP O B-NP I-NP O O B-NP I-NP O".

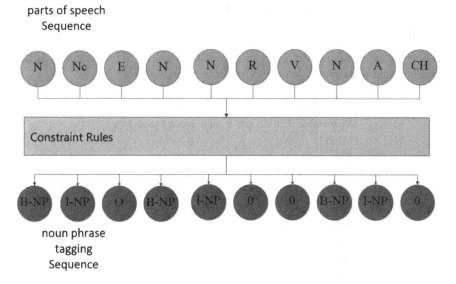

Fig. 1. Constraint rule model of Vietnamese noun phrase.

3 Construction of Noun Phrase Chunking Model Based on BiLSTM-CRF

In this paper, the framework of the BiLSTM-CRF model is shown in Fig. 2. The model is mainly divided into input layer, BiLSTM layer, CRF layer and output layer. The following is a description of each layer in the model.

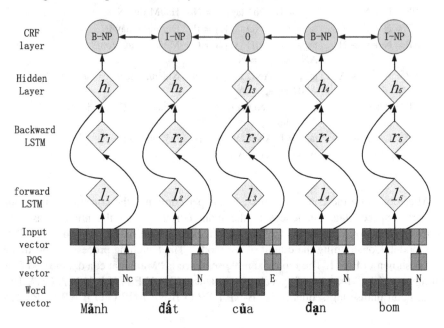

Fig. 2. The framework of the BiLSTM-CRF model.

3.1 Input layer

In this paper, the training text input by the model consists of Vietnamese, part-of-speech tags and noun phrase tags. The format is shown in Table 2.

Table 2. Examples of training texts.

Vietnamese	Part-of-speech tags	Noun phrase tags
Mảnh	Nc	B-NP
đất	N	I-NP
của	E	O
đạn	N	B-NP
bom	N	I-NP
không	R	O
còn	V	O
người	N	B-NP
nghèo	A	I-NP
.	CH	O

The input layer of the model mainly includes the training of word vectors and the generation of part-of-speech vectors. Then the input vectors are spliced and complemented. The process is shown in Fig. 3.

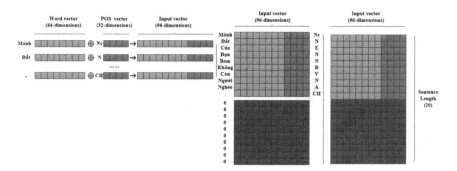

Fig. 3. Stitching and completion of input vectors.

The input word vectors and part-of-speech vectors in the model are queried from the vector dictionary. The word vector dictionary is trained by skip-gram model in Word2vec. There are 6642 words vectors in the dictionary, and the dimension of each word vector is 64. The part-of-speech vector dictionary is derived from the normal distributed random numbers generated by the program from-1 to 1. The generated dictionary is a vector matrix of 15 rows and 32 columns. 15 rows represent 15 kinds of part-of-speech tags, and 32 lists represent 32-dimensions of each part-of-speech vector.

The input vectors in the model are composed of word vectors and part-of-speech feature vectors. The dimension of a single input vector is 96. The input of the model is in sentences. The sentence length of this model is 20. If the length of a sentence is less than 20, the method of complement 0 is used to complete it. Among them, 0 represents empty character words, whose vectors are initialized by the model according to the normal distribution.

In this model, the minimum number of processing samples per time is set to 50. The dimension of the input tensor of the model is (20, 50, 96) when the word runs. As shown in Fig. 4.

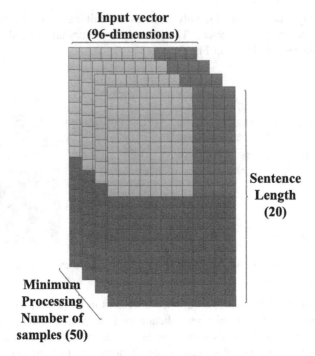

Fig. 4. The input tensor of the model.

3.2 BiLSTM Layer

In this model, the input of BiLSTM layer is the output of input layer, which is, the input tensor with dimension (20, 50, 96). The output of the BiLSTM layer is the hidden state HT. BiLSTM layer can overcome the gradient, solve the problem of long-distance dependence, and effectively utilize the context information. It is suitable for sequence annotation. So this paper chooses BiLSTM model.

Some principles of the BiLSTM model are described as follows:

1. LSTM model

LSTM model is specially designed to solve long-standing problems. It inherits the structure of RNN. Its main feature is that it has a special structure. This structure, called Cell, is mainly used to determine whether information is useful or not. In the Cell, information is added or deleted through a result called a gate. Three types of gates are placed in one Cell, namely the input gate, the forgotten gate, and the output gate. When the information enters the Cell, according to the rules, only the information that complies with the algorithm authentication will be left, and the non-compliant information will be forgotten through the forgotten gate. The internal structure of the Cell is shown in Fig. 5.

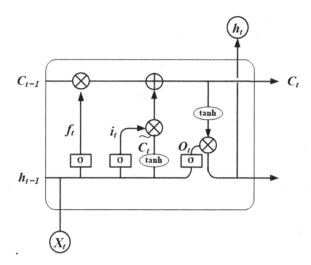

Fig. 5. The internal structure of the Cell.

As shown in Fig. 5, the first step in the Cell is the forgotten gate, which determines what information is discarded from the previous cell. The gate reads ht-1 and xt, then outputs a value between 0 and 1 for each cell state Ct−1. Among them, 1 means complete reservation, 0 means completely abandoned. In Eq. 1, ht-1 represents the output of the previous Cell, and xt represents the input of the current cell.

$$f_t = \sigma\big(W_f \cdot [h_{t-1}, x_t] + b_f\big) \tag{1}$$

The second step is the input gate, which determines how much new information is added to the Cell state. First, the sigmoid layer is used to determine which information needs to be updated, and then the tanh layer is used to generate an alternative to update the content. Finally, the two parts are combined. Update the status of the Cell.

$$i_t = \sigma(W_i \cdot [h_{t-1}, x_t] + b_i) \tag{2}$$

$$\tilde{C}_t = tanh(W_C \cdot [h_{t-1}, x_t] + b_C) \tag{3}$$

The latest state of Cell Ct, is calculated according to Eq. 4.

$$C_t = f_t \times C_{t-1} + i_t \times \tilde{C}_t \tag{4}$$

Finally, the output gate is used to determine what value we want to output. This value is in our cell state. The sigmoid layer is used to determine which part of the cell state will be output, and the tanh layer is used to obtain a value of −1 to 1, multiplied by the output of sigmiod to get the output part.

$$o_t = \sigma(W_O[h_{t-1}, x_t] + b_o) \tag{5}$$

$$h_t = o_t \times tanh(C_t) \tag{6}$$

2. BiLSTM model.

The BiLSTM model is built by two LSTM models, forward LSTM model and backward LSTM model. The input vectors are input into two LSTM models, the forward ($\vec{H}$) and backward ($\overleftarrow{H}$)sequences are calculated, and then the forward and backward sequences are combined $H_T = \left[\vec{H}_T, \overleftarrow{H}_T\right]$. Based on the LSTM model, the BiLSTM model turns the flow of information into two-way, and uses context information to predict the current output and improve the performance of the model.

3.3   CRF Layer

The input to the CRF layer is the hidden state HT, the output of the BiLSTM layer. The CRF layer borrows the basic idea of the CRF model. The CRF model is a typical discriminant model. The joint probability can be written in the form of a number of functions, so its output has a strong dependence. In the BiLSTM-CRF model, the CRF layer connects the output tags of BiLSTM through the transfer matrix, so that the order of these output tags can be integrated into the model. The outputs will not be independent of each other, but the best tags sequence.

The principle of the CRF layer is described as follows:

The parameter of the CRF layer in this paper is a 22 * 22 matrix A, where 22 is the sentence length 20 plus the start state and the end state. The value Aij in the matrix A represents the transfer score from the i-th label to the j-th label, and when the label is marked for one position, the label that has been previously marked can be utilized. If for the input sentence: $X = (x_1, x_2, \ldots, x_n)$, its corresponding tag sequence: $y = (y_1, y_2, \ldots, y_n)$. The score of the label y of sentence X as follows.

$$score(X, y) = \sum_{i=0}^{n+1} A_{y_i, y_{i+1}} + \sum_{i=1}^{n} P_{i, y_i} \tag{7}$$

Then the tag sequence is calculated by the softmax.

$$p(y|X, A) = \frac{\exp(score(X, y))}{\sum y' \in Y \exp(score(X, y'))} \tag{8}$$

Among them, the set Y is all the possibilities of the output sequence y. By Maximizing logarithmic likelihood function during model training.

$$L = \sum_{i=1}^{N} \log(y^i | x^i; A) \tag{9}$$

Among them, N is the number of training samples.

Finally, Viterbi algorithm is used to find the optimal path for annotation sequence in model prediction.

$$y^* = \underset{y \in Y}{argmax}\, p(y|x, A) \tag{10}$$

4 Construction of Vietnamese Noun Phrase Chunking Model Based on BiLSTM-CRF Model and Constraint Rules

In this paper, the Vietnamese noun phrase chunking model combines BiLSTM-CRF model and constraint rule model in the output layer, and optimizes BiLSTM-CRF model by using constraint rules. The process is shown in Fig. 6.

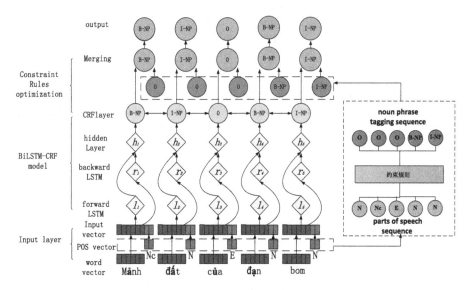

Fig. 6. The framework of Vietnamese noun phrase recognition model.

In the implementation of the model, the output of the constraints rules is transformed into a rule vector. Since the output of the model is BIO tag, the transformed rule vector is a 3-dimensions vector (Table 3).

Table 3. Rule vector conversion method.

Tag	Rule vector
B-NP	(1, 0, 0)
I-NP	(0, 1, 0)
O	(0, 0, 0)

Using the above conversion method, the rule vector of the example "Mảnh đất của đạn bom không còn người nghèo." is shown in Table 4.

Table 4. Constraint rule model output vector.

Vietnamese	Output vector		
Mảnh	0	0	0
đất	0	0	0
của	0	0	0
đạn	1	0	0
bom	0	1	0
không	0	0	0
còn	0	0	0
người	1	0	0
nghèo	0	1	0
.	0	0	0

The output of the BiLSTM-CRF model is fully connected and normalized to obtain the probability vector for each word pair label. The output of "Mảnh đất của đạn bom không còn người nghèo." is shown in Table 5.

Table 5. BiLSTM-CRF model output vector.

Vietnamese	Output vector		
Mảnh	0.5	0.16402113	0.33597887
đất	0.15605748	0.47642902	0.3675135
của	0.1636074	0.40497634	0.43141624
đạn	0.43000063	0.1623798	0.40761957
bom	0.16273756	0.43278787	0.40447456
không	0.1676929	0.38068765	0.45161948
còn	0.16951482	0.3649794	0.46550578
người	0.42115545	0.16566454	0.41318
nghèo	0.16447201	0.426197	0.40933102
.	0.16895267	0.37846977	0.45257756

Add the output vectors of the two models to get the final output vector. As shown in Table 6.

Table 6. Final output vector.

Vietnamese	Output vector		
Mảnh	0.5	0.16402113	0.33597887
đất	0.15605748	0.47642902	0.3675135
của	0.1636074	0.40497634	0.43141624
đạn	1.43000063	0.1623798	0.40761957
bom	0.16273756	1.43278787	0.40447456
không	0.1676929	0.38068765	0.45161948
còn	0.16951482	0.3649794	0.46550578
người	1.42115545	0.16566454	0.41318
nghèo	0.16447201	1.426197	0.40933102
.	0.16895267	0.37846977	0.45257756

After the output vector is obtained, the program finds the corresponding list index of the maximum value of the output vector, and converts it into a noun phrase tag for output. As shown in Fig. 7.

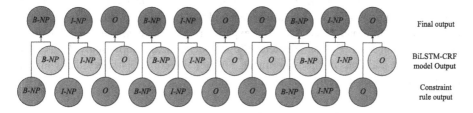

Fig. 7. The output of model

5 Experiments and Results Analysis

5.1 Experimental Data

The experimental data used in this paper come from two main sources. One of them is the Vietnamese Language Processing (VLSP) website, whose corpus is chunk corpus. It also contains other chunk identifiers, including verbs, adjectives and other eight types. Therefore, this paper preprocesses the corpus, including data cleaning tasks such as manual proofreading, marking, duplication and deletion. Then we get a corpus of noun phrases of about 193.95 million words. It contains noun phrase (represented by B-NP, I-NP, accounting for 40% of the total corpus) and non-noun phrase (represented by "O", accounting for 60% of the total corpus). The other is the corpus we built ourselves. This part of the corpus is mainly obtained by extracting Vietnamese text from the Vietnamese website, identifying it through the CRFS model, and then proofreading by Vietnamese experts to form a Vietnamese noun phrase corpus. The total size of the two corpora is about 574.68 million, and the saved text format of all data sets is UTF-8.

5.2   Evaluation Index

In this paper, precision (P), recall (R) and F-value are used to evaluate the effect of the experiment.

$$precision(\text{P}) = \frac{\text{Accurate number of noun phrase tagging}}{\text{Number of tagged noun phrase}} \qquad (11)$$

$$recall(\text{R}) = \frac{\text{Accurate number of noun phrase tagging}}{\text{Total number of noun phrase in corpus}} \qquad (12)$$

$$\text{F} - \text{value} = \frac{2 * \text{P*R}}{\text{P} + \text{R}} * 100\% \qquad (13)$$

5.3   Experimental Environment

This paper mainly carries out experiments based on noun phrase, namely non-nested noun phrase. It includes a single noun, a noun phrase without modifiers, a series of nouns which are difficult to determine the modifier relationship, juxtaposed nouns, proper nouns, time and place, etc. The experiment is written in win7 with python, which is version 3.6.1. Python uses tensorflow, undershesea, numpy, scipy, keras, nltk and so on. In addition, in the experiment, 80% of the corpus is selected for model training and 20% of the corpus is used for model testing.

5.4   Experimental Results and Analysis

In order to verify the performance of the proposed Vietnamese noun phrase recognition model, we designed the following four groups of experiments to verify. The experimental data are all noun phrase corpus in this chapter. The marked form is shown in Table 2. The dimension of word vector and part-of-speech vector is 64 and 32, and the maximum number of iterations of LSTM model, BiLSTM model and BiLSTM-CRF model is 50.

Experiment 1: The comparative experiment of the dimension of part-of-speech vectors in this model.

In order to effectively integrate the part-of-speech vector into the input vector, the part-of-speech vectors with different dimensions are respectively spliced into the word vector. Conduct a comparative experiment and select the optimal experimental protocol as the final solution. Among them, the word vector is a 64-dimensions vector trained by the word2vec model, and the part-of-speech vectors are selected as 128-dimensions,

64-dimensions, 32-dimensions, 16-dimensions and 8-dimensions, respectively. The model used in the experiment is the BiLSTM-CRF model, 80% of the corpus is selected for model training and 20% of the corpus is used for model testing. The experimental results are shown in Table 7.

Table 7. Experimental results.

The dimension of part of speech vector	P	R	F
8-dimensions	84.94%	83.63%	84.28%
16-dimensions	85.32%	85.54%	85.43%
32-dimensions	86.34%	87.11%	86.72%
64-dimensions	86.17%	86.98%	86.57%
128-dimensions	85.11%	84.91%	85.01%

From Table 7, we can see that the change of the dimension of the part of speech vector has a certain influence on the experimental results. The results of the experiments on the dimension of part of speech vectors from 8, 16 to 32 dimensions gradually improved, but when the dimension of part of speech vectors reached 64 and 128, the experimental results were worse than those of 32-dimensions. The experimental results of 64-dimensions and 32-dimensions are basically the same, but the operation speed of 64-dimensions is much slower than that of 32-dimensions. Therefore, the part-of-speech vector dimension selected in this experiment is 32-dimensions. According to the analysis, the reason for this phenomenon is that when the part-of-speech vector dimension is small, the proportion of the part-of-speech vector in the output vector is too small, and the improvement of the model does not reach the highest; When the part-of-speech vector dimension is large, the proportion of the part-of-speech vector in the output vector is too large, so that the proportion of the word vector in the output vector is too small, which affects the performance of the model.

Experiment 2: The comparison between the model and others

In order to verify that the proposed model is compared with the basic model BiLSTM-CRF model and the traditional machine learning model CRF and ME model, all of which incorporate part-of-speech features. The experimental results are shown in Table 8. The corpus used in the experiment is divided into five parts on average. The first one is chosen as the test corpus and the other four as the training corpus.

Table 8. Model contrast experiments.

Model	P	R	F
LSTM	83.88%	84.07%	83.97%
BiLSTM	85.72%	85.62%	85.67%
BiLSTM-CRF	86.34%	87.11%	86.72%
The proposed method	87.89%	88.74%	88.31%

From Table 8, we can see that the P, R and F values of the proposed method are the highest compared with the other three models, and the performance of the BiLSTM-CRF model is the second. This shows that the baseline model, the BiLSTM-CRF model, used in this paper can better integrate feature and context information into the model, and are more suitable for Vietnamese noun phrase chunking. Since the proposed method is better than the BiLSTM-CRF model, it can be seen that increasing the constraint rules can improve the precision and recall rate of Vietnamese noun phrase chunking.

Experiment 3: Comparing with references.

In order to verify the performance of the proposed method is better than that of the existing methods, the proposed method is compared with the CRF model in reference [15], which combines word features and part-of-speech features. The experimental results are shown in Table 9. The corpus used in the experiment is divided into five parts on average. The first one is chosen as the test corpus and the other four as the training corpus. The CRF model in the comparative reference [15] is implemented by using CRF++ toolkit, incorporating part-of-speech features and word features.

Table 9. Contrastive experiment of references.

Model	P	R	F
Model in reference [15]	85.68%	86.13%	85.90%
The proposed method	87.89%	88.74%	88.31%

From Table 9, we can see that compared with the model in reference [15], the experimental performance and effect of the proposed method are better. It can be seen that the BiLSTM-CRF model can improve the experimental results in Vietnamese noun phrase chunk recognition tasks, and the proposed method can achieve better results.

Experiment 4: Five-fold cross-validation experiment

In order to evaluate the impact of corpus on the proposed model, the experimental corpus is divided into five parts, one of which is chosen as the test corpus, the other four as the training corpus. Five-fold cross-validation experiments are conducted, and then the average values of the five experiments are used as the experimental results of the proposed model. The experimental results are shown in Table 10.

Table 10. Five-fold cross-validation experiment.

Test corpus	P	R	F
Fold-1	87.89%	88.74%	88.31%
Fold-2	88.43%	88.56%	88.49%
Fold-3	88.26%	89.01%	88.63%
Fold-4	87.68%	88.87%	88.27%
Fold-5	88.14%	88.46%	88.30%
Average	88.08%	88.73%	88.40%

From Table 10, it can be seen that different training corpus and test expectations will have a certain impact on the experimental results, making each group of experimental results different. The second group can achieve the highest precision of 88.43%, the third group can achieve the highest recall rate of 89.01% and the highest F-value of 88.63%. Finally, the average precision, recall and F-value of the model are 88.08%, 88.73% and 88.40 respectively.

6 Summary

Aiming at the problem of Vietnamese noun phrase chunking, a method based on BiLSTM-CRF model and constraint rule is proposed. This paper uses BiLSTM-CRF model to identify noun phrase, and uses the rules summarized in Vietnamese language characteristics to restrict the output layer. This paper preprocesses and expands the existing corpus. According to the statistics of corpus, nine main forms of noun phrase chunks are found. The experimental results show that the method can solve the problem of Vietnamese noun phrase chunking.

Acknowledgements. This work was supported by National Natural Science Foundation of China (Grant Nos. 61866019, 61972186, 61732005, 61672271, 61761026, 61762056), National Key Research and Development Plan (Grant Nos. 2018YFC0830105, 2018YFC0830101, 2018YFC0830100), Science and Technology Leading Talents in Yunnan, and Yunnan High and New Technology Industry Project (Grant No. 201606), Natural Science Foundation of Yunnan Province (Grant No. 2018FB104), and Talent Fund for Kunming University of Science and Technology (Grant No. KKSY201703005).

References

1. Zong, C.: Statistical Natural Language Processing. Tsinghua University Press, Beijing (2013)
2. Abney, S.P.: Parsing by chunks. In: Berwick, R.C., Abney, S.P., Tenny, C. (eds.) Principle-Based Parsing, pp. 257–278. Springer, Dordrecht (1991). https://doi.org/10.1007/978-94-011-3474-3_10
3. Bourigault, D.: Surface grammatical analysis for the extraction of terminological noun phrases. In: Proceedings of the 14th Association for Computational Linguistics, ACL, pp. 977–981. Association for Computational Linguistics, Stroudsburg (1992)
4. Ngai, G., Florian, R.: Transformation-based learning in the fast lane. In: Proceedings of the 2nd Meeting of the North American Chapter of the Association for Computational Linguistics, NAACL, Pittsburgh, USA, pp. 1–8 (2001)
5. Schmid, H., Im Walde, S.S.: Robust German noun chunking with a probabilistic context-free grammar. In: Proceedings of the 18th Association for Computational Linguistics, ACL, pp. 726–732. Association for Computational Linguistics (2000)
6. Sarkar, K., Gayen, V.: Bengali noun phrase chunking based on conditional random fields. In: Proceedings of the 2nd International Conference on Business and Information Management, ICBIM, Durgapur, West Bengal, India, pp. 148–153 (2014)

7. Sha, F., Pereira, F.: Shallow parsing with conditional random fields. In: Proceedings of the Human Language Technology Conference of the North American Chapter of the Association for Computational Linguistics, HLT-NAACL, Edmonton, Canada, pp. 134–141 (2003)
8. Ali, W., Malik, M.K., Hussain, S., et al.: Urdu noun phrase chunking: HMM based approach. In: Proceedings of International Conference on Educational and Information Technology, ICEIT, Chongqing, China, vol. 2, pp. V2-494–V2-497 (2010)
9. Yuan, W., Ling-yu, Z., Ya-xuan, Z., et al.: Combining support vector machines, border revised rules and transformation-based error-driven earning for Chinese chunking. In: Proceedings of Artificial Intelligence and Computational Intelligence, AICI, Sanya, China, vol. 1, pp. 383–387 (2010)
10. Gan, R., Shi, S., Wang, M., et al.: Chinese base noun phrase based on multi-class support vector machines and rules of post-processing. In: Proceedings of the 2nd International Workshop on Database Technology and Applications, DBTA, Wuhan, Hubei, China, pp. 1–4 (2010)
11. Zhai, F., Potdar, S., Xiang, B., et al.: Neural models for sequence chunking. In: Proceedings of the Thirty-First AAAI Conference on Artificial Intelligence, San Francisco, California, USA, pp. 3365–3371 (2017)
12. Zou, X.: Sequence labeling of chinese text based on bidirectional Gru-Cnn-Crf model. In: Proceedings of the 15th International Computer Conference on Wavelet Active Media Technology and Information Processing, ICCWAMTIP, Sichuan, pp. 31–34 (2018)
13. Thao, N.T.H., Thai, N.P., Le Minh, N., et al.: Vietnamese noun phrase chunking based on conditional random fields. In: Proceedings of the 1st International Conference on Knowledge and Systems Engineering, KSE, Hanoi, Vietnam, pp. 172–178 (2009)
14. Yanchao, L., Jianyi, G., Yantuan, X., et al.: A novel hybrid approach incorporating entity characteristics for vietnamese chunking. Int. J. Simul.-Syst. Sci. Technol. 17, 25–29 (2016)
15. Nguyen, L.M., Nguyen, H.T., Nguyen, P.T., et al.: An empirical study of Vietnamese noun phrase chunking with discriminative sequence models. In: Proceedings of the 7th Workshop on Asian Language Resources, ALR, Singapore, pp. 9–16 (2009)

Big Data Support and Architecture

A Distributed Big Data Discretization Algorithm Under Spark

Yeung Chan, Xia Jie Zhang, and Jing Hua Zhu$^{(\boxtimes)}$

Heilongjiang University, Harbin 150000, China
zhujinghua@hlju.edu.cn

Abstract. Data discretization is one of the important steps of data preprocessing in data mining, which can improve the data quality and thus improve the
accuracy and time performance of the subsequent learning process. In the era of
big data, the traditional discretization method is no longer applicable and distributed discretization algorithms need to be designed. Hellinger-entropy as an
important distance measurement method in information theory is context-
sensitive and feature-sensitive and thus are abundant of useful information.
Therefore, in this paper we implement a Hellinger-entropy based distributed
discretization algorithm under Apache Spark. We first measure the divergence
of discrete intervals using Hellinger-entropy. Then we select top-k boundary
points according to the information provided by the divergence value of discrete
intervals. Finally, we divide the continuous variable range into k discrete
intervals. We verficate the distributed discretization performance in the pre-
processing of random forest, Bayes and multilayer perceptron classification on
real sensor big data sets. Experimental results show that the time performance
and classification accuracy of the distributed discretization algorithm based on
Hellinger-entropy proposed in this paper are better than the existing algorithms.

Keywords: Big data mining · Discretization · Apache Spark · Hellinger
entropy

1 Introduction

Nowadays, with the wide application of Internet of things, data information perception
approach is diverse, but the density of valuable information is relatively low. Therefore,
in the era of big data, how to extract valuable information from massive data quickly
using learning and mining algorithms to purify the value of data is one of the key issues
of data mining and also a difficult problem to be solved.

Data preprocessing [1] is one of the important steps in the process of data mining
and analysis. Real data may contain a lot of noise, abnormal points, missing values and
repeated values, which is not conducive to the training of algorithm model. The result
of data preprocessing is to deal with various dirty data in a corresponding way, to
obtain standard and clean data for data statistics and data mining. Therefore, data
preprocessing is an essential stage for most data mining algorithms before modeling,
aiming to eliminate negative factors (missing, noise, inconsistency and redundant data)
existing in dataset. Data reduction simplifies data and its inherent complexity while

© Springer Nature Singapore Pte Ltd. 2019
H. Jin et al. (Eds.): BigData 2019, CCIS 1120, pp. 107–119, 2019.
https://doi.org/10.1007/978-981-15-1899-7_8

maintaining the original structure. Discretization as a method of data reduction has received extensive attention [2]. Discretization of numerical data is to set multiple segmentation points within the range of domain values of continuous attributes, divide the values in the range of domain values into discrete subdomains and use different symbols or integers to represent different subdomains of continuous attributes. Learning tasks preserve as much information as possible in the original continuous attributes, improve classification accuracy, and facilitate the underlying learning process.

The Hellinger divergence can be interpreted as a distance metric that measures the similarity between two probability distributions, where the distance corresponds to the divergence between the prior distribution and the posterior distribution. Its range of values is limited to the interval $[0, \sqrt{2}]$, and the distance is 0 if and only if the prior distribution and the posterior distribution are identical, so it is adapted to the case where other metrics are undefined. In other words, it applies to any probability distribution [10].

With the continuous development of the era of big data, distributed technologies and frameworks have become increasingly mature, bringing fast and reliable solutions for big data processing tasks. We have come up with a distributed version of the discretization method, which uses the Hellinger divergence proposed by Beran [3] to measure the amount of information contained in each interval, to select the boundary points. It solves the problems that traditional algorithms can't be applied to massive data or low computational efficiency, non-scalability, low classification accuracy, slow discrete speed, etc., which provides a new feasible solution for data preprocessing. The proposed scheme is applied to the real big data set for discretization, which is used for random forest, Bayesian and multi-layer perceptron classification.

This paper mainly contributes from two aspects. Firstly, it proves that the traditional discretization algorithm has greatly improved the time performance and discrete quality after optimization and reconstruction in the big data processing framework, especially in terms of time performance. The second is to improve the existing discretization algorithm that in some cases is inefficient and unreasonable. The experimental results show that the proposed distributed discretization algorithm based on Hellinger entropy leads the existing distributed discretization algorithm in discrete speed and discrete precision.

2   Related Work

Discretization technology can be divided into dynamic discretization and static discretization according to the implementation mode [4]. Dynamic discretization technology discretizes data in the process of constructing classification or prediction model, such as ID3 [5] and ITFP [6]. Such algorithms have high discretization quality, but high computational overhead and poor scalability. Static discretization technology is to discretize data before model construction. For example, AOA [7] is a discretization method adapted according to the distribution characteristics of datasets, which avoids the problem of unbalanced data distribution caused by fixed segmentation. However, its internal discretization adaptation methods are all traditional algorithms, and the performance is low when the data is very large.

Discretization technology can be divided into supervised and unsupervised according to the learning model. Unsupervised discretization algorithm abandons class information and the discretization quality is not as good as that of supervised method when there are more feature attributes. Such as Lee [8] proposed a numerical attribute discretization scheme based on Hellinger divergence, which uses the information measure in information theory to discretize continuous attributes. Wu et al. [9] proposed a discretization algorithm DAGMM based on Gaussian mixture model, which preserves the original mode of data by considering the multimodal distribution of continuous variables.

Traditional static and dynamic algorithms have high discretization accuracy, but their scalability is low, and the discretization time is slow and the discretization quality is low when used for large amounts of data. Therefore, in the era of big data, the above traditional methods cannot be directly used, and it is necessary to design a distributed data discretization algorithm. Sergio et al. [10] proposed a discrete DMDLP based on the entropy minimum description length of the Apache Spark platform, based on the principle of entropy minimum length proposed by Fayyad and Irani. Wang et al. [11] proposed a distributed attribute discretization based on the likelihood ratio hypothesis test for distributed processing using Hadoop framework, which overcomes the problem that the independence hypothesis test condition is unreliable. The above distributed discretization technology still has a heavy dependence on data distribution, resulting in low classification accuracy in some cases.

In this paper, we design a distributed discretization algorithm DHLG (Distributed Hellinger) based on Hellinger divergence under Apache Spark platform. it was proved that the discretization technology based on hellinger divergence has more advantages in the selection of boundary points in the reference [8]. The new version for the distributed environment is not a simple parallelization, but the original algorithm is deeply decomposed and reconstructed, providing an approximate, scalable, and flexible solution to handle the discreteness of big data. The problem is very challenging.

3 Prerequisite Knowledge

3.1 Discretization

Discretization is a data preprocessing technique that produces discontinuous intervals from continuous features. The resulting interval is associated with a set of discrete values, replacing the original data with nominal data. Given a data set D contains n instances, F features, C class labels. Where the feature $f \in F$, the class label $c \in C$, the discretization process divides the range of a continuous feature f into K_f disjoint subfields, as follows:

$$D_f = \left\{ [d_0, d_1], \ (d_1, d_2), \ \ldots, \left(d_{K_f-1}, d_{K_f}\right) \right\} \tag{1}$$

Where d_0 and d_{K_f} are the maximum and minimum values in the range of characteristic f, and all instances in data set D are sorted in ascending order. The segmentation point of feature f is defined as follows:

$$CP_f = \{d_1, d_2, \ldots, d_{K_f-1}\} \tag{2}$$

The CP is all the split points of all consecutive features in data set D. In general, in the traditional discretization method, unless the user is an expert in the problem domain, it is rarely possible to know the correct or optimal discretization partition. Although it is not possible to have an optimal discretization baseline, in order to design and evaluate an efficient discretization algorithm, some concepts of quality assessment are needed. Discretization aims to simplify and reduce the continuous value data in the dataset and find a concise data representation as a category. In supervised learning, this form can be expressed as dividing continuous values into multiple intervals, essentially the induction of the relative frequencies of the class labels within the interval. So, the definition of a high-quality discretization can be summarized as: maximization of the internal uniformity of the interval and minimization of the uniformity between the intervals.

3.2   Hellinger Entropy

The method proposed in this paper uses the Hellinger entropy that applicable to any probability distribution proposed by Beran [8] to realize high quality discretization, the difference between the class frequency of the target feature and the class frequency of the given interval is defined as the amount of information provided by the interval to the target feature. The more different the frequencies of the two classes, the more information is carried by the defined intervals, and the more information that can be provided to the target features. Therefore, it is important to measure the divergence between two class frequencies by the entropy function.

Definition 1: Interval Entropy.

$$E(I) = \left| \sum_i \left(\sqrt{P(x_i)} - \sqrt{P(x_i|I)} \right)^2 \right|^2 \tag{3}$$

Formula 3 defines the calculation formula of interval entropy. Hellinger-entropy measures the similarity between two probability distributions. $P(x_i|I)$ is the probability distribution of **label** of data in interval I in the current interval, and $P(x_i)$ is the probability distribution of **label** in total data set. The information contribution of the current interval to the total data set can be calculated by the formula.

Definition 2: Potential Tangential Entropy.

$$E(C) = E(a) - E(b) \tag{4}$$

The tangent point is associated with the corresponding two intervals. $E(a)$ and $E(b)$ are the entropies of the associated interval, respectively. Therefore, the entropy of the tangent point is defined as the entropy difference of the associated interval, and the importance of the tangent point is measured by this value, to choose the boundary point.

3.3   Parallel Computing Framework

The Apache Spark [12] platform has an outstanding performance in big data processing. It is designed based on the idea of MapReduce and uses the most advanced DAG scheduler, query optimizer and physical execution engine to achieve high performance of batch and stream data. Spark persists data in memory for high-speed read, write and operation, and is suitable for iterative and online jobs, so its computing performance is better than Hadoop. Meanwhile, Spark provides more than 80 advanced operations, enabling users to implement many distributed programs that are much more complex than MapReduce. Therefore, this paper uses Spark platform for algorithm design.

4   Distributed Discretization Algorithm (DHLG)

In this section, we mainly introduce the core ideas of this algorithm, which requires readers to understand the basic Spark primitive operation and Scala language, and give the pseudo-code of Spark implementation of the core algorithm.

4.1   Algorithm Framework

Figure 1 shows the main framework of DHLG, a decentralized algorithm proposed in this paper. Firstly, the source data stored in HDFS is preprocessed and cached in memory, to reduce the delay of data IO and network transmission and improve the efficiency of algorithm execution. Secondly, the pre-processed data is divided into two types of data according to the sample size of features: large features and small features. The boundary points of all large and small features are calculated iteratively. Sample data of each *bigFeature* will be divided into $n(n = dataSize/partitions)$ partitions. In these n partitions, n local solutions will be obtained in parallel according to the calculation criteria of small features, and the boundary points of *bigFeature* will be obtained after merging and selecting the best. The solution of small features is its final solution. Finally, all the boundary points of large and small features are combined to complete the discretization process.

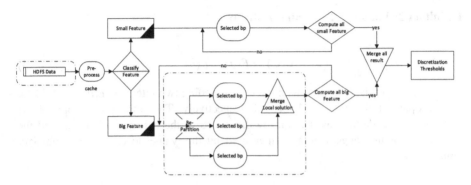

Fig. 1. Frame diagram of distributed discrete DHLG

4.2   Data Pre-processing

This article's experimental data is stored in HDFS, and to make it faster for task nodes to access it, we use DataFrame in Spark SQL to encapsulate the data, which allows us to manipulate it like a database table, with much higher query efficiency. The data preprocessing process is performed independently on each partition. In each task thread, the input data is converted into a triple form $<(f, th), label>$, where f represents the feature, th is the value of instance X under feature f, and label is instance X in feature f, the class label corresponding to the value. After the format conversion is completed, the dataset is subjected to sparse processing and deduplication. Sparse processing (to fill in the value of 0 when the instance X has no value in feature f) is performed if the input data set is sparse, otherwise the process is skipped. Deduplication is to accumulate the label value of the same th value under the same feature f, where **label** is an n-dimensional vector value and n is the number of categories of **label** labels. Finally, the datasets are sorted in ascending order by feature.

4.3   Selected Threshold Points

Algorithm 1: FindAllThresholds. This algorithm is the process encapsulation of computing feature boundary points. Step 1 counts the number of instances each feature contains by building in the countByKey function on spark (counting the number of occurrences of the value pair where the key is specified). Step 2 classifies features into two categories through a filtering function: small features and big features. Small features contain fewer instances than the maximum number of instances mp in a single partition, otherwise they are classified as big features. Step 3 broadcasts the indexes of the large features to the cluster so that some sizing characteristics can be determined in subsequent tasks. In step 3–4, large and small features to perform different discretization scheme, small features called algorithm 2 (FindSmallThresholds) computing the discrete boundary point, while the big features called algorithm 3 (FindBigThresholds) calculating the discrete boundary point, all the big and small features in their respective calculation scheme iterative calculate boundary point of each feature. Finally, step 7 combines the boundary points obtained by all features to obtain the final discrete result.

Algorithm 1: FindAllThresholds

Input: initialCands Initial candidate point
  mb Maximum number of feature intervals
  mp Maximum number of instances in a single partition
Output: All the boundary points
1: CountRDD ←countByKey(initialCands)
2: bigIndexes←filter(countRDD)
3: bBigIndexes ←broadcast(bigIndexes)
4: sThresholds← findSmallThresholds(mb, initialCands,bBigIndexes)
5: bThresholds←findBigThresholds(mb, mp, initialCands,bBigIndexes)
6: allThresholds ← union(sThresholds, bThresholds)
7: return allThresholds

Algorithm 2: FindSmallThresholds

Input: candidates Initial candidate point
  bigIndexes Index subscripts of large features
  mb Maximum number of feature intervals
Output: Small feature boundary points
1: fbCandidates ← filter(bigIndexes, candidates) #get small features
2: gCandidates ←groupByKey(fCandidates)
3: sThresholds ←
 mapValues ⟨fea, candidates⟩ ∈ gCandidates
  labelProbablity ← getProbablity(candidates)
  cutPoints ←head(candidates)∷ candidates
  for i ← 1 until length(cutPoints)−1 do
   cutPoints(i)←generateCutPoint((candidates(i−1),candidates(i))
  end for
  numIntvls ← length(cutPoints)−1
  ivtalEntropy ←
  for i ← 0 until numIntvls) do
   yield getIntervalEntropy(i, cutPoints, labelProbablity)
  end for
  cutEntropy←
  for cp ← 1 until numIntvls) do
   yield abs(ivtalEntropy(cp − 1) − ivtalEntropy(cp))
  end for
  EMIT takeMaxTop(cutEntropy, mb)
 end mapValues
4: return sThresholds

Algorithm 2: FindSmallThresholds. This algorithm is the calculation process of boundary points of small features. In the scheme proposed by Lee [8], the entropy of potential tangent points of each feature is calculated, and the points with the lowest entropy are merged continuously until the maximum interval number is satisfied. In the scheme of KEEL4 [13] library, the maximum entropy MB (maximum number of boundary points) potential tangency points are selected as the boundary points of the feature. After experimental comparison, the results of keel and keel are not different, and they have their own advantages when dealing with data of different structures. The former bottom-up iterative process is too complex to process large amounts of data. Therefore, keel is adopted in this paper.

First, steps 1–2 obtain indexes of big features from cluster broadcast, filter out all small feature data from RDD, and use spark RDD's primitive to operate groupByKey to group data by feature subscripts. in step 3, small feature's boundary point calculated by the following the calculation process is as follows: Firstly, the probability of the distribution of all data points under this feature is calculated. Secondly, the potential pointcut (each data is an independent candidate pointcut, and the midpoint of each adjacent two points is called potential pointcut) is generated, which relates two adjacent intervals (or potential pointcut). Secondly, the entropy of the potential tangent point can be obtained by Algorithm 4 (GetIntervalEntropy). The entropy difference of the correlation interval is the entropy of the potential tangent point. The greater the difference of the relative class distribution of the potential tangent point, the greater the entropy value, the more information the potential tangent point carries. Therefore, the Top K potential tangent points with maximum entropy are selected from all potential tangent points as the segmentation boundary points of this feature.

Algorithm 3: FindBigThresholds. This algorithm describes the characteristics of the boundary point of the selection process, steps 1 and 2 from the same cluster take out all the characteristics of data grouping and according to the characteristics of the index. Step 3 according to the sample upper limit in a single partition specified by the user, we customize a new partition to ensure the balanced distribution of data in each partition. In a distributed environment, the cost of network transmission of data in each node is very high. Controlling the distribution of data to obtain the minimum network transmission and IO read-write can greatly improve the overall performance and reduce the network overhead. Therefore, we use a custom partitioner to evenly distribute the data on each node. Step 4–6 traverse each big feature, partition and sort the data, through the Spark built-in function MapPartitions in each partition, finding a local solution incorporating all the local solution of the characteristics of a merit, select the optimal Top - K potential tangent point as its ultimate boundary point. Repeat the process until all the major features of the solution are obtained

```
Algorithm 3: FindBigThresholds
Input: candidates  Initial candidate point
    bigIndexes  Index subscripts of large features
    mb  Maximum number of feature intervals
Output:   Big feature boundary points
1:   fbCandidates ← filter(bigIndexes, candidates)    #get big features
2:   gCandidates ←groupByKey(fCandidates)
3:   partitioner ← createPartitioner(mp)
4:   for each BigFeature in gCandidates
5:     repartCandidates←partitionByWithSort(bigFeature, partitioner)
     bThresholds ←
     mapPartitions partition ∈ repartCandidates
     thresholds ← findSmallThresholds(partition)
     EMIT thresholds
     end mapPartitions
     end for
6:   bestThresholds ←getBestThresholds(bThresholds)
7:   return bestThresholds
```

Algorithm 4: GetIntervalEntropy. This algorithm describes the interval entropy calculation process between two potential tangent points. Step 1 find the upper and lower boundary points of the index interval in the boundary points according to the given interval index. Step 2 filters all instances from the ordered data set in the interval obtained in step 1. Then step 3 calculates the probability distribution of each label of the data in this boundary interval. Steps 4 and 5 substitute the probability distribution of each label in the range and the probability distribution of each label in the global data into the entropy calculation formula given in definition 1 to get the entropy in the range and return it as the result.

```
Algorithm 4: GetIntervalEntropy
Input: bounds   Boundary points
    initialDatas   Initial distribution of data points
    ivtalIndex   Interval index subscript
    labelFreqs   The frequency of each label in the feature
Output: The Entropy of specified interval
1: (botB, topB) ← calBounds(bounds , ivtalIndex)
2: inItval←
    filter <th, label> ∈ initialDatas
    if(botB <= th <= topB) EMIT <th, label>
    end filter
3: labelInItvalFreqs← getProbablity (inItval)
4: intvlEnt ← 0
    for labelIndex ← 0 until nLabels do
    labelProbablity= labelFreqs(labelIndex)
    labelInItvalProbablity= labelInItvalFreqs (labelIndex)
    intvlEnt ← intvlEnt + pow(sqrt(labelProbablity) −
sqrt(labelInItvalProbablity(labelIndex)),2.0)
    end for
5: return sqrt(abs(intvlEnt))
```

5 Experiment and Analysis

Discretization belongs to the pre-processing stage of data mining, which cannot measure the quality of discretization. Therefore, we use distributed classification algorithms in spark ml library (such as MLPC, naive bayesian and random deep forest) to evaluate the method proposed in this paper and classify data. Experimental results show that the time performance and classification accuracy of the distributed discretization algorithm based on hellinger-entropy proposed in this paper are better than the existing algorithms.

In order to reduce the error caused by random sampling data, each set of experiments contains 5 independent experiments and the average of the experimental results is taken as the final experimental result. The experiment uses the Spark 2.3.0, which is completed by a cluster of two computing nodes and one master node and two store nodes. The compute node is configured as Intel I core i7-4770, 3.4 GHz quad core, 16 GB RAM. The configuration of the master node is Intel core i5-7300HQ, 2.5 GHz, quad core, 8 GB RAM. The store nodes are configured as Intel core i5-7300HQ, 2.5 GHz, quad core, 4 GB RAM, 128 GB SSD. Experimental parameters are shown in Table 1.

Table 1. Parameters settings

Method	Parameter
DHLG	MaxBins = 12, sr = (0.7,0.3) maxCans = 20000,
DMDLP	MaxBins = 12, sr = (0.7,0.3) maxCans = 20000,
NB	Lambda = 1.0
RF	Impurity = gini, maxDeepth = 10
MLPC	HideLayers = 50, stepSize = 0.05

In the experiment, the heterogeneous human activity identification dataset [14] in the UCI machine learning repository was used, which is referred to as "HHAR", which is collected by the mobile device's accelerometer and gyroscope under random activity of the user. The experiment samples 30% of the sample size of the original data set as the experimental data set, and some brief information about the data set is given in Table 2. The data captured by the sensor has negative data and cannot be directly used in the Bayesian algorithm. The classification model is concerned with the change of the relative position of the target, not the actual value of the data. Therefore, we have performed data on the x, y, and z axes. Relative translation operation, it has no effect on the final prediction results. In addition, in order to reduce the number of initial candidate boundary points, all sample data is rounded to four decimal places.

Table 2. Data set properties

#Total instances	#Train	#Test	#Con	#Fea	#Class
14,548,945	10,186,262	4,364,683	5	10	6

The classification time consumption of the data using the distributed discretization algorithms DHLG and DMDLP and the unused discretization version NO_DIS (no discretion) in a random forest, Bayesian, multi-layer perceptron, etc. is given in Table 3. Discretized data using a discretizer has fewer data points, so training the classification takes less time, so the NO_DIS version has the longest classification time. Compared with the distributed discretization algorithm DMDLP, they are controlled by the maximum boundary point parameters, so the difference in classification is small. It can be seen from the table that DMDLP is slightly ahead of the Bayesian classification, and in other cases, our algorithm DHLG performance is leading.

Table 3. Classification time (seconds)

	DHLG	DMDLP	NO_DIS
MLPC	**846.62**	858.33	934.32
NB	104.83	**104.25**	106.21
RF	**34.48**	36.30	40.53

Figure 2 shows the discrete speed comparison of the distributed discretization algorithm DHLG (distributed Hellinger) proposed in this paper and the DMDLP (Distribute Minimum Description Length Principle). The horizontal axis represents the number of five sets of experiments, and the vertical axis represents the discrete velocity. After the discretization of the data set by the discretization algorithm, the data is divided into two parts in a ratio of 7:3: the training set and the test set, and the data of the test set is used for the classification model training, and the test set is used to test the discretization The classification accuracy of the data. Obviously, from the data of 5 sets of experiments, the discrete speed of the DHLG discretization algorithm is obvious.

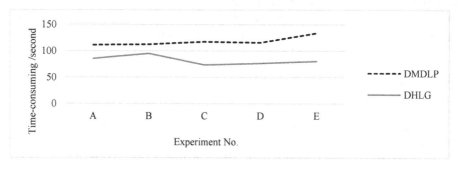

Fig. 2. Velocity comparison of discretion

Figure 3 shows the accuracy values when using the distributed discretization algorithms DHLG and DMDLP and the data without the discretization version NO_DIS in a classifier such as random forest, Bayesian, multi-layer perceptron, etc. From these results, we can see that the discrete DHLG proposed in this paper is superior to the distributed discretization algorithm DMDLP and the non-discrete version, indicating that our scheme is effective and the performance is improved.

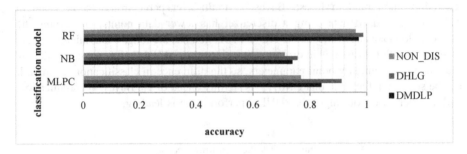

Fig. 3. Comparison of classification accuracy

6 Conclusion

This paper proposes a distributed discretization algorithm DHLG based on Apache Spark design, which optimizes the selection problem of boundary points by Hellinger-entropy. In this distributed version, relatively better interval boundary points are selected by measuring the amount of information for each segment with respect to the target feature. The algorithm follows the spark ml pipeline idea and seamlessly combines with the components in the spark ml library. It can maximize the efficiency and scalability of the algorithm when using the spark platform for big data mining and analysis. A new solution for the big data discretization technology that is seriously lacking in the research of big data is proposed. The experimental results show that the proposed scheme is superior to other distributed versions in terms of discrete precision and discrete speed compared with the existing distributed discretization algorithm.

In the future work, we will consider clustering the initial potential cut points according to the real distribution, and measure the amount of information in each class to select more boundary points. At the same time, we also consider whether we can apply this solution to the streaming big data environment.

References

1. García, S., Luengo, J., Herrera, F.: Tutorial on practical tips of the most influential data preprocessing algorithms in data mining. Knowl.-Based Syst. **98**, 1–29 (2016)
2. Ramírez-Gallego, S., García, S., Mouriño-Talín, H., et al.: Data discretization: taxonomy and big data challenge. Wiley Interdisc. Rev.: Data Min. Knowl. Discovery **6**(1), 5–21 (2016)

3. Beran, R.: Minimum hellinger distance estimates for parametric models. Ann. Stat. **5**(3), 445–463 (1977)
4. Ramírez-Gallego, S., et al.: Data discretization: taxonomy and big data challenge. Wiley Interdiscip. Rev.: Data Min. Knowl. Discov. **6**(1), 5–21 (2016)
5. Salzberg, S.L.: C4.5: programs for machine learning by J. Ross Quinlan. Morgan Kaufmann Publishers, Inc. 1993. Mach. Learn. **16**(3), 235–240 (1994)
6. Au, W.H., Chan, K.C., Wong, A.K.C.: A fuzzy approach to partitioning continuous attributes for classification. IEEE Educational Activities Department (2006)
7. Liu, Y.: Parallel discrete data preparation optimization in data mining. J. Sichuan Univ. (Nat. Sci. Ed.) **55**(05), 103–109 (2018)
8. Lee, C.H.: A Hellinger-based discretization method for numeric attributes in classification learning. Knowl.-Based Syst. **20**(4), 419–425 (2007)
9. Wu, C., Guo, S., Li, C.: Research on discretization algorithm based on gaussian mixture model. Small Microcomput. Syst. (4), 21 (2018)
10. Ramírez-Gallego, S., García, S., Mouriño-Talín, H., et al.: Distributed entropy minimization discretizer for big data analysis under apache spark. In: 2015 IEEE Trustcom/BigDataSE/ISPA, vol. 2, pp. 33–40. IEEE (2015)
11. Wang, L.: Power big data attribute discretization method based on cloud computing technology. Digit. Technol. Appl. (1), 56–58 (2015)
12. Zaharia, M., Xin, R.S., Wendell, P., et al.: Apache spark: a unified engine for big data processing. Commun. ACM **59**(11), 56–65 (2016)
13. Alcalá-Fdez, J., Fernández, A., Luengo, J., et al.: Keel data-mining software tool: data set repository, integration of algorithms and experimental analysis framework. J. Multiple-Valued Logic Soft Comput. **17**, 2–3 (2011)
14. UCI Machine Learning Repository: Heterogeneity Activity Recognition data. http://archive. ics.uci.edu/ml/datasets/Heterogeneity+Activity+Recognition

A Novel Distributed Duration-Aware LSTM for Large Scale Sequential Data Analysis

Dejiao Niu[✉][iD], Yawen Liu, Tao Cai[✉][iD], Xia Zheng,
Tianquan Liu, and Shijie Zhou

The School of Computer Science and Communication Engineering of Jiangsu
University, Zhenjiang 212013, Jiangsu, China
{djniu, caitao}@ujs.edu.cn

Abstract. Long short-term memory (LSTM) is an important model for sequential data processing. However, large amounts of matrix computations in LSTM unit seriously aggravate the training when model grows larger and deeper as well as more data become available. In this work, we propose an efficient distributed duration-aware LSTM(D-LSTM) for large scale sequential data analysis. We improve LSTM's training performance from two aspects. First, the duration of sequence item is explored in order to design a computationally efficient cell, called duration-aware LSTM(D-LSTM) unit. With an additional mask gate, the D-LSTM cell is able to perceive the duration of sequence item and adopt an adaptive memory update accordingly. Secondly, on the basis of D-LSTM unit, a novel distributed training algorithm is proposed, where D-LSTM network is divided logically and multiple distributed neurons are introduced to perform the easier and concurrent linear calculations in parallel. Different from the physical division in model parallelism, the logical split based on hidden neurons can greatly reduce the communication overhead which is a major bottleneck in distributed training. We evaluate the effectiveness of the proposed method on two video datasets. The experimental results shown our distributed D-LSTM greatly reduces the training time and can improve the training efficiency for large scale sequence analysis.

Keywords: Recurrent neural networks · Long short-term memory · Duration-aware LSTM · Sequential data analysis · Distributed neuron

1 Introduction

In recent years, deep learning method has grained remarkable successes. LSTM is a variant of recurrent neural networks and has revealed powerful capability for modeling sequential data [1–3]. However, the training of LSTM network is very challenging. The reason may come from two aspects. First, the computations in LSTM cell are very complicated and time consuming. Although LSTM perfectly addressed the vanishing gradient problem in recurrent neural networks, the involvement of gate mechanism substantially intricates more computations instead. Secondly, from the perspective of entire network, the training of LSTM is extremely difficult due to large amounts of parameters and increasingly deeper and larger model size. Large-scaled LSTM model usually leads to a preferable performance while makes it even harder to train and poorer to scale. Moreover, LSTM has difficulty in parallelizing its memory computations due

© Springer Nature Singapore Pte Ltd. 2019
H. Jin et al. (Eds.): BigData 2019, CCIS 1120, pp. 120–134, 2019.
https://doi.org/10.1007/978-981-15-1899-7_9

to intrinsic temporal dependencies among training steps. The computation in each step depends on the completion of previous step. Therefore, recurrent computations are less amenable to parallelization.

Generally, recurrent networks deal with temporal sequences one item at a time. At each step, the model read in an item and calculate the current memory using history information and current input. This unified memory update is carried out on each step all the way. However, if a sequential item does not change for some time, the repeated calculations on these time steps are unnecessary and of big wastes because no new information is flowing into LSTM. So, the standard and unified memory update may hold back the training on this occasion. Intuitively, a more efficient memory update and an optimized cell unit can be realized by taking the characteristics of the sequential data into account.

The training of large-scaled neural network with huge amounts of data are often carried out on distributed systems [4–6], where multiple computing nodes are allowed to work in parallel and thus speed up the learning. Currently, the most popular implementations of distributed DNNs are model parallelism and data parallelism [7]. Although they can improve the training efficiency, a high communication overhead remains a bottleneck and greatly impede the scalability of distributed DNNs. The reason for huge communication cost is that the models are typically split physically in current model parallelism, which destroys the logical structure of computations involved in the training.

Compared with artificial neural network, biological neuron system has billions of cells but still exhibit high efficiency [8]. Different from the artificial neuron, each biological cell acts as an autonomous agent and is able to work independently and parallelly with billions of others at the same time. Highly-performanced biological neuron system enlightens us to develop an efficient neural network with the idea of autonomous neurons which collaborating and working in parallel with one another.

In this work, we focus on an efficient LSTM model for large scale sequential data analysis. Towards this end, we first introduce a novel duration-aware LSTM(D-LSTM) unit. The duration of sequence items is leveraged to adaptively calculate LSTM memory and simplify redundant calculations when the input remains unchanged. Next, we propose a distributed training strategy for large-scaled D-LSTM model. Inspired by the biological neural system, multiple distributed neurons are introduced to imitate the activities of hidden neurons in D-LSTM. Each distributed neuron acts as an autonomous agent and can independently and parallelly perform respective operations, which avoids massive parameter transmission to the most extent. This neuron-centered strategy decomposes the network from a logicality perspective. Massive parameter calculations are thus reorganized and degraded to linear operations which are easier to implement in parallel.

The main contributions of the work can be concluded as follows:

(1) Inspired by the parallelism of biological neurons, a novel distributed D-LSTM is proposed to improve the training efficiency of D-LSTM.
(2) The distributed neuron and the interaction neuron are introduced for parameter-paralleled training of D-LSTM, where the computations are reorganized and assigned to multiple autonomous distributed neurons. This neuron-based decomposition greatly reduces the communication cost in distributed training of D-LSTM.

(3) The prototype of distributed D-LSTM is implemented on Spark and extensive experimental results on Charades and COIN datasets show the proposed method can improve the training efficiency and reduce time overhead for large-scale D-LSTM training.

2 Related Work

One challenge for deep neural networks (DNNs) is that they are difficult to train. Researchers have explored many techniques to accelerate training of DNNs, such as model compression, sparse and quantization, distributed implementations, etc. Due to the scope of the paper, we will mainly introduce some efforts on acceleration of RNNs.

In contrast to operations such as convolution, recurrent computations are less amenable to parallelization because the computation in each step depends on completing the previous step. In order to expose more parallelism in RNNs, Lei et al. [9] proposed the simple recurrent unit (SRU) architecture, which simplifies the computation and can be easily parallelized. SRU is as fast as a convolutional layer and 5–10× faster than an optimized LSTM. Khomenko et al. [10] proposed an RNN training acceleration algorithm based on sequence and multi-GPU data parallelization. The algorithm improved the training speed by grouping the sequences into buckets by length. Huang et al. [11] proposed a two-stage RNN and parallel training to accelerate RNN training. Two-stage RNN can be accelerated to 2 times and parallel RNN training can accelerate RNN model by 10 times through data parallelization. Ji et al. [12] proposed an approximate algorithm, BlackOut, which can effectively train a large-scale recurrent neural network model with a million words vocabulary. Keuper et al. [13] analyzed the factors affecting the efficiency of distributed large-scale neural network training, and improved the efficiency of neural network training by limiting the number of nodes participating in the training. Gholami et al. [14] designed asynchronous gossiping SGD, but the experimental results showed that asynchronous SGD was suitable for training neural networks with small clusters of computers.

Existing works on RNN acceleration are typically from the model view and take various optimization techniques to simplify the recurrent unit or make the network easier to parallel. However, we try to address the issue from the data view, that is, to leverage the properties of sequential data for a more efficient LSTM unit.

The training of large-scale models with huge amounts of data are often carried on distributed systems, where data parallelism and model parallelism are most popular styles. Dean et al. [4] developed a software framework called DistBelief that can utilize computing clusters with thousands of machines to train large models. Li et al. [7] proposed a parameter server for distributed machine learning. Current model parallelism divides the model mechanically and distributed model slices on multiple computing nodes. However, the physical division cannot take the logical dependencies of operations into account, which leads to a higher communication cost between the server and the workers. In recurrent neural network, there are recurrent connections between hidden layers. All training parameters are related to the neurons in the hidden layer. So, we aim to study a novel distributed system where the model is decomposed based on the neurons in hidden layer.

3 Distributed Duration-Aware LSTM for Large-Scale Sequential Data Analysis

3.1 Duration-Aware LSTM

In our prior work [15], we proposed an adaptive LSTM for durative sequential data. We adopt the same idea in this work and implement it on Spark. To avoid confusion, we name it duration-aware LSTM(D-LSTM). Here we show the architecture of D-LSTM in Fig. 1 and give the main procedures of the model.

Given an input x_t at time t, D-LSTM computes cell state and hidden output using the following equations.

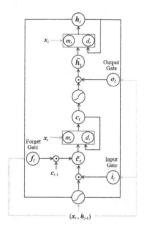

Fig. 1. Duration-aware LSTM(D-LSTM) unit.

$$i_t = \sigma(W_{xi}x_t + W_{hi}h_{t-1} + b_i) \tag{1}$$

$$f_t = \sigma(W_{xf}x_t + W_{hf}h_{t-1} + b_f) \tag{2}$$

$$o_t = \sigma(W_{xo}x_t + W_{ho}h_{t-1} + b_o) \tag{3}$$

$$c\_in_t = tanh(W_{xc}x_t + W_{hc}h_{t-1} + b_{c_{in}}) \tag{4}$$

$$c_t = f_t \odot c_{t-1} + i_t \odot c\_in_t \tag{5}$$

$$h_t = o_t \odot tanh(c_t) \tag{6}$$

$$\tilde{c}_t = f_t \odot c_{t-1} + i_t \odot c\_in_t \tag{7}$$

$$c_t = m_t \odot \tilde{c}_t + (1 - m_t) \odot d_t c_{t-d_t+1} \tag{8}$$

$$\tilde{h}_t = o_t \odot tanh(c_t) \tag{9}$$

$$h_t = m_t \odot \tilde{h}_t + (1 - m_t) \odot d_t h_{t-d_t+1} \tag{10}$$

Where f_t, i_t, o_t are forget gate, input gate and output gate, respectively. c_t and h_t are cell state and hidden state as in standard LSTM. In D-LSTM, m_t is a mask gate determining whether the unit should update the cell memory as standard LSTM or maintain the memory of previous time step. The open and close of m_t is controlled by the input x_t. Consequently, the memory update of LSTM unit is not carried out in a unified manner, but is more flexible to exploit the duration of sequential data. If the input remains constant for some time, D-LSTM will not repeatedly calculate the gate values using matrix operations but update its memory in an optimized way, which reduces the training time as well as complexity.

3.2   Distributed Duration-Aware LSTM

D-LSTM is able to provide more efficient memory update and avoid abundant calculations when the sequential input remains unchanged. Based on it, we further propose a distributed training algorithm for D-LSTM in order to improve the training efficiency of large-scale D-LSTM network.

The distributed D-LSTM consists of an interaction node and multiple distributed neuron nodes. The distributed neuron node (also called autonomous neuron) simulates the activities of hidden neurons in D-LSTM. Multiple distributed neurons are able to work parallelly and jointly complete the matrix operations in hidden layer. The interaction node is used to aggregate the training parameters and accomplish the computations which are unsuitable to be performed on the distributed neurons. Figure 2 shows the architecture of distributed D-LSTM. In order to accelerate the parameter transmission between the distributed neuron and the interaction node, high speed Infiniband and RDMA protocol are used.

Different from the traditional distributed implementations, our distributed D-LSTM adopts a neuron-centered strategy and the network is split by hidden neurons. Because we find the most complicated and time-consuming computations in D-LSTM are all associated with the hidden neurons. If decomposing and paralleling by neurons, the operation independence will be maintained to the most extent, and thus greatly reduce the communication cost and improve scalability. Moreover, from the neuron's perspective, matrix calculations are degraded into multiple concurrent linear operations, which speeds up the training and benefits the raising of efficiency.

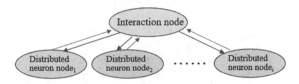

Fig. 2. The architecture of the distributed D-LSTM.

Next, we propose the autonomous training strategy for the distributed neuron and the coordination strategy for the interaction node, respectively.

Autonomous Training Strategy for the Distributed Neuron. The distributed neuron mainly includes the following phases:

Receiving Parameters from the Interaction Node. At initialization, each distributed neuron receives its respective parameters from the interaction node, including $w_{ix,t}^j$, $w_{ih,t}^j$, $w_{fx,t}^j$, $w_{fh,t}^j$, $w_{ox,t}^j$, $w_{oh,t}^j$, $w_{cx,t}^j$, $w_{ch,t}^j$, h_{t-1}^j, c_{t-1}^j, m_t, and d_t.

Calculating the Neuron Activation. Each individual neuron calculates its input gate i_t^j, output gate o_t^j, forget gate f_t^j, cell state c_t^j, and hidden state h_t^j.

$$i_t^j = \sigma\left(\sum\nolimits_{k=1}^{M} w_{ix,t}^{k,j} x_t^k + \sum\nolimits_{k=1}^{H} w_{ih,t}^{k,j} h_{t-1}^j\right) \tag{11}$$

$$f_t^j = \sigma\left(\sum\nolimits_{k=1}^{M} w_{fx,t}^{k,j} x_t^k + \sum\nolimits_{k=1}^{H} w_{fh,t}^{k,j} h_{t-1}^j\right) \tag{12}$$

$$o_t^j = \sigma\left(\sum\nolimits_{k=1}^{M} w_{ox,t}^{k,j} x_t^k + \sum\nolimits_{k=1}^{H} w_{oh,t}^{k,j} h_{t-1}^j\right) \tag{13}$$

$$\tilde{c}_t^j = \tan h\left(\sum\nolimits_{k=1}^{M} w_{cx,t}^{k,j} x_t^k + \sum\nolimits_{k=1}^{H} w_{ch,t}^{k,j} h_{t-1}^j\right) \tag{14}$$

$$\hat{c}_t^j = f_t^j \cdot c_{t-1}^j + i_t^j \cdot \tilde{c}_t^j \tag{15}$$

$$c_t^j = m_t \cdot \hat{c}_t^j + (1 - m_t) \cdot d_t \cdot c_{t-d_t+1}^j \tag{16}$$

$$\hat{h}_t^j = o_t^j \cdot \tan h(c_t^j) \tag{17}$$

$$h_t^j = m_t \cdot \hat{h}_t^j + (1 - m_t) \cdot d_t \cdot h_{t-d_t+1}^j \tag{18}$$

$$y_t^{(j)} = h_t^j V^{(j)} \tag{19}$$

where $V \in \Re^{H \times N}$ is the weight connecting the hidden layer and the output layer, N is the number of neurons in output layer, $V^{(j)}$ is the jth row of V. $y_t^{(j)}$ is the candidate output calculated by the distributed neuron j. The superscript k, j means the connection from the kth input neuron to the jth hidden neuron.

Calculating the Hidden Error. After the forward pass. The calculation of output error requires all distributed neurons' activations, thus carried out on the interaction node and then sent back to the distributed neurons. Then, the distributed neuron will calculate the hidden error.

$$v_t^j = v_{t-1}^j - \alpha h_t^j \delta y_t \tag{20}$$

$$\delta \hat{h}_t^j = \sum_{i=1}^{N} v_t^{ij} \delta y_t^j \tag{21}$$

$$\delta h_t^j = m_t \cdot \delta \hat{h}_t^j + (1 - m_t) \cdot d_t \cdot \delta h_{t-d_t+1}^j \tag{22}$$

$$\delta_{o,t}^j = \delta h_t^j \cdot [m_t \cdot \tanh(c_t^j) + (1 - m_t) \cdot d_t \cdot \tanh(c_{t-d_t+1}^j)] \cdot o_t^j \cdot (1 - o_t^j) \tag{23}$$

$$\delta_{f,t}^j = \delta h_t^j \cdot [m_t \cdot o_t^j \cdot (1 - \tanh(c_t^j)^2) + (1 - m_t) \cdot d_t \cdot o_{t-d_t+1}^j \cdot (1 - \tanh(c_{t-d_t+1}^j)^2)] \cdot [m_t \cdot c_{t-1}^j +$$
$$(1 - m_t \cdot d_t \cdot c_{t-d_t}^j)] \cdot f_t^j \cdot (1 - f_t^j) \tag{24}$$

$$\delta_{i,t}^j = \delta h_t^j \cdot [m_t \cdot o_t^j \cdot (1 - \tanh(c_t^j)^2) + (1 - m_t) \cdot d_t \cdot o_{t-d_t+1}^j \cdot (1 - \tanh(c_{t-d_t+1}^j)^2)] \cdot [m_t \cdot \tilde{c}_t^j +$$
$$(1 - m_t) \cdot d_t \cdot \tilde{c}_{t-d_t+1}^j] \cdot i_t^j \cdot (1 - i_t^j) \tag{25}$$

$$\delta_{\tilde{c},t}^j = \delta h_t^j \cdot [m_t \cdot o_t \cdot (1 - \tanh(c_t^j)^2) + (1 - m_t) \cdot d_t \cdot o_{t-d_t+1}^j \cdot (1 - \tanh(c_{t-d_t+1}^j)^2)] \cdot [m_t \cdot i_t^j +$$
$$(1 - m_t) \cdot d_t \cdot i_{t-d_{t+1}}^j] \cdot (1 - \tilde{c}_{t-d_t+1}^j)^2) \tag{26}$$

Updating Weight Parameters. Once the hidden errors are all available, the distributed neuron can compute the gradients and update the weight using BPTT (Back Propagation Though Time).

Coordination Strategy for the Interaction Node

Besides the autonomous calculations on the distributed neurons, there are still some computations which need the parameters from all distributed neurons and thus are carried out using an interaction node. We design a coordination strategy for the interaction node to complete initialization, output update and parameter concatenation.

In initialization, the interaction node calculates m_t and d_t according to the input x_t, and then release the parameters to the distributed neurons.

In output update phase, the interaction node will first compute the output activation of D-LSTM:

$$\tilde{y}_t = f(\sum_{j=1}^{H} y_t^{(j)}) \tag{27}$$

where f is the softmax function. Then calculate the error of output layer:

$$E = \sum_{t=1}^{T} -y_t \log \tilde{y}_t \tag{28}$$

$$\delta \mathbf{y}_t' = \widetilde{\mathbf{y}}_t - \mathbf{y}_t \tag{29}$$

where T is the length of sequence, $\mathbf{y}_t$ and $\widetilde{\mathbf{y}}_t$ are the true label and predicted label, respectively.

In concatenation phase, the interaction node collects all parameters from the distributed neurons, and generate new parameters for later steps.

The whole calculation procedure is shown in Fig. 3. Most complicated matrix operations are decomposed into linear operations and employed on multiple distributed neuron nodes. Each distributed neuron completes its individual learning task, thus greatly reduces the communication overhead and improves the training efficiency.

4 Experiments

In this section, we will evaluate the effectiveness of the distributed D-LSTM for sequential data analysis. We use video semantic recognition task for evaluation, because video analysis is more challenging than other types of sequential data. We two video datasets, Charades dataset and COIN dataset, which are recently published for video semantic analysis. The sub-actions have been labeled in these two datasets, so we will use them to recognize the high-level semantic of video sequence. The standard LSTM and the distributed LSTM are used as baseline architecture, and compared with the proposed distributed D-LSTM.

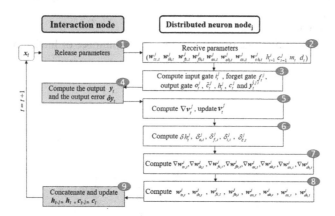

Fig. 3. The calculation procedure in the distributed D-LSTM.

We implement the distributed D-LSTM on a Spark cluster, where the driver node imitates the interaction node of distributed D-LSTM, while the worker node imitates the distributed neuron node. High-speed Infiniband and RDMA are used to provide higher data transmission and lower latency. In order to improve system scalability, multiple distributed neuron nodes will share a common worker node. In our experiments, the number of worker nodes is 4. The configuration of distribute D-LSTM is shown in Table 1.

Table 1. The configuration of experimental environment.

Configuration	Master node server	Slave node server×4
CPU	2×Intel®Xeon® CPU E5-2630	2×Intel®Xeon® CPU E5-2630
CPU cores	40	40
RAM	128 GB	128 GB
Disk	Intel S3520 800G SSD	Intel S3520 800G SSD
Spark	RDMA-Spark-0.9.1	RDMA-Spark-0.9.1
Network	56 GB Infiniband	56 GB Infiniband
Protocol	RDMA	RDMA

4.1 Charades Dataset

Charades dataset is a large-scale video dataset published by Allen Institute in 2016 [16], which includes 9848 annotated videos and 66,500 annotations in total. The dataset is collected in 15 types of indoor scenes, involves interactions with 46 object classes and 157 action classes. We choose five representative scenes (bedroom, bathroom, kitchen, living room, home office) and use the proposed distributed D-LSTM network to work on the scene recognition. The training set and test set have 4336 and 1003 video data, respectively. The sampling time for the video sequence is 1 s and we set the sequence length 50. The input at each time step is a 157-dimension vector where each dimension represents one of 157 action classes.

First, we evaluate the time overhead with various number of hidden neurons. The epoch is set 10, and the result is given in Fig. 4.

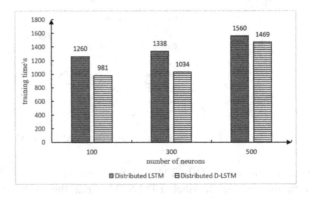

Fig. 4. The training time comparison on Charades.

From Fig. 4, we can see the training time of distributed D-LSTM is shorter that of distributed LSTM. The training efficiency is improved by 22.1%, 22.7% and 5.8% when the number of hidden neurons grows from 100, 300, to 500. So, the duration-aware LSTM needs less time to complete a fixed epoch compared with standard LSTM. The duration of sub-semantic is helpful for the improvement of training efficiency. When the hidden neuron reaches 500, the gap between the distributed LSTM and D-LSTM is narrowing, reducing by 16%. This may because more hidden neurons lead to higher computational complexity, which makes the communication cost and memory footprint of Spark cluster raised dramatically, and thus affects to the time overhead.

Then, we will compare the proposed method with the popular GPU acceleration method. We use a NVDIA Tesla P100 16G GPU and set epochs 10. Figure 5 gives the training time with various number of hidden neurons.

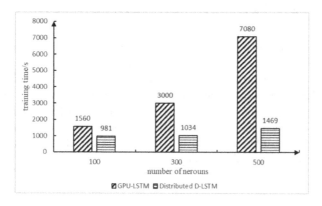

Fig. 5. The training time comparison with GPU.

Compared with the LSTM accelerated on GPU(GPU-LSTM), the improvement on training time is remarkable. When the number of hidden neurons is 100, the gap between two methods is 579 s, with the distributed LSTM 37.1% faster than GPU-LSTM. while the hidden neurons raise to 500, the distributed LSTM is 79.3% faster than GPU-LSTM. With the increase of hidden neuron, the performance improvement becomes increasingly obvious. It means the distributed D-LSTM can provide superior acceleration compared with GPU counterpart and has better scalability.

Figure 6 gives the result with different epochs. The number of hidden neurons is 100. If executes more epochs, the training time will increase accordingly, while distributed D-LSTM need less time compared with the distributed LSTM. When the epochs raise from 10, 20 to 30, the gaps between two methods are 279 s, 590 s and 787 s, respectively. This also validates the effectiveness of our method.

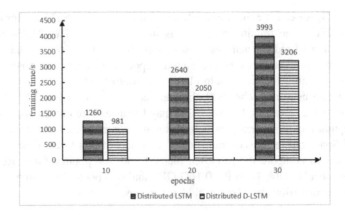

Fig. 6. The training time with various epochs on Charades.

Next, we will calculate the time when reaching the same error. We set various error thresholds, and check how long it will take for both models to arrive at the errors. Figure 7 gives the comparisons on Charades dataset. We can see, the training time for distributed D-LSTM is slightly higher than distributed LSTM when error is 0.8. However, continuously decreasing the error, the distributed D-LSTM outperforms the distributed LSTM. When the error is 0.4, its convergence speed is improved by 8.7% compared with the distributed LSTM, which indicates the effectiveness of semantic duration for speeding up LSTM training.

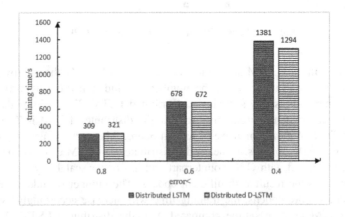

Fig. 7. The time overhead under different error thresholds.

Finally, we will evaluate the semantic recognition accuracy. Figure 8 shows the accuracy with different number of hidden neurons.

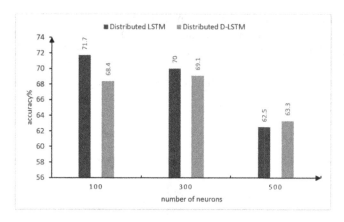

Fig. 8. The accuracy on Charades.

We can see the accuracy of distributed D-LSTM is a little lower than that of the distributed LSTM. But with the increase of hidden neurons, the accuracy loss is gradually decreased, which means the proposed method can achieve high efficiency with a closer or comparative accuracy compared to standard LSTM baseline.

4.2   COIN Dataset

COIN is a large video dataset for comprehensive instructional video analysis, which covers an extensive range of everyday tasks with explicit steps [17]. The dataset consists of 11,827 videos related to 180 different tasks. Each video is labelled with 3.91 step segments, where each segment lasts 14.91 s on average. In our test, we use six categories from three domains, which are "Use Analytical Balance", "Paste Car Sticker", "Assemble Bed", "Assemble Sofa", "Assemble Cabinet", and "Assemble Desktop PC". The training set and test set are 302 and 105 videos, respectively. The length of each video sequence is 200, where a null action is used to fill up a sequence.

First, we evaluate the time overhead with various number of hidden neurons. The result is shown in Fig. 9.

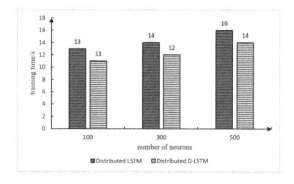

Fig. 9. The accuracy on COIN.

Because the training set is relatively small, the time overhead is shorter. Compared with the distributed LSTM, the distributed D-LSTM can improve the efficiency by 12.5%–15.4%, which indicates the effectiveness of modeling with the duration of sub-semantic.

Then we increase the iteration count and test the training time. The result is shown in Fig. 10.

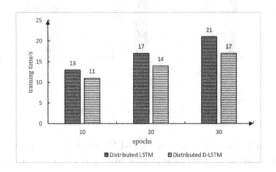

Fig. 10. The training time with various epochs.

We find the result is very similar to that of Charades, where the training time increases with the epoch number. When more epochs are involved, the distributed D-LSTM shows a higher promotion in training efficiency.

Finally, we test the accuracy on COIN. Compared with the distributed LSTM, the proposed method only shows 1.9%, 1.8% and 1% accuracy loss when the number of hidden neurons is 100, 300 and 500, respectively. The result is shown in Fig. 11.

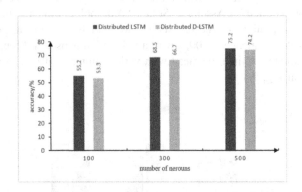

Fig. 11. The accuracy on COIN.

From all the above experimental results, we find the distributed D-LSTM can greatly improve the training efficiency while still achieve a comparative accuracy. The reason why the accuracy of distributed D-LSTM is slightly lower than that of distributed LSTM is that

5 Conclusions

In this work, we propose an efficient distributed LSTM for sequential data analysis. In order to fully utilize the properties of sequential data, a duration-aware LSTM unit is first proposed. The improved unit is capable of perceiving the duration of sequence item and adaptively update LSTM memory, thus leads to efficient memory computation within the unit. Then, a distributed training algorithm is proposed for D-LSTM, where the autonomous neurons are introduced and can work in parallel to accomplish parameter update. The prototype of distributed D-LSTM is implemented on a Spark cluster. Experimental results on two video datasets indicate the proposed method can greatly reduce the time overhead and improve the training efficiency for sequential data analysis, while with only a minor accuracy deterioration.

Acknowledgment. This work was partly supported by the National Natural Science Foundation of China No. 61806086, and the China Postdoctoral Science Foundation No. 2016M601737.

References

1. Hochreiter, S., Schmidhuber, J.: Long short-term memory. Neural Comput. **9**(8), 1735–1780 (1997)
2. Graves, A., Mohamed, A., Hinton, G.: Speech recognition with deep recurrent neural networks. In: Proceedings of IEEE International Conference on Acoustics, Speech and Signal processing, pp. 6645–6649. IEEE, Piscataway (2013)
3. Salehinejad, H., Sankar, S, Barfett, J., Colak, E., Valaee, S.: Recent advances in recurrent neural networks. arXiv preprint arXiv:1801.01078 (2018)
4. Dean, J., Corrado, G.S., Monga, R., Chen, K., Ng, A.Y.: Large scale distributed deep network. In: Proceedings of Advances in Neural Information Processing Systems, pp. 1223–1231. Neural Information Processing Systems Foundation, San Diego (2012)
5. Abadi, M., Agarwal, A., Barham, P., Brevdo, E., Devin, M.: Tensorflow: large-scale machine learning on heterogeneous distributed systems. arXiv preprint arXiv:1603.04467 (2016)
6. Xing, E.P., Ho, Q.R., Dai, W., Kim, J.K., Yu, Y.L.: Petuum: a new platform for distributed machine learning on big data. IEEE Trans. Big Data **1**(2), 49–67 (2015)
7. Li, M., Andersen, D.G., Park, J.W., Smola, A.J., Su, B.Y.: Scaling distributed machine learning with the parameter server. In: Proceedings of International Conference on Big Data Science & Computing, p. 1. ACM, New York (2014)
8. Eluyode, O.S., Akomolafe, D.T.: Comparative study of biological and artificial neural networks. Eur. J. Appl. Eng. Sci. Res. **2**(1), 36–46 (2013)
9. Lei, T., Zhang, Y.: Training RNNs as fast as CNNs. arXiv preprint arXiv:1709.02755 (2017)
10. Khomenko, V., Shyshkov, O., Radyvonenko, O., Bokhan, K.: Accelerating recurrent neural network training using sequence bucketing and multi-GPU data parallelization. In: Proceedings of International Conference on Data Stream Mining and Processing, pp. 561–570 IEEE, Piscataway (2016)
11. Huang, Z., Zweig, G., Levit, M., Dumoulin, B., Chang, S.: Accelerating recurrent neural network training via two stage classes and parallelization. In: Proceedings of IEEE Workshop on Automatic Speech Recognition and Understanding (ASRU), Olomouc, Czech Republic, pp. 326–331. IEEE, Piscataway, 8–12 December 2013

12. Ji, S., Vishwanathan, S.V.N., Satish, N., Anderson, M.J., Dubey, P.: BlackOut: speeding up recurrent neural network language models with very large vocabularies. Comput. Sci. **115** (8), 59–68 (2015)
13. Keuper, J., Preundt, F.J.: Distributed training of deep neural networks: theoretical and practical limits of parallel scalability. In Proceedings of the Workshop on Machine Learning in High Performance Computing Environments, pp. 19–26. IEEE, Piscataway (2017)
14. Gholami, A., Azad, A., Jin, P., Keutzer, K., Buluc, A.: Integrated model, batch, and domain parallelism in training neural networks. arXiv preprint arXiv:1712.04432 (2017)
15. Niu, D.J., Xia, Z., Liu, Y.W., Cai, T., Zhan, Y.Z.: ALSTM: adaptive LSTM for durative sequential data. In: Proceedings of IEEE 30th International Conference on Tools with Artificial Intelligence, pp. 1018–1026. IEEE, Piscataway (2018)
16. Sigurdsson, G.A., Varol, G., Wang, X., Farhadi, A., Laptev, I., Gupta, A.: Hollywood in homes: crowdsourcing data collection for activity understanding. In: Leibe, B., Matas, J., Sebe, N., Welling, M. (eds.) ECCV 2016. LNCS, vol. 9905, pp. 510–526. Springer, Cham (2016). https://doi.org/10.1007/978-3-319-46448-0_31
17. Tang, Y., Ding, D., Rao, Y., Zheng, Y., Zhang, D., Zhao, L.: COIN: a large-scale dataset for comprehensive instructional video analysis. In: Proceedings of International Conference on Computer Vision and Pattern Recognition (accepted, 2019)

Considering User Distribution and Cost Awareness to Optimize Server Deployment

Yanling Shao[1,3] and Wenyong Dong[1,2(✉)]

[1] Nanyang Institute of Technology, Nanyang 473000, China
shaoyl1204@163.com, dwy@whu.edu.cn
[2] Computer School, Wuhan University, Wuhan 430072, China
[3] Department of Computer and Science, Wuhan University of Technology,
Wuhan 430063, China

Abstract. In edge computing systems, it is crucial issue to select suitable placement sites and quantity of servers so as to realize the low latency of Internet of Things (IoT) applications and balance the sever utilization. Hence, this paper proposes a cost-aware edge server optimization deployment method. Firstly, we model the edge server placement problem as a Mixed Integer Nonlinear Programming problem (MNIP), which comprehensively considers the resource allocation ratio, regional average load, and access delay. And then, the Benders decomposition algorithm is employed to solve it. The simulation results show that the proposed method can find better solution to place the edge micro datacenter (MDC) compared with the state-of-art server deployment strategies in terms of latency for applications and utilization of resources.

Keywords: Edge computing · Server deployment · Benders decomposition

1 Introduction

Recently, some Internet of Things (IoT) applications, such as industrial intelligence control, virtual reality/augmented reality, and online games, require cloud computing platforms to provide delay-sensitive service. But, the long-distance network communication will result in poor user satisfaction because of the long data transfer times. In order to improve the user's satisfaction and reduce the traffic pressure on the backbone network, the cloud computing service provider delivers the service content and part of the service processing to the network edge server close to the access side for localized real-time edge processing. Therefore, the deployment of edge servers has become a key issue for edge infrastructure service providers.

There are three key factors to consider when deploying an edge server: end-to-end latency, resource utilization, and deployment costs. There are two key factors that affect server deployment: the locations and quantity of the edge servers. For example, when a live user watches video live in a large stadium such as a sports ground, if the edge server is deployed at a far location, the user device may need to go through multiple hops to access the nearest edge resource. Therefore, the location of edge servers can have a significant impact on end-to-end latency. In addition, the price of leasing varies according to the location of server deployment. Therefore, location has a large impact

© Springer Nature Singapore Pte Ltd. 2019
H. Jin et al. (Eds.): BigData 2019, CCIS 1120, pp. 135–147, 2019.
https://doi.org/10.1007/978-981-15-1899-7_10

on deployment costs. Therefore, to ensure the quality of service for low-latency applications, edge computing infrastructure service providers need to plan to deploy edge servers in close proximity to user equipment. Due to the different density of user distribution in each service region, the number of edge servers to be deployed in each region is different. Edge computing infrastructure providers need to determine the number of servers needed in a service area covered by an access point based on the density of user distribution in that service area. When user-submitted IoT business requests need to handle a large amount of data flow, too few or too many edge servers may cause some edge servers to be overloaded or under loaded. Heavy load may increase network latency and reduce user satisfaction. Meanwhile, insufficient load for a long time will lead to low resource utilization and high deployment cost. Therefore, a reasonable deployment strategy has become an important research content.

This paper will analyze and model the edge server optimization deployment problem, and then propose a cost-aware edge server optimization edge server placement method. There are two main contributions of this study as below.

(1) The edge server deployment problem is modeled as mixed integer nonlinear programming problem (MINP).
(2) In order to achieve the purpose of low latency demand and minimum deployment cost, a sparse deployment optimization strategy of edge server based on Benders decomposition is proposed.

2   Related Work

In edge computing, the server is deployed on the edge of the network between the terminal device and the cloud data center. User equipment accesses the edge server remotely in a wireless or wired manner to meet the needs of low latency applications. At present, industry and academia have already studied the issue of edge server placement. ETSI MEC working group has collected representative application cases and deployment scenarios [1] to improve MEC standardization. Heavy Reading also published a white paper on mobile edge computing cases and deployment options [2] and pointed out that the location selection of MEC infrastructure should be based on specific network environment and business requirements. It can be deployed in LTE macro base stations, multi-standard base station convergence points or 3G wireless controllers. For edge computing system, edge servers can be deployed in the above network locations to flexibly adapt to different business requirements.

Fan et al. [3] considered mobile user density and Cloudlet location to study the Cloudlet deployment strategy based on cost and end-to-end latency trade-off optimization in mobile edge computing environments. This paper used mixed integer programming (MIP) tool in CPLEX solver to find the suboptimal solution. Furthermore, this study also designed a dynamic workload distribution scheme for each access point coverage area in slot t to minimize the total E2E delay requested in each slot, and concluded this problem as a linear programming problem. Xu et al. [4] studied capacity-constrained Cloudlet placement in large-scale wireless metropolitan area networks (WMANs) with the goal of minimizing the average access latency between mobile users and Cloudlets. The problem was modeled as integral linear programming (ILP) problem, and a solution with exact solution was proposed. Furthermore, in view of the shortcomings of the integer linear programming method in solving large-scale problems, the paper designed a heuristic greedy algorithm to solve the problem. Jia et al. [5] studied cloudlet sparse deployment and user attribution in WMAN. The goal of this work was to minimize the average delay time for job offloading. They proposed a multi-user multi-cloudlet task offloading model (M/M/c) based on queuing theory, then designed the load-heavy access point first (HAF) placement algorithm and density-based clustering (DBC) placement algorithm. Yin et al. [6] proposed the Tentacle decision support framework for online edge services. The framework leveraged edge computing platforms, Cloudlet and network function virtualization technologies to implement a flexible edge server deployment approach that optimized the performance and cost of edge facilities. Xiang et al. [7] studied the mobile adaptive Cloudlet deployment problem. As users move frequently, because of the limited coverage of Cloudlet, the number of mobile devices covered by Cloudlet at location a decreases, resulting in a large amount of waste of resources. To solve this problem, they proposed an adaptive mobile Cloudlet deployment method based on mobile application geo-location information big data. Lee et al. [8] proposed a simple fog server deployment algorithm based on SDN control considering the service completion time, network traffic and fog computing resources in the fog computing environment. Wu et al. [9] studied the optimization of server placement in parallel distributed systems. Considering the potential revenue and the construction cost of each location, the additional servers are deployed to maximize the revenue while the original servers compete. Chen et al. [10] proposed PacketCloud, an open platform based on Cloudlet. The platform helps Internet service providers (ISPs) and content providers select appropriate outlets to deploy Cloudlet services. Cloudlet server resources can be shared among different services to achieve flexible resource allocation.

Through the above analysis, it can be seen that Cloudlet deployment covers a small range. In this work, we study the large range of wireless metropolitan area network and comprehensively consider the factors such as its utilization rate, deployment economic cost and low latency for each strategic site. Most of the existing work aims to ensure low latency of edge computing, or optimize job completion time, or improve resource utilization, or increase the number of mobile devices in the coverage area. In the actual edge computing environment, it is necessary to meet the low latency of users' demands. Based on stable user distribution density and edge computing infrastructure provider's resource deployment costs, this work aims to improve resource utilization and reduce service budget costs. In addition, this paper adopts Benders algorithm to solve the problem of edge server deployment for the first time, which can effectively find the optimal solution of edge server deployment with economic cost and low delay balance.

3   Problem Description and Formulation

3.1   Problem Description

The server deployment scope of this study is the wireless metropolitan area network (WMAN) area. The candidate deployment location is the existing router device location near the base station location and data convergence point of user equipment. We describe the problem of edge server deployment as follows:

Within the scope of the wireless metropolitan area network, given the potential edge server deployment location set I and service coverage area set J, where the coverage area is the service range of the base station or the router's one-hop service distance. User devices are connected to edge micro data centers via service base stations. Edge micro data centers can handle task requests and data offloaded by user terminals. Due to different user distribution density, different load in each coverage area, the rental cost of each potential location and the number of edge servers to be deployed are not the same. The main goal of this paper is to deploy edge servers from these potential edge locations according to the low latency requirements of the application, and to determine the number of nodes in each edge micro data center according to the user distribution density, so as to minimize the overall cost under the restriction of low latency the application.

3.2   Problem Formulation

In a Wireless Metropolitan Area Network (WMAN) range, $WMAN = (V, E)$, where $V = I \cup S$, I is the network access point, S edge server deployment location, and E is the link set between the access point and a potential location of the edge server. $U = \{u_1, u_2, ...U_N\}$ represents the set of all user devices. Geographically distributed user devices access edge computing resources through network access points they serve. Let u_l be the lth user, and $l \in \{1, 2, ..., n\}$.

Suppose the edge server is collocated with the base station or the network aggregation device (router, switch). Set J as the set of service coverage areas of different network access points. According to the density of user distribution, the server clusters $\{s_1, s_2, ...,s_k\}$ is deployed at the selected edge location, where $k \geq 1$. And set $j \in J$ as the jth coverage area of the network access point.

Figure 1 depicts an example of base stations, users, and edge server deployments in a wireless metropolitan area network. In the figure, there are 11 base stations as candidate locations, where each base station covers a service area, and user equipment in the service area sends requests to access the network from the base station, and then the local manager performs load distribution according to the available resource context in the edge computing cluster. As can be seen from the figure, the deployment location of edge server A is relatively remote in geographical location, and the rental price is relatively low compared with the central area which the service range is within one hop and the user distribution density is large. Similarly, locations A, B, and C have a high density of user equipment within one hop of the service area, where multiple edge servers can be deployed. It can be seen that in the actual wireless metropolitan area network, due to the relatively large network access point size and the large number of candidate locations, thousands of candidate edge sites are available under multiple constraints such as delay and lease cost. It is a tricky problem to find the best feasible solution.

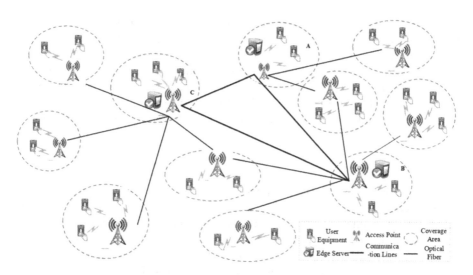

Fig. 1. Edge server deployment in a wireless metropolitan area network

In order to illustrate the deployment of edge server at candidate edge site, when the decision variable $y_i = 1$, it means that the edge server is deployed at the ith candidate edge site. Otherwise it means that the edge server is not deployed at the ith edge location. Since requests from users in one region may be distributed to different edge MDC's for processing, a continuous variable x_{ij} $(0 \leq x_{ij} \leq 1)$ is declared here, which represents the ratio of load allocated in region j to edge MDC at edge location i. And further, we define the non-negative integer variable χ_i as the number of edge servers located at edge site i.

Overall Cost of Edge Server Deployment. The overall cost of edge server deployment consists of two parts: total user access delay and economic cost of edge server deployment.

Total User Access Delay. Set d_{ij} is the unit delay caused by logical link access objects between the access points in coverage area j to edge site i. Then the total delay T_j of requesting access in the region j is shown in Eq. (1).

$$T_j = \sum_{i \in I} \omega_j x_{ij} d_{ij} \tag{1}$$

Where the average number of user requirements $\omega_j = \sum_{l \in \{1,2,...,n\}} p_{lj} \lambda_l \overline{\psi_j}, \psi_j$ as average user density in the area j. The request arrival rate λ_l of user l and the proportion p_{lj} of time that the user stays in service coverage area. The number of user terminals in the time slot can be collected. ψ_t is the number of user terminals in the covered area of the time slot t, and $\overline{\psi_j} = \sum_{\tau=1}^{\delta} \psi_\tau / \delta$, here, δ as the total time slots.

Economic cost of edge server deployment. When the edge computing infrastructure provider prepares to deploy edge servers, it needs to select a fixed location and equip the infrastructure. The economic cost of edge server deployment consists of fixed construction costs such as location rental price and the price and quantity of deployed servers. Let's set f_i is the fixed construction cost of edge site i (including the cost of leasing and other basic hardware), g_i as the unit price of server and χ_i as the number of server nodes. Then the definition of edge server deployment cost is shown in Eq. (2).

$$Cost_1 = \sum_{i \in I} (f_i + g_i \chi_i) y_i \tag{2}$$

Modeling of Edge Server Deployment. On the one hand, in order to achieve the minimum end-to-end access latency for processing user requests, it needs to be allocated to the edge computing resource with the minimum proximity to the user. As the number of edge micro data centers deployed increases, end-to-end latency for user request processing decreases. On the other hand, edge computing infrastructure providers must spend more money on edge server deployments. So from the point of view of edge facilities provider, its goal is to use the lowest cost of deployment for handling user requests the minimum end-to-end network latency, in order to realize the deployment cost and trade-off between user requests total delay optimization, therefore, this chapter studies the server deployment optimization strategy to balance consider deployment cost and low latency of user request processing. The deployment cost of edge server includes the sum of the deployment cost of edge computing resources and the total access delay cost. Therefore, the comprehensive cost minimization model for edge server deployment can be described as Eq. (3).

$$\min\left(\sum_{i \in I} (f_i + g_i \chi_i) y_i + \varsigma \sum_{i \in I} \sum_{j \in J} \omega_j x_{ij} d_{ij}\right) \tag{3}$$

s.t.

$$C1 : \sum_{j \in J} \omega_j x_{ij} \leq s\chi_i, \ \forall i \in I$$
$$C2 : \sum_{i \in I} x_{ij} = 1, \ \forall j \in J$$
$$C3 : \chi_i \leq cy_i, \ \forall i \in I$$
$$C4 : \sum_i y_i \leq k, \ \forall i \in I$$
$$C5 : x_{ij} \in [0,1] \forall i \in I, \forall j \in J$$
$$C6 : y_i \in \{0,1\}, \ \forall i \in I$$
$$C7 : \chi_i \in Z_0^+, \ \forall i \in I$$

Among them, ζ is the adjustment constant, which is used to adjust the proportion of the total access delay cost and the edge server deployment cost. $\varsigma = \left\lceil \theta_2 \sum_{i=1}^{k} (f_i^{\max} + g_i c) \right\rceil / \left\lceil \theta_1 \sum_j \omega_j d_j^{\max} \right\rceil$, $\varsigma > 0$. And $d_j^{\max}$ is the maximum delay of the farthest edge server in coverage area j, c is the maximum number of servers in the edge site, $\sum_{i=1}^{k} f_i^{max} + g_i c$ represents the highest deployment cost, $\sum_j \omega_j d_j^{max}$ represents the maximum total delay of all user requests in coverage area j, θ_1 and θ_2 is the equalization parameter, $\theta_1 + \theta_2 = 1$, $\theta_1, \theta_2 \in [0, 1]$.

It can be seen that the objective function (3) guarantees that the deployment cost and total delay cost of deploying the edge server in the WMAN are minimized. Constraint (C1) guarantees that the task cannot exceed the maximum load on the edge server cluster at that site. Constraint (C2) ensures that the load is distributed to different edge server clusters across all areas. Constraints (C3) and (C4) ensure that the number and total number of locations deployed in the metropolitan area network do not exceed the defined maximum value, and the constraint (C5–C7) defines the range of values of the variables. It can be seen that the variable of the number of servers at a certain location is an Integer variable, the variable of load distribution at a certain edge location in the region is a continuous variable, and the model of this edge server deployment is formulated as Mixed Integer Nonlinear Programming problem (MINP).

4   Edge Server Deployment Algorithm Based on Benders Decomposition

Benders decomposition algorithm is suitable for solving mixed integer programming problems [11–13], so this study uses Benders decomposition algorithm to solve the problem of edge server deployment. Its basic idea is to decompose the original problem into the main problem containing complex integer decision variables and the sub-problem containing only continuous variables according to the different types of variables. In the process of solving, the main problem and the sub-problem are iteratively performed. The main problem provides the lower bound of the original problem, and the obtained integer solution is passed to the sub-problem. The sub-problem provides the upper bound for the original problem, and returns the Benders cut, the main sub-problem to the main problem. Solve alternately, until the upper and lower bounds are equal, the algorithm stops, and the optimal solution of the original problem is obtained.

The edge server deployment algorithm based on Benders decomposition proposed in this chapter is shown in Algorithm 1. In line 2, the maximum upper limit UB and minimum lower limit LB are initialized, and the feasible initial position is selected. In the iterative process of using Benders algorithm to solve the deployment of edge servers, line 7 shows that the dual sub-problem C_k provides an upper bound for the original problem, returns Benders to the MP problem to constrain the main problem, and updates UB to form a new main problem to solve the configuration of edge servers. In line 4, the optimal solution for the primary problem MP is to provide a lower bound for the original problem. To prevent the MP problem from becoming unbounded in previous iterations, many cuts were initially added to the viable MP solution.

ALGORITHM 1. THE EDGE SERVER DEPLOYMENT ALGORITHM BASED ON BENDERS

DECOMPOSITION（Benders_SD）

Input: I: the AP set of base station; U:user set; J: the area set

Output: Y: the deploy site set; χ_i :the server number of the deploy site i

Begin

1: initialize g_i, the server price; f_i, the price of an edge position ; s, the maximum load of a server; c, the largest server number in a single edge position which accommodates

2: initialize $UB = +\infty$, $LB = -\infty$, $k=0$;

3: do{

4:   Select the initial server deployment scenario

$$\overline{y_i} = \begin{cases} 1, & \text{if } d_{ij} \leq D, \ \forall i,j \\ 0, & \text{else} \end{cases}$$

5:   In the first step, all nodes in the region j that satisfy the delay condition are selected for initial deployment

6:   Initialize the main problem model MP

7:   Compute $C_k = \sum_j \beta_j + \sum_i c\overline{y_i}\gamma_i$

8:   if $(C_k < UB) UB = C_k$

9:   Solving MP to get the lower bound Lk by Benders cut constraints

$\varsigma \leq \sum_j \beta_j^* + \sum_i c y_i \gamma_i^*$

10: if$(lk > LB)$ $LB = lk$;

11:      if the MP problem has no solution, the original problem has no solution and the algorithm ends.

12: update $\overline{y_i}$ and χ_i by MP's solution

13: $k=k+1$

14: }while$((UB - LB) / UB > 0.001 \ || \ k < 100)$

End

---

5   Performance Evaluation

This simulations were tested on a personal laptop with Inter(R) Core(TM) i7-3770 CPU@3.40 GHz processor, RAM 12.0 GB memory and 1T hard disk space. The Benders_SD algorithm was programmed in C++ language. The experimental results were obtained by performing 25 times independently under the same conditions and taking the average value.

5.1   Experiments Settings

This section evaluates the performance of the proposed algorithm based on real and synthetic network topology datasets according to the references [4, 5]. The amount of resources requested by each user is a random value with a range of [50, 200]. The maximum number of requests processed per server is 50. And the delay between edge sites is generated randomly between 5 ms and 50 ms. The main system parameters of this experiment are set as shown in Table 1.

Table 1. System parameter setting

Parameter	Value
Number of user requests per access point	[50, 200]
Slot length (min)	10
Latency between edge sites and client access points (ms)	[5, 50]
Number of candidate edge sites	{200,400,600,800,1000}
price of a server ($)	1000
The maximum number of servers in an edge MDC	10
Maximum of requests processed per server	50
Rental cost per edge site ($)	[10000, 80000]
Controlled parameter θ_1 and θ_2	{0.1,0.2,0.3,0.4,0.5,0.6,0.7,0.8,0.9}

Datasets. According to the references [4–6], the real data set of this experiment comes from the network topology of Hong Kong MTR, including 18 potential marginal locations corresponding to 18 regions in Hong Kong.

Comparison Algorithm. In order to evaluate the performance of the proposed Benders decomposition-based server deployment strategy (Benders_SD), this paper chooses the load-first priority placement (HAF [5]), the greedy algorithm, and the CPLEX algorithm for comparative analysis.

Metrics. There are three metrics for this experiment: the edge MDC deployment economic cost ($), the total user access delay (s), and the overall cost, where the overall cost is derived from Eqs. (5–6), and the unit of the overall cost is jointly determined by delay (s) and creation cost ($), which is represented by s ($) in this study.

5.2   Experimental Results and Analysis

Sensitivity Analysis on Controlled parameter θ_1. The values of the delay sensitivity parameters are set to {0.1, 0.2, 0.3, 0.4, 0.5, 0.6, 0.7, 0.8, 0.9}, and θ_1 gradually increases, indicating that the more sensitive the delay, the candidate edge position of this experiment is 200. As can be seen from Fig. 2, with the gradual increase of the controlled parameters θ_1, the deployment economic cost of the edge MDC gradually increases. Figure 3 illustrates how the total end-to-end latency cost decreases with the increase of parameter θ_1, which makes edge server deployments more sensitive to end-to-end latency. As θ_1 increases, the algorithm proposed in this chapter focuses more on latency costs, so more edge servers need to be deployed to reduce latency, which corresponds to an increase in edge MDC deployment economic costs. It can be seen that controlled parameters θ_1 has a great influence on the result of the algorithm. Figure 4 depicts the overall cost as a function of parameters θ_1. At $\theta_1 = 0.2$, the overall cost is minimal.

Therefore, this chapter comprehensively considers the comprehensive interests of edge computing service providers and users, and sets the system adjustment parameter $\theta_1 = 0.2$ with the goal of minimizing overall cost.

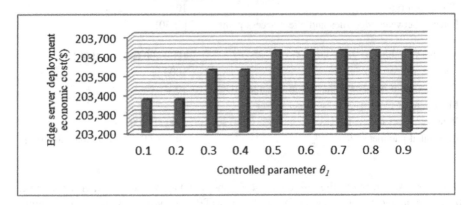

Fig. 2. Edge server deployment economic cost under different controlled parameter values

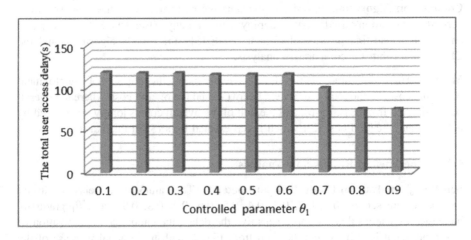

Fig. 3. The total user access delay under different controlled parameter values.

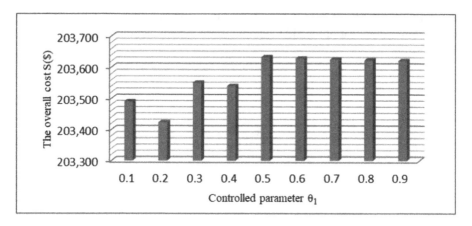

Fig. 4. The overall cost under different controlled parameter values.

Performance Evaluation for Small Network Service Area. This group of experiments uses the real Hong Kong subway network HKMTR data set to evaluate the edge server deployment algorithm proposed in this chapter, including 18 AP access points, and choose 3 of them as deployment locations. This group of experiments relies on the results of parameter sensitivity experiments and sets the system adjustment parameter $\theta_1 = 0.2$.

Figure 5 shows the edge MDC deployment economic cost under the four algorithms Benders_SD, HAF, Greedy, and CPLEX. It can be seen that the Benders_SD algorithm proposed in this work has the lowest creation cost, which is superior to the other three comparison algorithms, and the HAF algorithm has the highest server deployment cost.

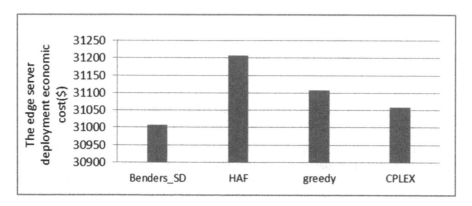

Fig. 5. Edge server deployment economic cost

Figure 6 depicts the total end-to-end access delay cost under the four algorithms. Benders_SD proposed in this paper is also superior to the other three comparison algorithms. AS can be seen from Fig. 7, the integrated cost of the Benders_SD algorithm is lower than that of the HAF algorithm, the Greedy algorithm and the CPLEX algorithm.

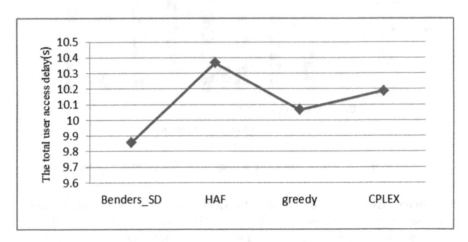

Fig. 6. The total end-to-end access delay cost.

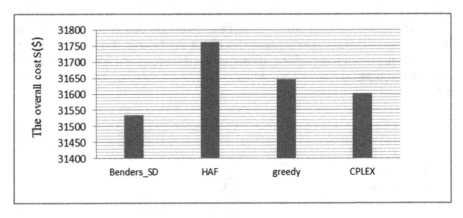

Fig. 7. The overall cost of edge server deployment.

Based on the above comparison results and the evaluation results of the four algorithms by HKMTR data set, Benders_SD proposed in this paper has the best performance in the three aspects of edge MDC creation cost, end-to-end delay and comprehensive cost, which can minimize the cost of edge computing infrastructure provider and end-to-end delay of user access.

6 Conclusion

This paper proposes a cost-aware edge server optimization deployment method, which establishes an objective function based on edge server deployment and access cost minimization through resource allocation ratio, regional average load, and access delay between users and the edge node locations served by them. The Benders decomposition algorithm is used to solve the mixed integer nonlinear programming problem. The simulation results show that the sparse deployment optimization strategy based on Benders decomposition can guarantee users' QoE for low-latency applications and improve the utilization rate of resources so as to lower the overall deployment cost of edge servers.

References

1. ETSI, White Paper No. 11: Mobile Edge Computing: A key technology towards 5G. http://www.etsi.org/images/files/ETSIWhitepapers/etsiwp11mecakeytechnologytowards5g. pdf. Accessed 14 Aug 2016
2. Gabriel, B.: Mobile edge computing use cases & deployment options. https://www.juniper. net/assets/uk/en/local/pdf/whitepapers/2000642-en.pdf. Accessed 09 Mar 2017
3. Fan, Q., Ansari, N.: Cost aware Cloudlet placement for big data processing at the edge. In: IEEE International Conference on Communications, pp. 1–6. IEEE Computer Society Press, Washington (2017)
4. Xu, Z., Liang, W., Xu, W., Jia, M., Guo, S.: Efficient algorithms for capacitated Cloudlet placements. IEEE Trans. Parallel Distrib. Syst. 27(10), 2866–2880 (2016)
5. Jia, M., Cao, J., Liang, W.: Optimal Cloudlet placement and user to Cloudlet allocation in wireless metropolitan area networks. IEEE Trans. Cloud Comput. 5(4), 725–737 (2017)
6. Yin, H., Zhang, X., Liu, H., Luo, Y., Tian, C., Zhao, S., et al.: Edge provisioning with flexible server placement. IEEE Trans. Parallel Distrib. Syst. 28(4), 1031–1045 (2017)
7. Xiang, H., Xu, X., Zheng, H., Li, S., Wu, T., Dou, W., et al.: An adaptive cloudlet placement method for mobile applications over GPS big data. In: Global Communications Conference, pp: 1–6. IEEE Press, Piscataway (2017)
8. Lee, J.H., Chung, S.H.: Fog server deployment considering network topology and flow state in local area networks. In: IEEE International Conference on Ubiquitous and Future Networks, pp. 652–657. IEEE Computer Society Press, Washington (2017)
9. Wu, J.J., Shih, S.F., Liu, P., Chung, Y.M.: Optimizing server placement in distributed systems in the presence of competition. J. Parallel Distrib. Comput. 71(1), 62–76 (2011)
10. Chen, Y., Chen, Y., Cao, Q., Yang, X.: PacketCloud: a Cloudlet-based open platform for in-network services. IEEE Trans. Parallel Distrib. Syst. 27(4), 1146–1159 (2016)
11. Hooker, J.N.: Planning and scheduling by logic-based benders decomposition. Oper. Res. 55(3), 588–602 (2007)
12. Costa, A.M.: A survey on benders decomposition applied to fixed-charge network design problem, Elsevier Science Ltd. (2005)
13. Ma, L., Wu, J., Chen, L.: DOTA: delay bounded optimal cloudlet deployment and user association in WMANs. In: IEEE/ACM International Symposium on Cluster, Cloud and Grid Computing (CCGRID), pp. 196–203. IEEE Computer Society Press, Washington (2017)

Convolutional Neural Networks on EEG-Based Emotion Recognition

Chunbin Li[1], Xiao Sun[1(✉)], Yindong Dong[1], and Fuji Ren[1,2]

[1] School of Computer Science and Information Engineering,
Hefei University of Technology, Hefei 230601, China
`lichunbin@mail.hfut.edu.cn`, `sunx@hfut.edu.cn`, `dongyindong66@163.com`
[2] Faculty of Engineering, The University of Tokushima, Tokushima, Japan

Abstract. Human Computer Interaction (HCI) enables people to transfer and exchange information with computers. For the purpose of friendliness, integrating HCI with emotional factors has been intensively investigated. In this paper, an effective method is proposed to recognize human emotions by electroencephalogram (EEG) signals, which record electrical activities of the brain. First of all, the EEG signals are converted to the multispectral image that preserves the local distance between any two nearby electrodes. Notably, our method preserves the features of EEG signals in frequency and spatial dimensions, unlike standard EEG analysis techniques inaccurately interpreting the location of electrodes. And then a Convolutional Neural Network (CNN) model is performed to identify human emotions by virtue of the image containing EEG feature, for the reason of CNN's significant effect in image recognition. A publicly available dataset, DEAP dataset, is used to validate our algorithm. The results show that the mean classification accuracy is 81.64% for valence (low and high) and 80.25% for arousal (low and high) across 32 subjects.

Keywords: Emotion recognition · EEG signal · Image · CNN

1 Introduction

Recent developments of Human Computer Interaction (HCI) have attracted much attention. In particular, one thing heavily involved in social activities is the human emotion, and considerable research efforts have been devoted to integrate HCI with emotional factors. And in this field, searching for emotion recognition is a fundamental problem. It is well established that human emotions can be detected by using text, speech and facial expressions. But these things are easily masked by people's behavior. For this reason, more and more work has been carried out on classifying human emotions from electroencephalogram (EEG) signals, which can reflect the brain dynamics directly and provide information for emotional states reliably.

Deep Learning (DL) has been on the fast lane across the past ten years, it has made huge achievements of application to the recognition in the context of text,

H. Jin et al. (Eds.): BigData 2019, CCIS 1120, pp. 148–158, 2019.
https://doi.org/10.1007/978-981-15-1899-7_11

speech, images and videos. Among all deep learning architectures, the Convolutional Neural Network (or shortly, CNN) has been confirmed to be fit for coping with the variability of two-dimensional (2-D) shapes [10]. In this paper, we attach importance to the features of EEG signals in frequency and spatial dimensions, and transfer the EEG signals to the images capturing multiple frequencies and also preserving the local distance between nearby electrodes, according to Power Spectral Density (PSD) and the positions of EEG acquisition electrodes [3]. For every channel, we extract the PSDs of 4 frequency bands. And in each band, we project PSDs of all channels onto a 2-D matrix according to the positions of the electrodes, using the Azimuthal Equidistant Projection, borrowed from map projection, and Clough-Tocher scheme, borrowed from mathematics. Thus, we convert one-dimensional (1-D) chain-like EEG vectors to three-dimensional (3-D) arrays, whose shape is $32 \times 32 \times 4$ (height $\times$ width $\times$ frequency bands). For a colorful image, there are three channels in RGB color model, corresponding to three colors. So we regard each of our processed samples as an image having four "colors". Then, a CNN model is developed to finish binary classification (high and low) tasks for valence and arousal respectively. Afterwards, the performance of our method is tested on the DEAP, an openly accessible dataset. The respective average accuracies with respect to valence and arousal are 81.64% and 80.25%, which demonstrates the effectiveness of classifying human emotions by our method.

What remains to be presented in this current research are: we review relative literatures in the second section; the description of benchmark dataset and the introduction to our emotion classification model appear in the third and the experimental outcomes the fourth; finally we make conclusion in the fifth section.

2   Related Work

The EEG signal is packed with abundant information, but its complexity is one of huge challenges for researchers. There are lots of literature reviews coping with the question that which features are effective for EEG-based emotion recognition. Using Welch's Method, Horlings et al. [4] select the mobility, activity, and complexity out of the EEG signal then accomplish a software system capable to recognize emotions. Petrantonakis and Hadjileontiadis [14] implement an EEG-based, user-independent emotion recognition system through higher order crossing (HOC) analysis. And their system achieves a classification accuracy of 83.3% among six emotions. Lahane and Sangaiah [9] compute Density Estimate of EEG using Kernel Density Estimation (KDE) and high emotion classification accuracy is obtained. In [19], Thammasan et al. play music for subjects in order to stimulate their emotions and collect their EEG signals. Experimental results show that Fractal Dimension (FD) and PSD are both effective in classifying subjects' emotion states.

In 2012, DEAP dataset was released and quickly became an affective benchmark for testing the emotion recognition algorithm on the basis of EEG. Rozgic et al. [15] use Fast Fourier Transform (FFT) to get Spectral Power and Spectral Power Differences of EEG signals, and use segment level decision fusion.

They achieve 76.9% and 68.4% accuracy for 2 classes on valence and arousal respectively. Liu et al. [12] adopt Maximum Relevance Minimum Redundancy (mRMR) for feature selection after extracting several features on time and time-frequency domains. Classification is done by using k-Nearest Neighbor (KNN) and Random Forest (RF) and the highest accuracy achieved is 69.9% and 71.2% on valence and arousal. Considering stimulus familiarity, Thammasan et al. [20] first divide DEAP into two parts, low and high familiarity. Then they extract FD and PSD of EEG signals, and classify emotions using Support Vector Machine (SVM), Decision Tree C4.5, and also Multilayer Perceptron (MLP), getting the accuracy of 73.30% and 72.50% on valence and arousal respectively. The recent development of deep learning techniques gives researchers more ideas. For the same classification task, Li et al. [11] conduct on each-channel signal Continuous Wavelet Transform (CWT) and the outputs are transformed into scalograms, then the constructed frames serve as the input of CNN and Long Short-Term Memory, a renown recurrent neural network structure, which has average accuracy rates of 72.06% for valence and 74.12% for arousal. Mert and Akan [13] explore the property of MEMD, the full name of which is Multivariate Empirical Mode Decomposition, and achieve the $72.87\% \pm 4.68$ emotion recognition accuracy for high/low valence and $75\% \pm 7.48$ in the case of high/low arousal using Artificial Neural Network (ANN) classifier. Tripathi et al. [21] extract several statistic features of DEAP's EEG signals and represent them for each experiment as a 2-D array. Then the Deep Neural Network (DNN) and CNN applied to the binary classification of emotions are trained by this 2-D array. Experimental results show that CNN provides better accuracy rates, 81.41% and 73.36%, respectively with respect to valence and arousal.

3   Materials and Methods

3.1   Dataset

One of the most used multimodal data in the human emotion analysis is called DEAP, the Database for Emotion Analysis using Physiological signals, collected through an experiment and issued in 2012 [7]. In the experiment, 40 one-minute music videos were selected as stimuli after several selection steps and recruited were 32 healthy participants including 16 males and 16 females whose age were over 19 and below 37 (26.9 in average). Each of them watched 40 videos and during this period, his/her EEG and peripheral signals were recorded with 32 electrodes located by 10–20 System at sampling rate 512 Hz. As shown in Fig. 1. Signals recorded in a trial include 40 channels and other 8 are peripheral signals of physiology the example of which are temperature and Galvanic Skin Response (GSR). When one video ended, the experimenter self-assess the arousal level, valence level, liking level, and, dominance level, and score each of the four a continuous value between 1 and 9. We evaluate emotion by two indicators namely the valence and the arousal within our research, referring to the arousal-valence scale proposed by Russell [16]. Illustrated in Fig. 2, the emotional 2-D

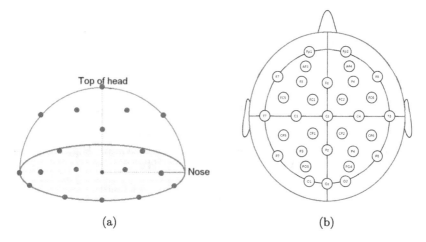

Fig. 1. The EEG cap layout for 32 channels according to the 10–20 System. (a) Locations of electrodes in the 3-D space. (The red points represent electrodes and we just show 18 of them because others are covered up by the front spherical surface.) (b) The vertical view of EEG cap. (Color figure online)

plane is divided into four quadrants, which are respectively high-valence high-arousal area (HVHA), high-valence high-arousal area (HVLA), low-arousal area (LVLA) and low-valence high-arousal area (LVHA).

In contrast to the original data, the preprocessed version also provided to researchers and used in our work is a collection of recordings downsampled from the original recordings to the rate of 128 Hz, and filtered with a bandpass of 4 to 45 Hz after the elimination of EOG artifacts. The dataset consists 40 trials of 32 subjects, each of which lasts for 60 s. In order to multiply the samples, in our work, each one-minute trial is segmented by windows of 8 s with 50% overlapping. Therefore, for every subject, we have 560 samples (14 samples × 40 trials) to train our classification model. As for labels, we divide valence and arousal scales into high/low levels (5 is the demarcation point), respectively. And the samples which are segmented from the same trial are attached the same label.

3.2 Converting EEG Signals to Images

In EEG tests or experiments, an worldwide accepted tool we use for the description and application of scalp electrode positions is 10–20 System. When the system was designed, the correlation of the electrode location to the corresponding cerebral cortex region is taken into full consideration. Thus, we transform chain-like EEG vectors into images, which can reveal the relative positions of the electrodes, to take advantage of the spatial features of EEG signals.

The PSD approach, based on FFT, is reported in many papers to be effective in indicating the brain activity and has been studied extensively in the scene of EEG-based emotion identification. The EEG signals in a trial include 32

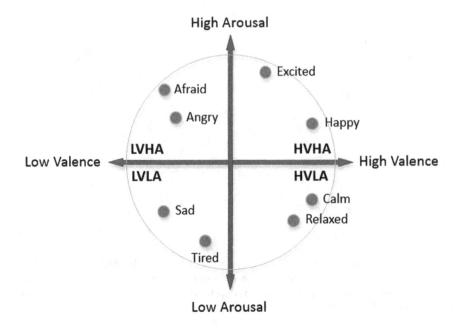

Fig. 2. The valence-arousal dimension emotion model.

channels, corresponding to 32 electrodes, and the raw data of each channel is a 1-D data vector. In our work, we decomposed the segmented samples to separate frequency bands, the alpha band of 8–13 Hz, the beta band of 13–30 Hz, the gamma band of 30–40 Hz, and the theta band of 4–8 Hz. And then averaged PSDs over four bands were extracted in every channel.

As mentioned above, we want to utilize the spatial features of EEG signals. But the EEG electrodes are distributed over the scalp in a 3-D space, so we need to find a method for projecting the locations of electrodes from a 3-D space onto a 2-D surface. Borrowing from map projection, we used the Azimuthal Equidistant Projection (AEP), as the cap worn on a human's head is spherical, so is the global. In mapping applications, this method ensures that all points on the map are at proportionally correct distances and correspondingly correct direction from the chosen center point [1]. The South Pole or the North Pole is usually regarded as the center point, and all meridians are shown as straight lines and all parallels are shown as concentric circles whose shared center is the pole point. As for our problem, Cz is chosen to be the center point and given the coordinate (φ_1, λ_0), with φ referring to latitude and λ referring to longitude in the Geographic Coordinate System. The points that have the same azimuth, which is defined that the line subtends from the vertical, will project along a straight line from the center. And the azimuth is marked by θ. We denote the distance between another point of projection and center point by ρ. So the Cartesian coordinate of the point (θ, ρ) is computed by the following set of equations:

$$x = \rho \sin \theta \tag{1}$$

$$y = -\rho \cos \theta \tag{2}$$

For a point, given the Geographic coordinate, we should transform φ and λ into θ and ρ, and the relationships between them are shown as Eqs. (3) and (4).

$$\cos \rho = \sin \varphi_1 \sin \varphi + \cos \varphi_1 \cos \varphi \cos (\lambda - \lambda_0) \tag{3}$$

$$\tan \theta = \frac{\cos \varphi \sin (\lambda - \lambda_0)}{\cos \varphi_1 \sin \varphi - \sin \varphi_1 \cos \varphi \cos (\lambda - \lambda_0)} \tag{4}$$

Using AEP to the Geographic coordinates of electrodes in 3-D space, we obtain their Cartesian coordinates in 2-D surface. Finally, Clough-Tocher scheme [2] is applied for interpolating the scattered PSDs corresponding to electrodes and estimating the values in-between the electrodes over a 32×32 mesh. Up to this point, we have a processed dataset shaped $32 \times 560 \times 32 \times 32 \times 4$ (subjects $\times$ samples/subject $\times$ height $\times$ width $\times$ frequency bands).

3.3 Model

Although CNN has already been used to solve various problems, it has always been considered remarkably effective in dealing with the variability of 2-D shapes since it appeared. CNN learns the features of images through the cooperation of convolutional layers and sub-sampling layers. During the learning process, it adjusts parameters and updates weights through the backpropagation algorithm. The algorithm finishes this task by Eq. (5):

$$w(t + 1) = w(t) + \eta \delta(t) x(t) \tag{5}$$

$x(t)$ being the output of a neuron and $\delta(t)$ represents its error, and η the step size.

When the feature map output by the last layer is input to the current convolution layer with several trainable filters, the filter will act on the feature map by convolution operation, the acting result then pass to an activation which subsequently produces another map that can be a combination of multiple input maps, as shown in Eq. (6):

$$x_\beta^\gamma = f \left(\sum_{\alpha \in M_\beta} x_\alpha^{\gamma-1} \cdot k_{\alpha\beta}^\gamma + b_\beta^\gamma \right) \tag{6}$$

where f is a predetermined activation function, M_β denotes a collection of inputs of the convolution layer, and b assigned to the output is an additive bias. Moreover, an output map could be the result of the input maps convoluting to different filters k.

Reducing the sizes of maps without reducing the number, sub-sampling layers also plays important roles in CNN networks. It can be formulated as

$$x_\beta^\gamma = f\left(B_\beta^\gamma \ sub\left(x_\beta^{\gamma-1}\right) + b_\beta^\gamma\right) \tag{7}$$

where $sub(\cdot)$ is a sub-sampling function, and with respect to a particular output map, b denotes additive bias, and B the multiplicative bias.

Now, we modify the VGG network, first proposed in [17] and proved to be good at solving various kinds of problems related to image recognition in different fields. For a colorful picture, RGB color model has three channels of red, green and blue. Similarity, each of our samples has four 2-D meshes, feature maps of theta, alpha, beta and gamma bands, so we can regard every sample as an image having four "colors". Therefore, CNN is exactly suitable for our problem.

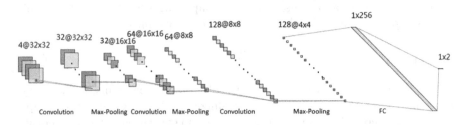

Fig. 3. Proposed CNN structure. Each plane is a feature map.

There are three convolutional layers in our model, each having a layer of maximum pooling tailed. For all convolutional layers, the shape of filters is 3×3 and the stride is 1 "pixel". And the max-pooling layers are over 2×2 blocks, with stride 2. Passed through these six layers, the tensor becomes thinner. It then passes to the Fully Connected (FC) layer of 256 units with dropout rate 0.5 [18]. The final output is a 2-dimension vector, representing the number of classes for valence and arousal, respectively. All hidden layers use Rectified Linear Units (ReLU) [8] and the output layer is equipped with softmax as non-linear activation function. Our model is detailed in Fig. 3.

Our network is implemented with the TensorFlow framework. The loss function for the binary classification is the binary cross entropy, optimized by Adam [6]. The network parameters are updated at learning rate that equals 0.001. In addition, the model performance will be measured by the average of 10-fold cross validations.

4   Results and Discussion

Our model was evaluated by using the publicly available DEAP dataset. In the binary classification, the state of indicators in consideration, the valence and the arousal, was only high or low, which was identified by virtue of the features from

Table 1. Accuracy in the comparison. All of the methods are performed on DEAP dataset and for binary classification (high and low).

Study	Valence classification accuracy	Arousal classification accuracy
Li et al. [11]	72.06%	74.12%
Thammasan et al. [20]	73.30%	72.50%
Tripathi et al. [21]	81.41%	73.36%
Mert and Akan [13]	72.87%	75%
Our method	**81.64%**	**80.25%**

EEG signals in frequency and spatial dimensions and a CNN discriminator. We compare our model with others on the same dataset.

Table 1 summarizes the accuracy of all methods in comparison. We observe that our method achieves accuracy of 81.64% and 80.25% for valence and arousal respectively, outperforming the previous studies mentioned above, in which little attention has been devoted to spatial features. Even though features of EEG signals in time and frequency dimensions are indispensable, spatial features play significant roles in EEG-based emotion recognition. In [21], Tripathi et al. convert DEAP data into 2-D arrays and used a CNN model for classification, but they only extracted several statistic features and arranged them from top to bottom in 2-D mesh-like arrays according to the serial numbers of channels, neglecting the position relationships of electrodes. On the contrary, the spatial features of EEG signals are attached great importance in our work. We shifted the data into an image that has features from different frequencies and preserves the topology, instead of vectorizing the time feature or frequency feature as done by traditional EEG analyses. In this way, the data structure could be retained throughout the process.

Table 2. Accuracy (%) of the binary classification under different window sizes.

Dimension	1s	2s	3s	4s	5s	6s	8s	10s	12s	15s	20s	30s	60s
Valence	68.52	70.04	72.46	75.30	77.73	79.48	81.64	80.45	78.47	78.13	76.41	77.34	75.39
Arousal	69.91	71.88	73.20	75.69	76.92	78.70	80.25	79.62	77.78	78.24	75.62	74.22	72.57

Time window size also has an influence on the classification accuracy. The samples segmented by size-varying time window were put into our model for training and testing, and the mean accuracy is illustrated in Table 2. And for intuition, the line chart of the results is also plotted as Fig. 4. As is shown, when the size of time window is less than 8s, the accuracy for both valence and arousal increases as the window becomes larger. And the highest accuracy is obtained when the size is 8s, 81.64% for valence and 80.25% for arousal. So we select 8s as the time length of samples in our work. In practical applications of HCI, this length is both real-time and high accuracy when the data collection and emotion recognition are performed. Although user's emotion can be recognized more quickly if we undertake the emotion recognition every 1s, 2s or 3s, the accuracy is lower. After 8s, the accuracy is generally declining, except for 30s for valence and 15s for arousal. For this phenomenon, our inference is that too small time window has too little information to present a subject's emotion, while his/her emotion may change several times in a window for too long time.

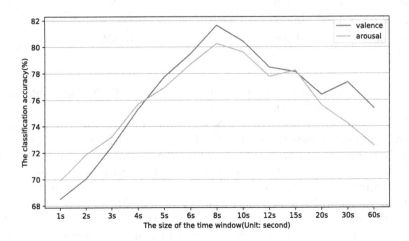

Fig. 4. The variation of the classification accuracy with different time window sizes.

To gain a better understanding of the functionality of separate frequency bands, we identified the level of the indicators only using feature maps from every single band, where the CNN (Fig. 3) structure was revised for receiving the feature map shaped in 2-D matrix as the input. The results of binary classification for valence and arousal respectively are shown in Table 3. Obviously, it is the fact that all frequency bands together generated better results than using a single band. And giving accuracy at more than 74% in both of the two dimensions, higher bands of beta or gamma has superior performance to the lower ones, which are theta and alpha, implying higher frequency features from EEG signals are more related to valence and arousal than the lower, a finding uniform to conclusion of [5].

Table 3. Mean classification accuracy for valence and arousal in different frequency bands.

Dimension	Frequency band				
	Theta	Alpha	Beta	Gamma	All
Valence	65.68%	68.36%	76.84%	79.58%	81.64%
Arousal	61.83%	64.34%	74.16%	77.96%	80.25%

5   Conclusion

From this research, a method of EEG-based emotion recognition is proposed. We shift EEG signals to the multi-spectral images that preserve the features of EEG signals in frequency and spatial dimensions. And then for identifying human emotion from the images containing EEG features, we construct a CNN model. Evaluated by the publicly available DEAP dataset, the results show that the mean classification accuracy is 81.64% for valence (low and high) and 80.25% for arousal (low and high), respectively, across 32 subjects, which outperform other methods in comparison.

Later the influences of four separate frequency bands on EEG-based emotion recognition are studied through the experiment. The findings are that better results can be gotten through using the combination of four frequency bands than a single band and that high frequency bands are more related to valence and arousal than low frequency bands.

References

1. Wikipedia.   http://en.m.wikipedia.org/wiki/Azimuthal_equidistant_projection. Accessed 10 Jul 2019
2. Alfeld, P.: A trivariate clough–tocher scheme for tetrahedral data. Comput. Aided Geom. Des. **1**(2), 169–181 (1984)
3. Bashivan, P., Rish, I., Yeasin, M., Codella, N.: Learning representations from EEG with deep recurrent-convolutional neural networks. arXiv preprint arXiv:1511.06448 (2015)
4. Horlings, R., Datcu, D., Rothkrantz, L.J.: Emotion recognition using brain activity. In: Proceedings of the 9th International Conference on Computer Systems and Technologies and Workshop for PhD Students in Computing, p. 6. ACM (2008)
5. Jatupaiboon, N., Pan-ngum, S., Israsena, P.: Emotion classification using minimal EEG channels and frequency bands, pp. 21–24. IEEE (2013)
6. Kingma, D.P., Ba, J.: Adam: a method for stochastic optimization. arXiv preprint arXiv:1412.6980 (2014)
7. Koelstra, S., et al.: Deap: a database for emotion analysis; using physiological signals. IEEE Trans. Affect. Comput. **3**(1), 18–31 (2012). https://doi.org/10.1109/T-AFFC.2011.15
8. Krizhevsky, A., Sutskever, I., Hinton, G.E.: Imagenet classification with deep convolutional neural networks. In: Advances in Neural Information Processing Systems, pp. 1097–1105 (2012)
9. Lahane, P., Sangaiah, A.K.: An approach to EEG based emotion recognition and classification using kernel density estimation. Procedia Comput. Sci. **48**, 574–581 (2015). https://doi.org/10.1016/j.procs.2015.04.138. item_number: S187705091500647X
10. LeCun, Y., Bottou, L., Bengio, Y., Haffner, P.: Gradient-based learning applied to document recognition. Proc. IEEE **86**(11), 2278–2324 (1998)
11. Li, X., Song, D., Zhang, P., Guangliang, Y., Hou, Y., Hu, B.: Emotion recognition from multi-channel EEG data through convolutional recurrent neural network, pp. 352–359. IEEE (2016)

12. Liu, J., Meng, H., Nandi, A., Li, M.: Emotion detection from EEG recordings. In: 2016 12th International Conference on Natural Computation, Fuzzy Systems and Knowledge Discovery (ICNC-FSKD), pp. 1722–1727. IEEE (2016)
13. Mert, A., Akan, A.: Emotion recognition from EEG signals by using multivariate empirical mode decomposition. Pattern Anal. Appl. **21**(1), 81–89 (2018). https://doi.org/10.1007/s10044-016-0567-6. identifier: 567
14. Petrantonakis, P.C., Hadjileontiadis, L.J.: Emotion recognition from EEG using higher order crossings. IEEE Trans. Inf. Technol. Biomed. **14**(2), 186–197 (2010). https://doi.org/10.1109/TITB.2009.2034649. item_number: 5291724
15. Rozgić, V., Vitaladevuni, S.N., Prasad, R.: Robust EEG emotion classification using segment level decision fusion. In: 2013 IEEE International Conference on Acoustics, Speech and Signal Processing, pp. 1286–1290. IEEE (2013)
16. Russell, J.A.: A circumplex model of affect. J. Pers. Soc. Psychol. **39**(6), 1161–1178 (1980). https://doi.org/10.1037/h0077714. identifier: 1981-25062-001
17. Simonyan, K., Zisserman, A.: Very deep convolutional networks for large-scale image recognition. arXiv preprint arXiv:1409.1556 (2014)
18. Srivastava, N., Hinton, G., Krizhevsky, A., Sutskever, I., Salakhutdinov, R.: Dropout: a simple way to prevent neural networks from overfitting. J. Mach. Learn. Res. **15**(1), 1929–1958 (2014)
19. Thammasan, N., Moriyama, K., Fukui, K.I., Numao, M.: Continuous music-emotion recognition based on electroencephalogram. IEICE Trans. Inf. Syst. **E99.D**(4), 1234–1241 (2016). https://doi.org/10.1587/transinf.2015EDP7251
20. Thammasan, N., Moriyama, K., Fukui, K.I., Numao, M.: Familiarity effects in EEG-based emotion recognition. Brain Inform. **4**(1), 39–50 (2017). https://doi.org/10.1007/s40708-016-0051-5. identifier: 51
21. Tripathi, S., Acharya, S., Sharma, R.D., Mittal, S., Bhattacharya, S.: Using deep and convolutional neural networks for accurate emotion classification on deap dataset. In: Twenty-Ninth IAAI Conference (2017)

Distributed Subgraph Matching Privacy Preserving Method for Dynamic Social Network

Xiao-Lin Zhang$^{(\boxtimes)}$, Hao-chen Yuan, Zhuo-lin Li, Huan-xiang Zhang, and Jian Li

Inner Mongolia University of Science and Technology, Baotou, P. R. China
zhangxl@imust.cn, mryuanhaochen@163.com

Abstract. The growing popularity of cloud platforms store and process large-scale social network data, if we do not pay attention to the method of using a cloud platform, privacy leakage will become a serious problem. In this paper, we propose a distributed k-automorphism algorithm and a distributed subgraph matching method, the distributed k-automorphism algorithm can efficiently protect the privacy of the social networks in the cloud platform by adding noise edges to ensure k-automorphism and the distributed subgraph matching method can quickly obtain temp subgraph matching results. After temp results are joined, we can obtain correct results by recovering and filtering temp results according to the symmetry of the k-automorphism graph and k-automorphism functions in the client. We also propose a modified method that utilizing incremental thought to solve the problem of dynamic subgraph matching. The experiments show that the above methods are effective in dealing with large scale social network graph problem and these methods can effectively solve the problem of privacy leakage of subgraph matching.

Keywords: Cloud platform · Subgraph matching · Protecting privacy · Distributed

1 Introduction

With the rapid development of social networks, social networks contain richer sensitive information, for example, Facebook, Twitter, and Weibo, etc. As a consequence, we need to process data in the cloud platform to reduce time costs. However, social network data contains user's sensitive information, sensitive information will leak when we handle data in error way. Many scholars have studied how to protect private information. The paper [1] proposes three ways of attack privacy information based on background information like structural recognition. The paper [2] assumes that the attacker only uses one structural attack method, but the attacker may use multiple structural attack methods to obtain private information. The paper [7] proposes an algorithm called KM that ensures k-automorphism, the algorithm can protect the social network data well. So we decide to use the k-automorphism algorithm as basic privacy preserving algorithm and the reason of using distributed k-automorphism social network privacy protection algorithm as an algorithm to anonymize social network is as

H. Jin et al. (Eds.): BigData 2019, CCIS 1120, pp. 159–173, 2019.
https://doi.org/10.1007/978-981-15-1899-7_12

follows: Because the algorithm does not delete any edges so it will not lose any data, but traditional KM algorithm processes big data graph slowly. So we use an algorithm called DKA (distributed k-automorphism) to protect social network privacy quickly.

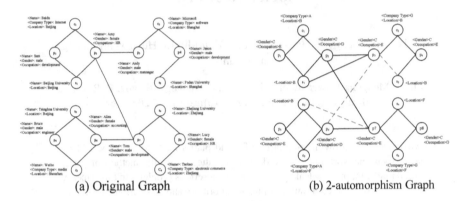

(a) Original Graph   (b) 2-automorphism Graph

Fig. 1. Data graph

Table 1. Label generalization group table

Label group	Labels
A	{Internet, Media}
B	{Shanghai, Beijing}
C	{male, female}
D	{HR, accounting}
E	{development, manager, engineer}
F	{Shenzhen, Zhejiang}
G	{electronic commerce, software}

Meanwhile, more people utilize the cloud platform as their tools to analysis data. We mainly focus problem of subgraph matching in this paper. When you get the k-automorphism anonymous graph G^k, if you only upload the graph to cloud, the data is so large that the subgraph matching will run slowly and you could not get the right results. It is very important to protect privacy information when someone needs to run matching algorithms in cloud platforms [3–6]. The k-automorphism algorithm [7] is a traditional social network privacy protection algorithm, the traditional k-automorphism algorithm adds noise edges to the original graph [8]. Figure 1(a) is an original social network graph. As shown in Fig. 1(b), k-automorphism algorithm can resist structural attacks well. However, this method adds three noise edges, which leads to false matching results and a large amount of space and time consumption in the cloud.

So we propose a solution:

1. We use a distributed k-automorphism social network privacy protection algorithm called DKA (distributed k-automorphism) to protect social network privacy

quickly. In order to reduce the cost of subgraph matching time and space, we only upload a part of k-automorphism graph and k-automorphism function to the cloud platform.

2. We propose a distributed subgraph matching method for performing subgraph matching in the cloud platform, it can get temp matching results about k-automorphism graph quickly and we can use modified algorithms to solve problems about temp results is wrong for original date graph and make matching method fit in dynamic graph data.

3. Experiments prove that the above methods ensure that the social network does not leak privacy when uploading data to the cloud for subgraph matching and reduce costs of space and time.

Organization: Section 2 introduces related work and the background information of the research of this paper. Section 3 presents a privacy protection algorithm and the rule of how to upload anonymized graph into the cloud platform. The method of distributed subgraph matching in dynamic graph is proposed in Sect. 4. Section 5 presents our experimental results, followed by conclusion for this paper in Sect. 6.

2 Related Work

Many scholars have studied the method of protecting structural privacy information on the social network. These studies can be used as the basic method to protect the private information of social networks. In papers [9–12] proposed privacy protection algorithms based on deleting edges, the deletion of the edge may cause the uploaded graphs to lose important information. The basic method of using distributed k-automorphism social network privacy protection algorithm as an algorithm to anonymize social network is as follows: Because the algorithm does not delete an edge, so it keeps important information in the anoymized graph; we utilize the symmetry of k-automorphism to restore subgraph matching results in the client; distributed k-automorphism social network privacy protection algorithm is suitable for a large scale social network environment.

Paper [13] proposed that the LP model can anonymize the social network graphs contain weight and preserve the shortest path between nodes in the graph, but an attacker can reidentify an anonymous graph when the attacker has arbitrary topological background information and can't protect the accuracy of the subgraph matching results in the client. The paper [14] proposed a social network privacy protection method to protect the correctness of subgraph matching results. The paper [6] propose a distributed subgraph matching method, the paper did not discussed the case of dynamic graphs.

So this paper proposes DKA algorithm to protect the privacy of the social network graph in cloud platforms and a distributed dynamic subgraph matching method suitable for large scale social networks is proposed to ensure efficient subgraph matching in the cloud platform and privacy is not leaked, at last, correct subgraph matching results are obtained by restoring and filtering the matching results.

3 Social Network Privacy Protect Method

3.1 Preliminary

Definition 3.1: Undirected attributed graph [2]. Representing a social network as an undirected attributed graph $G = (V(G), E(G), T, \Gamma, L)$. (1) $V(G)$ is a set of vertices of graph G; (2) $E(G) \subseteq V(G) \times V(G)$, $E(G)$ is a set of undirected edges of the graph G; (3) T is the type of vertex, there is only one type per vertex; (4) Γ is a set of attribute of a vertex, a vertex maybe have lots of attributes or only one attribute, and the types of vertices are different, the attributes of the vertices are different. (5) L is a set of labels of vertices of graph. For example, in Fig. 1, the type of vertex c_1 is c (company), the type of p_1 is p (person).

Definition 3.2: Graph Automorphism [6]. An automorphism of a graph $G = (V(G), E(G), T, \Gamma, L)$ is an automorphic function f of the vertex set V, such that for any edge $e = (u, v)$, $f(e) = (f(u), f(v))$ is also an edge in G, i.e., it is a graph automorphism from G to itself under function f. If there exist k-automorphisms in G, it means that there exists $k - 1$ different automorphic functions.

Definition 3.3: k-automorphism function [7]. If v is a vertex in a k-automorphism G^k, v and its corresponding $k - 1$ symmetric vertices form an alignment vertex instance AVI, all AVIs are stored in an alignment vertex table AVT, each row in the AVT table represents a AVI. Building k-automorphism function based on the AVT table of k-automorphism graph G^k: F_i ($i = 0,..., k - 1$).

$$F_0(v) = v, \ F_i(v) = F_{i-1}(v)._{\text{next}} \text{ for } 1 \leq i \leq k - 1. \tag{1}$$

A 2-automorphic graph G^k generated from G is given in Table 2. Each row of AVT shows k symmetric vertices. For example, in Table 2, the first row of AVT contains p1 and p5, meaning p1 and p5 are symmetric vertices. Each column of AVT shows all vertices in one block of the k-automorphic graph. For example, there are two blocks in the 2-automorphic graph in Fig. 1(b), based on AVT, k-automorphism function as follows: $F_0(p_1) = p_1$; $F_1(p_1) = p_5$.

Definition 3.4: Subgraph matching [6]. For a graph and a subgraph query q, the goal of subgraph matching is to find every subgraph $Q = (V(Q), E(Q), T, \Gamma, L)$ in $G = (V(G), E(G), T, \Gamma, L)$, if there is at least one bijection f: $V(Q) \rightarrow V(G)$ such that:

$$\forall q_i \in V(Q), g(q_i) \in V(G) \Rightarrow L(q_i) \subseteq L(g(q_i)). \tag{2}$$

$$\forall q_i, q_j \in V(Q), \text{edge}(q_i, q_j) \in E(Q) \Rightarrow \text{edge}\left(g(q_i), g(q_j)\right) \in E(G). \tag{3}$$

Then Q is an isomorphic subgraph of G. If Q is an isomorphic subgraph of G. Finding all subgraphs in the graph G that are homogeneous with Q is subgraph matching. The subgraph matching result of the query graph Q on the data graph G is denoted as $R(G)$.

We believe that the cloud platform will run exactly as required by the user. But the data storage and operation is in the cloud platform provided by the third party, so there is a danger of privacy leakage for data providers. If an attacker knows the structure information of the vertex t in the uploaded graph in the cloud platform, then attackers can identify the vertex in the upload graph with a higher probability. So the privacy of the vertex will leak.

3.2 Distributed K-Automorphism Algorithm

The goal of social network privacy for subgraph matching is to protect the privacy of social networks when making subgraph matching in the cloud platform and getting correct subgraph matching results in the client.

Note that for ease of presentation, we use terms "graph" and "social network" interchangeably. According to the symmetry of k-automorphism graph, a new uploading scheme is proposed to reduce the cost of subgraph matching in a cloud platform.

The DKA algorithm consists of two parts: First, we use the MLP algorithm [16] to make the original graph be parted to k blocks by multi-level label passing method, and obtain the AVT table according to k blocks. Then we add noise edges within the block and add noise edges between blocks by the marked vertices. Utilizing features of Trinity [17] framework that can send and receive messages between adjacent vertices, storing the vertices of the graph in the compute nodes. The addition of noise edges of a block is mainly based on sending marked information between adjacent vertices in the same column of the AVT table. For example, we divide graph such as Fig. 1 into 2 blocks, and we get AVT table like Table 2; vertices in compute nodes send their marked message according to table AVT. Completed algorithm as follows:

```
Algorithm 1: Adding noise edges within the block
Input: AVT; G°;
Output: Gᵏ, ;
int line=AVT.row.count;
for(i=0;i<line;i++)
    {
        for(j=i+1;j<line+1;j++)
            {
                i.vertex sendmess to the j.vertex;
                while(j.vertex be marked and not
all.j.vertex be marked)
                    addedge(i,j);
            }
    }
```

The addition of noise edges between blocks is mainly based on sending marked information between adjacent vertices in different columns of the AVT table. For example in Table 2, vertex p_2 send marked message to vertex p_7, when vertex p_7 receives the marked information, it marks itself, then add an edge $(F_2(p_2), F_2(p_7))$, F is the k-automorphism function. Completed algorithm as follows:

```
Algorithm 2: Adding noise edges between blocks
Input:  AVT; G_k, ;
Output: G^k;
int line=AVT.row.count;
for(i=0;i<line;i++)
{
   for(j=i+1;j<line+1;j++)
     {
       i.vertex sendmess to the j.vertex;
       while (j.vertex be marked and not all.j.vertex be
marked)
       addedge(F_i(i.vertex),F_i(j.vertex));
     }
}
```

Table 2. AVT table

k_1	k_2
p_1	p_5
p_2	p_6
p_3	p_7
p_4	p_8
c_1	c_3
c_2	c_4
s_1	s_3
s_2	s_4

3.3   Uploaded Graph

When we get the k-automorphism graph, if we upload it directly to the cloud, it will lead to huge storage costs and time costs. Therefore, we only upload a part of the graph and the k-automorphism function to the cloud. We will run the subgraph matching program in the cloud platform because the k-automorphism graph is symmetry, the accurate subgraph matching result can still be obtained in the client. The method can reduce the storage space, increase the speed of subgraph matching.

Definition 3.5: Uploaded graph [2]. For a k-automorphism graph $G^k = (V(G^k),$ $E(G^k)$ T, Γ, L), its uploaded graph is $G^u = (V(G^u), E(G^u), T, \Gamma, L)$, (1) $V(G^u)$ consists of a vertices set $V(B_1)$ of the first block of G^k and all one-hop neighbor vertices of $V(B_1)$; (2) $E(G^u)$ is a subset of the undirected edge $E(G^k)$, and it connects the vertices in

$V(B_1)$ and the vertices of $V(B_1)$ and all one-hop neighbor vertices of $V(B_1)$. For example, Fig. 2. is uploaded graph of Fig. 1(a).

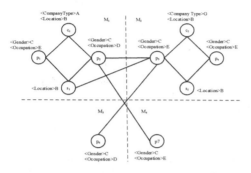

Fig. 2. Uploaded graph

4 Distributed Dynamic Subgraph Matching Method

The distributed dynamic subgraph matching method consists of three parts: Decomposing the query graphs, subgraph matching in a cloud platform, and subgraph matching result processing.

4.1 Decomposing the Query Graphs

Definition 4.1: Decomposition of query graphs: Let Q be the query graph and $S = \{QSG_1,..., QSG_n\}$, S is a set of QSG (query subgraph). Any edge of Q is only contained in one QSG_i. We call set S is a QSG coverage of searching graph Q.

For example, Fig. 3(a) is a query graph, it means to search for two people who are involved in the school in Beijing, and their work in the Internet and software company, P, S, and C, respectively represent the types of vertices. Figure 3(b) and (c) is QSGs of the original query graph Fig. 3(a). A QSG is a tree that has 2 layers. Recorded as QSG = {r, L}, r represents the label of the root node, L is a set of labels of the child node of r.

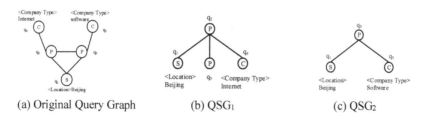

(a) Original Query Graph   (b) QSG₁   (c) QSG₂

Fig. 3. Query graph

Theorem 4.1: The minimum QSG coverage problem is equivalent to the minimum node coverage problem.

Proof: Let S is a minimum coverage of QSG, and V is a set of root vertices of all QSG$_i$. Any edge is connected to at least one vertex in the set V, so V is coverage of vertices of Q. If there exist other coverage of V, make $|V'| < |V|$, we build a set S, consists of vertices of V' and $v \in V'$, making the vertices of searching graphs and edge associated with vertex V form a QSG. Randomly delete edges in QSGs until any edge belongs to one QSG, so QSGs of S' have at least one edge. Obviously, $|S'| \leq |V'| < |V| = |S|$, so S is not a minimum QSG coverage.

The minimum vertex coverage problem is an NP-hard problem, so the QSG coverage problem is an NP-hard problem too. The 2-approximate algorithm is a classical algorithm to deal with the minimum vertex coverage problem. The algorithm chooses an edge (u, v) in every step, then adding u, v to the result set and delete all edges that are connected to u, v. We repeat the above steps until all edges are deleted. We will modify the algorithm to make it suitable for the minimum QSG coverage problem.

The algorithm mainly modifies the method that chooses edges, we choose edge by their selectivity. Setting the function:

$$f(v) = \frac{deg(v)}{freq(v)} \tag{4}$$

$f(v)$ expresses selectivity of vertex v, $deg(v)$ expresses the degree of vertex v, $freq(v)$ expresses the number of occurrences of the label of vertex v. The first rule ensures that the root nodes of the QSGs are bound by previous QSGs. The second rule favor generating QSGs of the higher value of f(v) in order to reduce the size of intermediary join results. The complete algorithm as follows:

```
Algorithm 3: Decomposing the query graphs
Input: Q
Output: QSGs
D=∅
H=∅
while Q has edge
{
   if(D=∅)
pick an edge (v,u) that f(u)+f(v) is the largest;
else
pick an edge (v,u) that v∈S and f(u)+f(v) is the largest;
Hᵥ=the QSG rooted at v;
add Hᵥ to H;
   D=D ∪ neighbor(v);
remove edges in Hᵥ from Q;
if deg(u)>0
Hᵤ=the QSG rooted at u;
add Hᵤ to H;
remove edges in Hᵤ from Q;
D=D ∪ neighbor(u);
   remove u,v and all nodes with degree 0 from D;
}
```

For example, in Fig. 3(a), we choose the edge q_2q_3 first, because the value of $f(q_2) + f(q_3)$ is highest; then $QSG_1 = \{q_2, (q_1, q_3, q_4)\}$ and $QSG_2 = \{q_3, (q_1, q_5)\}$. After that, no edge is left and algorithm halts.

Regarding the time complexity of the algorithm, we note that computing f-values and sorting vertices by f-values have $O(n\log n)$ cost, where n is the number of vertices in the query graph. In each round, at least two vertices are removed from Q, so the iteration needs $O(n)$ steps. Thus, the time complexity of Algorithm is $O(n^2 \log n)$.

4.2 Subgraph Matching in Cloud Platform

With graphs stored in the cloud platform, and a string index mapping vertices label to vertex IDs. For prevent leakage of privacy, we replace each vertex of QSGs label with the corresponding label group according to the label generalization group table like Table 1. We run algorithm 4 in every compute node to get subgraph matching results. Specifically, a QSG is a two-level tree structure. We use $QSG = (r, L)$ to denote a QSG, where r is the label of the root vertex and L is the set of labels of its child vertices. The complete algorithm as follows:

```
Algorithm 4: Subgraph matching
Input: QSGs, Gu;
Output: R;
Sr ←Index.getID(r); // find vertex id labeled R
R= ∅;
for each n in Sr do
  c=Cloud.Load(n); //Load the vertex with id as n
  for each li in L do
      S_l_i={m|m∈c.children and Index.hasLabel(m,l)};
  R=R∪{{n}× S_l_i × S_l_2 × ...× S_l_k};
return R;
```

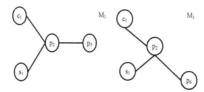

(a) Subgraph Matching Results of QSG_1

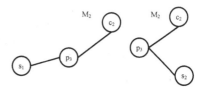

(b) Subgraph Matching Results of QSG_2

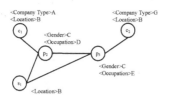

(c) Subgraph Matching Result R_{in}

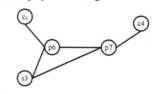

(d) Mapping Result R_{out}

Fig. 4. Subgraph matching result

4.3   Processing Subgraph Matching Result

Obviously the result of subgraph matching for uploading graph G^u is not equal to the result of subgraph matching of complete k-automorphism graph. We can utilize these results of subgraph matching for uploading graph G^u and k-automorphism function to obtain complete subgraph matching result. For example, we use result R_{in} of Fig. 4(c) and AVT table of Table 2, we can obtain F_1 $(c_1) = c_3$, F_1 $(p_2) = p_6$, F_1 $(p_3) = p_7$, F_1 $(c_2) = c_4$, F_1 $(s_1) = s_3$, get the mapping results like Fig. 4(d), R_{out}. $R(G_k) = R_{in} \cup R_{out}$ is subgraph matching result of k-automorphism graph. So Fig. 4(c) and (d) are subgraph matching results of the anoymized graph.

Subgraph matching results of k-automorphism graph $R(G^k)$ is different from the result of original graph $R(G^o)$. Due to the addition of edges in the original graph, the result of subgraph matching is likely to increase so the result will be inaccurate for the owner of the original data. Therefore, after the client gets the result of k-automorphism graph, the client should also filter the result of subgraph matching, delete the matching result that vertices or edge do not exist in the original graph; delete the label of vertices and results which is different from the corresponding labels of the vertex in the query graph. For example, edge p_7s_3 in Fig. 4(d) does not exist in the original graph, so the right subgraph matching result is Fig. 4(c).

4.4   Incremental Subgraph Matching

The above methods about subgraph matching only apply to the static state. When utilizing static subgraph matching methods to deal with the dynamic data graph, it will produce much redundant calculation. For example, static methods need to match complete data graph after the data graph is updated. So we will use a incremental algorithm [17] method proposed by the paper to modify the above methods.

Proposition 4.1 [18]: Given a normal query graph QSG and a data graph G, only the deletions of ss edges for some QSGs in G may reduce the subgraph matching results. In other hands, inserting edges into a data graph G may only add new matches to subgraph matching results, in other words, it may only add new vertices and edges to the subgraph matching graph G_r. There are two groups of edges that, when added to G, may yield new matches, referred to as cc edges and cs edges. A newly inserted edge (v', v) is a cs (resp. cc) edge for a QSG edge (u', u) if $v' \in cand(u')$ and $v \in match(u)$ (resp. $v \in cand(u)$). One can verify the following:

Proposition 4.2 [18]: (1) For a query graph G_q, only insertions of cs edges into a data graph G may increase matches of query graph G_q. (2) For a general query graph G_q, only insertions of cs or cc edges into G may add new matches of G_q. (3) Moreover, cc edges alone only add new matches for vertices of query graph in some strongly connected component of G_q.

Definition 4.3: Updated Unit. We used to denote the diameter of a QSG, the length of the longest shortest path in the QSG. If a edge $e = (u, v)$ in the data graph G is deleted or added, let $V(d, e)$ be the set of vertices in G that are within a distance d of both u and v. The subgraph of G induced by $V(d, e)$, the subgraph of G consisting of vertices in

V(d, e) along with edges of G connecting theses vertices. We call the subgraph is a updated unit.

Proposition 4.3: Given a QSG, a update unit and a data graph, the changed matching results are the difference between matching results of the updated unit and matching results of the data graph.

So we modify original subgraph matching algorithm with above methods, complete algorithm as follows: (1) find the diameter of the QSG; (2) extract the updated unit from the data graph; (3) compute subgraph matching results $R_d(.)$ in the data graph; (4) compute subgraph matching results $R_u(.)$ in the updated unit. The complete algorithm as follows:

```
Algorithm 5:  Subgraph matching in Dynamic graph with de-
letion
input: d; QSG; G₀; and a edge (u, v) to be deleted from
G₀.
output: Rd; Ru; Gt; R
find a subgraph Rd of u, v in the data within d in the
data graph;
run algorithm 4 get Ru;
R= Rd-(Rd\Ru);
```

When the edges are deleted from the data graph, the result of matching will not increase. The deleted edges, such as deleting (a, b), will result in the reduction of matching results. In this case, all matching results containing the edge (a, b) can be removed from the previous matching result set in the client.

```
Algorithm 6:  Subgraph matching in Dynamic graph with in-
sertion
input: d; QSG; G₀; and a edge (u, v) to be deleted from
G₀.
output: Rd; Ru; Gt; R
find a subgraph Rd of u, v in the data within d in the
data graph;
run algorithm 4 get Ru;
R= Rd + (Ru\Rd);
```

When the edges are added from the data graph, the result of matching will not decrease. The added edges, such as adding (a, b), will result in the increase of matching results within R_d. In this case, all subgraph matching results to be added can be found in the updated unit. Note that, the diameter of the QSG is typically small, so the time of computing R_d is obviously less than computing the subgraph matching results in the complete dynamic graph again.

5   Experiments and Results

5.1   Experimental Data Set

We performed experiments on two real data sets: roadNet-CA and roadNet-PA. Data sets roadNet-CA contains a total of 1965206 nodes and 2766607 edges; roadNet-PA contains a total of 1088092 nodes and 1541898 edges. The two datasets do not contain labels, so three labels are generated manually in each dataset.

5.2   Experimental Environment

The experiment uses a cluster of 8 computing nodes as the computing platform. The configuration of hardware configuration is CPU 1.80 Hz, RAM 16 GB. Experiments use Windows Server 2008 R2 system and Trinity computing framework.

5.3   Experimental Results

Figure 5 shows the time of using DKA algorithm to anonymize the original graph under 8 computing nodes. It can be seen that the anonymity time increases with the increase of k value because the noise edges need to be added to the anonymous graph increase with the increase of k value, so the running time of the algorithm increases with the increase of k value.

Figure 6 shows how long the DKA algorithm runs compared to the traditional k-automorphism algorithm. It is found that the cost of running time is greatly reduced compared with the traditional algorithm.

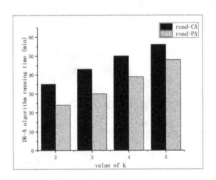
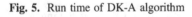
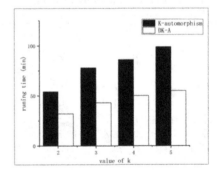

Fig. 5. Run time of DK-A algorithm   **Fig. 6.** Run time comparison of algorithm

Figure 7 compares the size of the space required for fully anonymous graphs and uploaded graphs. It is found that uploaded graphs can save a lot of space costs. At the same time, with the increase of k, uploaded graphs take up more and more space because the noise edges that anonymous graphs need to add become more and more, so the space cost of uploaded graphs becomes larger and larger.

Figure 8 shows the time required for subgraph matching when the uploaded graph is a 2-automorphic anonymous graph and the number of nodes in the search graph is 4. It can be seen that under the cloud platform, subgraph matching is very fast.

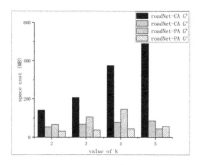

Fig. 7. Space cost of upload graph

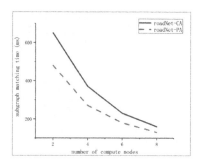

Fig. 8. Subgraph matching time

Figure 9 shows when the uploaded graph is an upload graph of a 2-automorphic anonymous graph, the matching time of subgraph increases with the number of nodes in the searching graph increasing. Because of the number of QSGs increases when the number of nodes in the searching graph increases, the matching result of QSGs increases, which requires more join operation time.

Figure 10 shows that the recovery subgraph matching results take very little time and almost do not affect the total time required for subgraph matching.

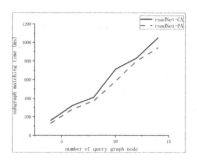

Fig. 9. Subgraph matching time about the number of query graph vertices

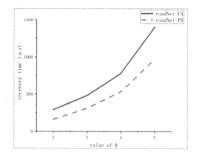

Fig. 10. Recovery of the subgraph matching result

Figure 11(a) and (b) shows time cost that we use subgraph matching method improved by IncSimMatch algorithm. We use dataset roadNet-CA as data graph and the number of query graph vertices is 4. The number of compute nodes is 6. The vertices of data set edges change by adding and deleting edges randomly. The number of edges change and matching time is used to measure algorithm efficiency. In the initial stage, since the original algorithm does not require auxiliary data structures, so the improved algorithm requires more time costs. When the number of edges changes, the modified algorithm will have less time cost, because there is no need to match the complete data graph again, especially in the case of deleting edges, the improved algorithm performs better.

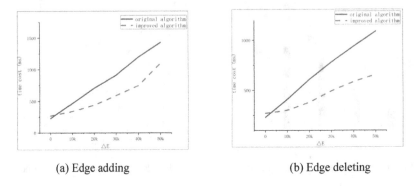

(a) Edge adding   (b) Edge deleting

Fig. 11. Subgraph matching time of updated data graph

6   Conclusions

We presented a distributed k-automorphism (DKA) anonymous algorithm to preserve privacy of the upload graph, the algorithm can increase the speed of traditional k-automorphism anonymous algorithm by distributed paralleling the traditional algorithm. We also use a new upload method for only uploading a part of the anonymous graph, by utilizing the symmetry of k-automorphism graph, we can overcome the shortcoming of not getting the right matching results on the anonymous and data graph have the large space cost. Our results show a improvement in performance on large scale datasets and DKM algorithm can save a lot of costs.

Then, we propose a distributed subgraph matching method, the method based on the incremental method can solve the problem that the time of tradition subgraph matching method running is long and the method could not fit in the dynamic graph.

Acknowledgments. This work is partially supported by Natural Science Foundation of China (No.61562065). The authors also gratefully acknowledge the helpful comments and suggestions of the reviewers, which have improved the presentation.

References

1. Hay, M., Miklau, G., Jensen, D., et al.: Resisting structural re-identification in anonymized social networks. VLDB J. **19**(6), 797–823 (2010)
2. Chang, Z., Zou, L., Li, F.: Privacy preserving subgraph matching on large graphs in cloud. In: International Conference on Management of Data, pp. 199–213. ACM (2016)
3. He, H., Singh, A.K.: Query language and access methods for graph databases. In: Aggarwal, C., Wang, H. (eds.) Managing and Mining Graph Data, pp 125–160. Springer, Boston (2010). https://doi.org/10.1007/978-1-4419-6045-0_4

4. Yan, X., Yu, P.S., Han, J.: Substructure similarity search in graph databases. In: ACM SIGMOD International Conference on Management of Data, Baltimore, Maryland, USA, June, pp. 766–777 (2005)
5. Yuan, Y., Wang, G., Chen, L., Wang, H.: Efficient subgraph similarity search on large probabilistic graph databases. Proc. VLDB Endowment **5**(9), 800–811 (2012)
6. Sun, Z., Wang, H., Wang, H., Shao, B., Li, J.: Efficient subgraph matching on billion node graphs. Proc. VLDB Endowment **5**(9), 788–799 (2012)
7. Zou, L., Chen, L.: k-automorphism: a general framework for privacy preserving network publication. Proc. VLDB Endowment **2**, 946–957 (2009)
8. Yuan, M., Chen, L., Yu, Philip S., Mei, H.: Privacy preserving graph publication in a distributed environment. In: Ishikawa, Y., Li, J., Wang, W., Zhang, R., Zhang, W. (eds.) APWeb 2013. LNCS, vol. 7808, pp. 75–87. Springer, Heidelberg (2013). https://doi.org/10.1007/978-3-642-37401-2_10
9. Tai, C.H., Yu, P.S., Yang, D. N., Chen, M.S.:. Privacy-preserving social network publication against friendship ttacks. In: ACM SIGKDD International Conference on Knowledge Discovery and Data Mining, vol. 7, pp. 1262–1270. ACM (2011)
10. Cheng, J., Fu, W.C., Liu, J.: K-isomorphism: privacy preserving network publication against structural attacks. In: ACM SIGMOD International Conference on Management of Data, SIGMOD 2010, Indianapolis, Indiana, USA, June, vol. 4, pp. 459–470. DBLP (2010)
11. Bhagat, S., Cormode, G., Krishnamurthy, B., Srivastava, D.: Class-based graph anonymization for social network data. Proc. VLDB Endowment **2**(1), 766–777 (2009)
12. Campan, A., Traian, M.: A clustering approach for data and structural anonymity in social networks. In: Privacy, Security, and Trust in KDD Workshop, PinKDD, pp. 33–54 (2008)
13. Cormode, G., Srivastava, D., Yu, T., Zhang, Q.: Anonymizing bipartite graph data using safe groupings. VLDB J. **19**(1), 115–139 (2010)
14. Das, S., Egecioglu, O., El Abbadi, A.: Anonymizing weighted social network graphs. In: IEEE, International Conference on Data Engineering, vol. 41, pp. 904–907. IEEE (2010)
15. Cao, N., Yang, Z., Wang, C., Ren, K., Lou, W.: Privacy-preserving query over encrypted graph-structured data in cloud computing. vol. 6567, no. 6, pp. 393–402 (2011)
16. Wang, L., Shao, B., Xiao, Y., Wang, H.: How to partition a billion-node graph (2014)
17. Shao, B., Wang, H., Li, Y.: Trinity: a distributed graph engine on a memory cloud. In: ACM SIGMOD International Conference on Management of Data, pp. 505–516. ACM (2013)
18. Fan, W., Wang, X., Wu, Y.: Incremental graph pattern matching. ACM Trans. Database Syst. **38**(3), 1–47 (2013)

Big Data Processing

Clustering-Anonymization-Based Differential Location Privacy Preserving Protocol in WSN

Ren-ji Huang$^{(\boxtimes)}$ ⬥, Qing Ye, and Mo-Ci Li

Department of Information Security, Naval University of Engineering,
Wuhan 430033, China
wuyingjituan@163.com

Abstract. Playing a vital role in the period of big data and intelligent life, wireless sensor networks (WSN) transmits a bulk of data. Location information as the vital data in transmission is widely used in detecting and routing for the network. With the big data mining and analysis, the security of location and data privacy in WSN faces great challenges. To the problem of active attacking like node capture in wireless sensor network node location privacy, existing location privacy preserving protocols are analyzed and Differential Location Privacy protocol based on Clustering Anonymization is proposed. By sensor nodes clustering using genetic clustering algorithm, the individual location is hidden in the statistical location information of the group. The Laplace Mechanism is also added to the protocol to realize differential location privacy. Node location privacy in WSN is preserved as well as privacy preserving budget is saved. The result of theoretical analysis and contrastive simulation experience shows that the protocol can be useful.

Keywords: Wireless Sensor Network · Location privacy preserving · Differential privacy · Clustering anonymization

1 Introduction

Our life has been deeply connected with the network in the age of big data. We can get much more information making us wiser and more convenient. Wireless Sensor Networks(WSN) are local area networks which consist of a large amount of sensor nodes through wireless communication to collect and analyze data [1]. WSN plays an important role in data interconnection and future intelligent life. Now, with oceans of data from monitored objects, it has various applications in many areas of our daily life such as transportation, environmental monitoring, health care and military reconnaissance [2]. Meanwhile the security of sensor nodes (objects) location information used in the routing and data analysis is facing great challenges by the big data analysis and data mining. The characteristic of opening, self-organization and wireless communication in WSN leads to its inherent security defects that makes it easy to be attacked. Once the sensor nodes location is obtained by an attacker, he can move forward to analyze sensitive information of the objects monitored by the WSN or destroy the sensor nodes and even the whole network. So it is vital to keep location privacy of wireless sensor nodes [3].

© Springer Nature Singapore Pte Ltd. 2019
H. Jin et al. (Eds.): BigData 2019, CCIS 1120, pp. 177–193, 2019.
https://doi.org/10.1007/978-981-15-1899-7_13

There are two main attack patterns in WSN location privacy protection which are ordinary attack mode and complex attack mode [4]. In ordinary attack mode, the attackers passively get sensitive information via eavesdropping, backtracking [5] and flow analyzing [6]. Because of wireless communication, these attacks receive the electromagnetic signal and analyze signal characteristics of nodes, which makes it difficult to be discovered. In complex attack mode, attackers can capture nodes [7] or tamper the data via active method. They can disguise themselves to be a normal node in the network to receive and forward packages. An active attacker can take control of some fragile nodes by cracking passwords or injecting viruses. It is very possible for an active attacker to obtain and analyze the actual content in the data and the location privacy of nodes. The exiting location privacy preserving schemes in WSN normally defend an ordinary attacker but consider less to complex attack mode. The limit energy and computing power of sensor nodes leads to serious security risk in forwarding sensitive data. So, it is meaningful to protect the node location privacy under the condition of complex attack mode.

To solve the problem of node location privacy preserving in WSN when attacked by an active attacker, we propose clustering-anonymization-based differential location privacy preserving protocol in WSN. The algorithm first clusters the network nodes and selects the clustering center by genetic algorithm. Then it uses Laplace Mechanism to add noise to location information of clustering centers. Finally, source nodes collect data and forward packages in the network by the disturbed cluster location.

2   Related Works

2.1   Overview

Traditional source node location privacy protection in WSN use controlled routing to change the distribution of electromagnetic signal to disturb the analyzing of attackers. The earliest scheme was proposed by Ozurk which is called phantom routing [8]. In a phantom routing scheme, the data package generated by source node first goes through a random walk progress, which means the data is forwarded randomly and the routing patterns will be changed to make it difficult to trackback for attackers. After that, a number of location privacy preserving schemes are proposed. PUSBRF [9] is an improved phantom routing algorithm to strengthen security. Circle routing [10] algorithm and SVCRM [11] algorithm construct a loop in routing that makes attackers unable to trace to the source. FitProbRate [12] and GFS [13] algorithm inject fake packages to the network to confuse the flow patterns. ADRing [14] scheme combines fake packages and phantom routing to protect location privacy. All these schemes aim at protecting node location privacy form backtracking and flow analyzing but they protect sensor nodes from node capturing and data analyzing just by simple encryption which is obviously not enough.

2.2   Clustering Anonymous Mechanism

Clustering refers to the process of dividing the set of objects (patterns) into multiple classes according to the similarity of their attribute [15]. The clustering anonymous mechanism in privacy preserving is to hide individual sensitive information in the whole while ensuring the accuracy. Based on DBSCAN clustering algorithm and encryption protocol, an effective clustering privacy preserving mechanism can be constructed [16]. Liu [17] improved DBSCAN algorithm and proposed a privacy preserving method of data dissemination based on clustering anonymous mechanism. Chai [18] clustered sensitive data attribute based on k-means clustering method and dividing data into different equivalence classes to protect privacy. Ma [19] improved optimized clustering anonymous algorithm using the method of clustering without the isolated points and optimizing the original k-mean clustering center.

2.3 Differential Privacy

Differential privacy [20] is a new privacy definition proposed by Dwork in 2006 for the privacy disclosure of statistical databases. Under this definition, the calculation and processing results of the data sets are insensitive to the changes of a specific record. The single record in or out of the data set has little impact on the calculation results and the attacker cannot analyze the individual privacy through the changes of the background knowledge of the data set.

Definition 1. Set D and D' have the same attribute structure, and the symmetric difference between them is denoted as $D\Delta D'$. $|D\Delta D'|$ means the data recorded in $D\Delta D'$. If $|D\Delta D'| = 1$, then, D and D' are said to be Adjacent Dataset.

Definition 2. Let the random algorithm M, P_M is the set composed of all possible outputs of M. For any two adjacent data sets D, D' and any subset S_M of P_M, if the algorithm M satisfies,

$$\Pr[M(D) \in S_M] \le \exp(\varepsilon) \times \Pr[M(D') \in S_M] \tag{1}$$

Then algorithm M provides ε-differential privacy, where, the parameter ε is privacy budget [21].

Definition 3. Set the input of function f as a data set D and the output as a d dimensional real vector. For any adjacent data set D and D',

$$GS_f = \max_{D,D'} \|f(D) - f(D')\| \tag{2}$$

Called sensitivity [22], which represents the degree of influence of introduced noise on query results of the data set.

Definition 4 Laplace Mechanism [23]. For Laplace distribution noise Lab(b) with the probability density,

$$p(x) = \frac{1}{2b} \exp\left(-\frac{|x|}{b}\right) \tag{3}$$

Where, b is the scale parameter, if the given data set D, function f is a query on D and Δf is its sensitivity, the random algorithm $M(D) = f(D) + Y$ provides ε-differential privacy [23]. And $Y \sim Lap(\Delta f / \varepsilon)$ is the random Laplace distribution noise of scale parameter $\Delta f / \varepsilon$.

Based on the Laplace mechanism, Hardt [24] et al. proposed a multiplicative weights mechanism for privacy-preserving data analysis. Zhang [25] introduced differential privacy into location protection and proposed a user-centric location privacy-preserving method with differential perturbation. You [26] proposed a Laplace noise adding method by the Markov properties of motion trail.

3 Differential Location Privacy Preserving Based on Clustering Anonymization

To the problem of WSN node location privacy preserving, we propose a differential location privacy preserving protocol based on clustering anonymization (DLPCA). Classifying approximate nodes into the same cluster, replacing the precise location of each node to the statistical cluster information, the protocol firstly clusters sensor nodes according to the collected data. Then based on the Laplace Mechanism in differential privacy, Laplace noise is added into the clustering anonymous location information in order to achieve differential privacy requirements. Finally packages are forwarded by the disturbed location information.

3.1 Clustering Anonymous

Definition 5. Let the n-dimensional vectors x and y be two vectors in the set of vectors $\{a_1, a_2, \cdots, a_m\}$. If

$$d^2(x,y) = (x-y)^T V^{-1}(x-y) \tag{4}$$

While

$$V = \frac{1}{m-1} \sum_{i=1}^{m} (a_i - \bar{a})(a_i - \bar{a})^T \tag{5}$$

$$\bar{a} = \frac{1}{m} \sum_{i=1}^{m} a_i \tag{6}$$

Let's call $d(x,y)$ the mahalanobis distance between the vectors x and y. The mahalanobis distance is constant for all non-singular linear transformations, so it is not affected by the feature dimension. Meanwhile, the feature correlation is also processed to reduce the value of strongly correlated feature. Therefore, mahalanobis distance is used as the scale of distance between the data in the process of clustering anonymity.

Completed the WSN node layout, the base station (sink node) first forms a data set according to the collected data of each node and perform clustering to determine clusters and clustering center of nodes. Then each sensor node anonymizes its location information according to its cluster and forwards data to the base station through the clustering center nodes.

Algorithm 1 Clustering Anonymous Algorithm.

Input: nodes data set D, threshold T and initial clustering center set C.
Output: nodes after clustering and anonymization.
Step1: to calculate the mean of node data set D ($\bar{a} = \frac{1}{m}\sum_{i=1}^{m} a_i$) and further calculate the covariance matrix of the data set ($V = \frac{1}{m-1}\sum_{i=1}^{n}(a_i - \bar{a})(a_i - \bar{a})^T$, in which $a_i = (a_{i1}, a_{i2}, a_{i3}, \cdots, a_{in})^T$ is the vector composed of data collected by a single node and information of the node itself).   **Step2**: select the data a_i of node i from the node data set. And calculate its distance from each point in clustering center set C. Select the clustering center j with the smallest distance and compare it with the threshold value T. If $d(c_j, a_i) < T$, then a_i is included in cluster j. Otherwise, take the node as the new clustering center point and add it into the clustering center set C.   **Step3**: select the next node $i+1$ and repeat the execution of step 2 - step 3 until all nodes in the data set have been classified.   **Step4**: calculate the barycenter of nodes location in each cluster and replace the location information of each node by its clustering center.   **Step5**: Take each clustering center as the clustering head and output clustering anonymous *Nodes*.

The barycenter of nodes for each cluster is calculated below.

$$P_c(x, y) = \sum_{i=1}^{|C|} P(node_{ix}, node_{iy}) \tag{7}$$

Where, $P_c(x, y)$ is the node location after clustering anonymization and P is the actual location of the node.

3.2 Clustering Center Based on Genetic Algorithm

Genetic algorithm is an intelligent algorithm based on the principle of biological genetics [27]. Imitating the evolution process of organisms in nature to select the best population, the algorithm use the method of multi-initial value parallel search to continuously search for the optimal solution. For the simple operation and global search ability, genetic algorithm is used here to determine the clustering center.

Algorithm 2 Determining Clustering center

Input: node data set D, threshold T, genetic cut-off difference threshold δ, population individual number S, mutation probability p.

Output: optimized clustering center set C'.

Step1: Individual coding. An individual in a population corresponds to a solution to the problem (a collection of clustering centers). Each individual is identified by a separate code called an individual gene. The coding of clustering center set is realized by node identification, and the identification of each node is concatenated in binary form to form individual chromosome.

Step2: Population initialization. The initial population should ensure diversity to improve the search efficiency. Therefore, we choose K nodes randomly in WSN to form a clustering center set as an individual and generate individual repeatedly to form initial $Chrom$.

Step3: Fitness function. Fitness function is used to evaluate the clustering effect to the clustering center set. In genetic algorithm, excellent individuals are selected to be passed on to the next generation through fitness function and low-strain individuals are eliminated. For each individual of the parent population, we cluster with algorithm 3.1 and calculate the average of adjacent distance sum of squares in the class.

Step4: Select. The individuals with high fitness were selected from the parent $Chrom$. The fitness of each individual is calculated and the individuals for reproduction are selected by selection method.

Step5: Cross. The crossover process is the exchange of chromosome fragments between two individuals. Inputting two individuals that need cross operation, they will be cut into two parts and crossed together to form two new individuals. Since it is impossible to contain duplicate nodes in a clustering center set, it is necessary to ensure that there are no duplicate nodes in each body after the crossover. If there are duplicate nodes, adjust the duplicate nodes and make them different.

Step6: Variations. According to the mutation probability P, a mutation operation is carried out on the generated new individuals to change one of their individual genes to increase the diversity of the population. It is also important to ensure that there are no duplicate nodes in the mutated body.

Step7: Insert the newly generated individuals into the parent population to form a new daughter population $Chrome'$.

Step8: Calculate the fitness of the daughter population and compare with the parent population. If the difference is less than the threshold δ, the algorithm ends and output the individual with the best fitness. Or go to step4 to produce the next generation.

The fitness of the population is calculated as below.

$$\bar{S}_W = \frac{1}{N}\sum_{j=1}^{K}\sum_{l=1}^{L_j} d^2\left(c_j, node_{jl}\right) \tag{8}$$

Where K is the number of clustering, L is the number of nodes in a cluster and N is the number of all nodes. The smaller the $\bar{S}_W$ is, the smaller the individual difference is, so that it is more in line with the clustering requirements. While the number of classifications also has an impact on the fitness. Too many classifications can easily lead to the reduction of class difference, and make the difference of chromes become smaller. Therefore, the number of classifications k is inversely proportional to the fitness. If the same fragment is produced in the individual gene, which means duplicate nodes are generated in the clustering center, we set the fitness degree 0 for not suitable for the actual demand. So the fitness function is.

$$fitness = \begin{cases} \frac{1}{k \cdot S_w}, & where\ no\ duplicate\ nodes\ in\ C \\ 0, & where\ duplicate\ nodes\ in\ C \end{cases} \tag{9}$$

3.3 Differential Location Privacy Preserving Algorithm

To preserve the privacy of cluster location and meet the requirement of differential privacy, add noise to the clustering anonymous nodes.

Algorithm 3 Differential Location Privacy

Input: clustering anonymous location $P_c(x, y)$, privacy budget ε.
Output: collection of disturbed nodes location $Nodes'$.
Step1: Generate $2N$ random number fs in line with Laplace distribution as noise according to the privacy budget ε.   **Step2**: Add the Laplace noise to the clustering anonymous node location.   **Step3**: Return the collection of disturbed nodes location $Nodes'$.

The noise is generated as below.

$$fs = Lap(\Delta f / \varepsilon) \tag{10}$$

Where, Δf is the sensitivity of the query function (the sensitivity of the center of gravity of cluster node location).

The noise is added as below.

$$P'_i(x', y') = P_c(x, y) + noise(fs_i, fs_{i+N}) \tag{11}$$

3.4 Communication Mechanism

The DLPCA algorithm can be summarized as the combination of the algorithms described above and a proper communication mechanism. It is shown as a whole in Fig. 1.

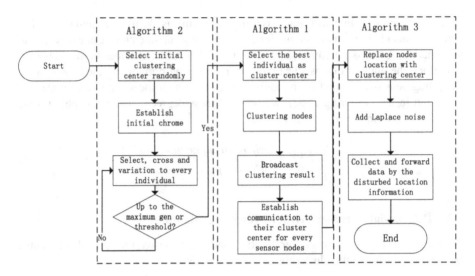

Fig. 1. The whole process of DLPCA

In WSN, the base station node can obtain the data information of all nodes in the network and has strong computing capacity and sufficient energy. On the other hand, the computing capacity and energy of other sensor nodes in WSN are limited, so clusters needs to be completed through the base station. In order to ensure the safety of the communication and reduce the energy consumption, it is necessary to establish a communication mechanism.

Before the deployment of WSN, all sensor nodes and base station are given a public-private key pair. After deployment, the base station collects the first data conveyed by source nodes and clusters the whole networks by algorithm proposed. Then the base station broadcast the ID of each clustering center and the clusters of nodes to every sensor nodes with its public key encrypted. Choose the clustering centers as head nodes and they make nodes in the clustering anonymous and noised. Next the head nodes broadcast the disturbed location information to every single node. And a source node sends the collected data to the base station through its head node. For the sake of security and the movement of sensor nodes, the head nodes recollect the location information in the cluster and update the disturbed nodes location in a period. The communication mechanism is shown in Fig. 2.

Wireless communication makes WSN open and easy be attacked. So there is the encryption in the initial communication between source nodes and the base station without location disturbed.

Step 1: The base station sends the data, timestamp and random number to the receiving node with the private key signed and public key encrypted.
Step 2: The sensor node decrypts the data with its private key and verifies it by the base station public key. Then it sends the collected data and its location information to the base station in the same way.

Step 3: The base station decrypts the information, calculates clusters and sends them to clustering heads with signed and encrypted.

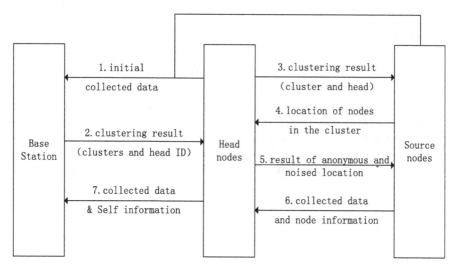

Fig. 2. Communication of DLPCA

4   Analysis and Verification

DLPCA provides WSN source location privacy preserving with differential privacy. The usability, security and efficiency of the algorithm is analyzed below.

4.1   Usability

On the one hand, the sensor node location information is mainly used to analyze the objects and environments and it requires information from multiple nodes in a zone, but not the accuracy of a single node location. On the other hand, node location can be used for routing, which needs more accurate location information. However, appropriately fuzzy location will not have too much impact on routing effects.

In clustering anonymous, supposing the data collection of source nodes be D, clustering anonymous data collection of nodes be D', the minimum size of cluster be k, the query function of node location be f_{loc} and its sensitivity be $\Delta(f_{loc})$, the difference of adjacent data set will be distributed to k nodes in a location privacy query. So the sensitivity of the query will be $\Delta(f_{loc})/k$. The sensitivity is decreased and the privacy budget is preserved to $1/k$ with the peer protection.

Let P_i be a node location in the cluster, P_c be the clustering center location and P'_c be the disturbed location. Then the error of location is.

$$\delta = d\left(P_i, P'_c\right) \le d(P_i, P_c) + fs = d(P_i, P_c) + \sqrt{fx^2 + fy^2} \tag{12}$$

Where, fs is the location offset comes from Laplace noise and fx, fy are the noise on x, y axis.

The expectation of error is.

$$E\delta = E(d(P_i, P_c)) + E\left(\sqrt{fx^2 + fy^2}\right) = ES + \sqrt{2}Ef \tag{13}$$

Where ES is the average distance from nodes to the center in the cluster. When ES is minimum, it will be the location of the center of gravity. In the algorithm, the clustering rule is the minimum intra-class proximity to ensure the optimal ES under the consideration of nodes location and data collection. The difference between intra-class nodes are small when a cluster is not very large. Hence, the clustering center is close to the center of gravity.

Ef is the expectation of a single noise.

$$E(fs) = \frac{\Delta(f_{loc})}{k\varepsilon} \tag{14}$$

As is said above, a large amount of nodes in a cluster and a small range of cluster will lead to a smaller error in disturbed location. So proper clustering reduces the error and improve the usability.

4.2   Security

In the process of clustering anonymous, the location of nodes is replaced by the clustering center. The location privacy is hidden in a zone of the cluster. The cluster comes from collected data of nodes and nodes in the same cluster has similar location and environment, which makes it undistinguishable. The attacker can only know a zone of nodes but not a particular node.

The anonymous location information is disturbed by Laplace noise to meet the requirement of differential privacy. There are no nodes to be in two clusters so they are non-intersect data set. According to the Laplace mechanism and the parallel theorem [28] in differential privacy, the disturbed nodes location is ε-differential privacy and every non-intersect data set is ε-differential privacy. So the whole network is ε-differential privacy.

The initial communication between source nodes and the base station is protected by public key system. There is a communication protocol to authentication and encryption. With the signature and encryption, the data is secret and integrity. The timestamp and random number protect data from replay attack. The updating of nodes and keys makes the effectivity and security of the network.

4.3   Efficiency

The clustering algorithm is quite simple, which only needs to traverse all sensor nodes and clustering centers for every node in classification. Therefore, the time complexity of the clustering process is $O(k \times n)$. In the process of optimizing the initial clustering center, the genetic algorithm is used to improve the global search ability with fast convergence speed at the early stage. So, it finds the better clustering center quickly. The algorithm is more efficient when choosing a proper optimum precision. Most calculations are in the base station or the sink node, so the cost for sensor nodes is acceptable and has little influence in collecting and forwarding data.

5   Simulation and Evaluation

To evaluate the effect of DLPCA, we simulate the algorithm in Matlab and compare with normal Laplace privacy preserving mechanism.

5.1   Simulation Environment and Method

Matlab language was used for simulation, and the experimental environment was Windows 7, 8 GB memory, core i7 CPU and Matlab 2017a software. DLPCA protocol is simulated in the range of 1000 m × 1000 m. Specific parameters used in the simulation are shown in Table 1.

Table 1. Simulation parameters

Variables	Meaning	Value
N	Nodes number	600
T	Clustering threshold	100
k	Clustering center number	50
S	Initial chrome scale	100
$MAXGEN$	Maximum generation	100
$GGAP$	Genetic gap	0.95
pm	Cross probability	0.7
p	Variation probability	0.01
δ	Genetic stop precision	10^{-3}
	Privacy budget	0.1

Firstly, generate 600 wireless sensor nodes randomly when simulating. Then randomly select 50 nodes as the initial clustering center and repeatedly 100 times as the initial chrome. Next optimize the clustering center through genetic algorithm and clustering nodes. Finally replace the nodes location with the closeting center location and add Laplace noise. The simulation process is shown in Fig. 3.

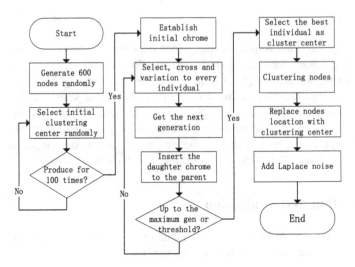

Fig. 3. Simulation process of DLPCA

Clustering with genetic algorithm, k-means, similarity threshold and max-minimum method respectively, the result of clustering in different cluster number and scale are simulated. The location error in different clustering methods are analyzed through comparison of average distance from clustering center. Simulating DLPCA and normal Laplace privacy preserving mechanism respectively, the average node location offset in different privacy budget are calculated and the relationship between privacy preserving and cluster number are analyzed.

5.2   Clustering Evaluation

Clustering results are evaluated by intra-class deviation matrix S_W, inter-class deviation matrix S_B and total deviation matrix S_T. They show the class structure from different perspectives [15]. S_W shows the similarity inner the class. S_B shows the difference between the classes. S_T only depends on the distribution of the data set. We use $\|S_W\|_2$ as the evaluation criterion and compare the clustering results from genetic algorithm, K-means [29], similarity threshold, and max-minimum method.

$$S_W = \sum_{j=1}^{c} \frac{n_j}{N} S_W^{(j)} \tag{15}$$

Where,

$$S_W^{(j)} = \frac{1}{n_j} \sum_{i=1}^{n_j} \left(x_i^{(j)} - m_j \right) \left(x_i^{(j)} - m_j \right)^T, j = 1, 2, \cdots, c \tag{16}$$

$$m_j = \frac{1}{n_j} \sum_{i=1}^{n_j} x_i^{(j)}, j = 1, 2, \cdots, c \tag{17}$$

Figure 4 shows the evolution in genetic algorithm.

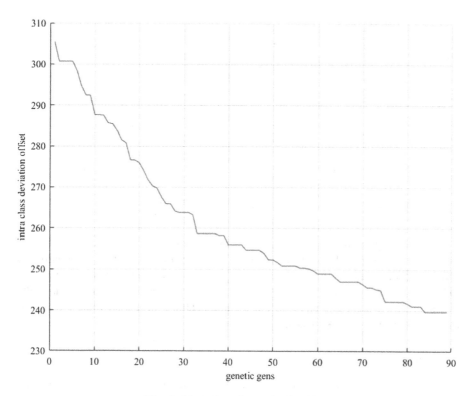

Fig. 4. Evolution of genetic algorithm

Similarity threshold clustering method is to cluster from an initial clustering center, using the closest distance and a distance threshold. Max-minimum clustering method is to use the maximum distance criterion to decide a clustering center and use minimum distance criterion to decide the class. K-means clustering method is to use average of every cluster as the clustering center and dynamically adjust it. Comparing the simulation these methods, the results are shown in Fig. 4.

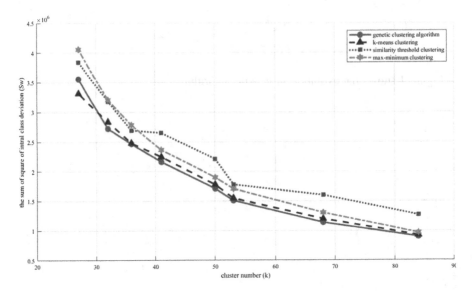

Fig. 5. Comparison of cluster algorithms in different clustering number

From the Fig. 5, we can figure out that the methods (k-means and genetic algorithm) with optimized clustering center get lower intra class deviation. Comparing to k-means method, genetic algorithm mostly gets lower deviation, which means the nodes in the same cluster have less difference. But with a small clustering number, the effects of genetic algorithm and k-means method is almost the same, while the genetic algorithm is slower than k-means.

5.3 Anonymous and Disturbing

Clustering anonymous algorithm decreases the sensitivity of query to save privacy budgets and improve data usability. Here we use the average offset distance from anonymous and actual location as the evaluation index.

$$\bar{d} = \frac{1}{N} \sum_{i=1}^{N} \sqrt{(x_i - fx_i)^2 + (y_i - fy_i)^2} \qquad (15)$$

Where, x, y are the real location of the nodes and fx, fy are the disturbed location.

Comparing the location privacy preserving from DLPCA and normal Laplace mechanism [23], the results are shown in Fig. 6.

In the Fig. 6, the solid line is the DLPCA and the dotted line is the ordinary Laplace mechanism. Since the Laplace mechanism has nothing to do with clustering, it shows a nearly horizontal line. For DLPCA, with the increase of cluster number, the average offset distance from nodes is small. When the privacy budget is small, the offset is

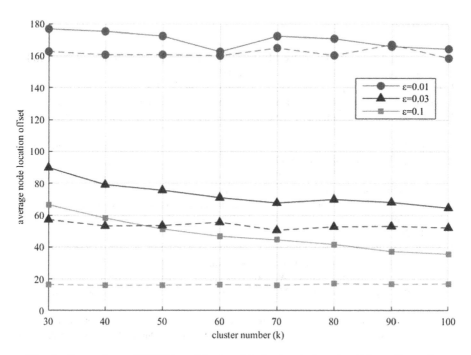

Fig. 6. Comparison of DLPCA and Laplace Mechanism in Location Privacy Protection

larger than normal Laplace mechanism. Hence, the clustering anonymous algorithm saves privacy budgets. With the larger privacy protection budgets, node location is less affected by regional anonymity and the noise becomes the main factor, which may cause excessive noise to affect the usage of location information. In conclusion, DLPCA improves the usability of location information with the same privacy budget, which is in the line with the theory.

6   Results and Overlook

The paper proposed a differential location privacy preserving algorithm based on clustering anonymous (DLPCA) towards the location privacy leak problems in the big data interconnection of WSN nodes. The location usability improved with the peering privacy budget and protection level. A security communication mechanism is designed to ensure the initial data exchange and nodes deployment. With the analysis and simulation, the DLPCA has advantages over other methods in location usability and privacy preserving.

Next works:

(a) Improve the efficiency of genetic algorithm in clustering anonymous.
(b) Study the proper parameters and cluster number in the algorithm to provide better protection.

References

1. Li-min, S., Yuang, Z., Qing-chao, L., et al.: Fundamentals of wireless sensor networks, Tsinghua University, Beijing, pp. 3–13 (2014)
2. Feng, H., Hong-wei, C., Zong-ke, J., Ti-jiang, S., et al.: Review of recent progress on wireless sensor network applications. J. Comput. Res. Dev. **47**(S2), 81–87 (2010)
3. Ynag, X., Ma, K.: Evolution of wireless sensor network security. In: World Automation Congress, pp. 1–5 (2016)
4. Peng, H., Chen, H., Zhang, X.-Y., et al.: Location privacy preservation in wireless sensor networks. J. Softw. **26**(3), 617–639 (2015)
5. Cheng, L., Wang, Y., Wu, H., et al.: Non-parametric location estimation in rough wireless environments for wireless sensor network. Sens. Actuators, A **224**, 57–64 (2015)
6. Groat, M., He, W., Forrest, S.: KIPDA: k-indistinguishable privacy-preserving data aggregation in wireless sensor networks. In: Proceedings of the INFOCOM, pp. 2024–2032. IEEE (2011)
7. Babar, S.A., Prasad, N., et al.: Proposed embedded security framework for Internet of Things (IoT). In: The Wireless Communication, Vehicular Technology, pp. 1–5 (2011)
8. Ozturk, C., Zhang, Y., Trappe, W.: Source-location privacy in energy-constrained sensor network routing. In: The 2nd ACM Workshop on Security of Ad Hoc and Sensor Networks, pp. 88–93 (2004)
9. Juan, C., Bin-xing, F., Li-hua, Y., et al.: A source-location privacy preservation protocol in wireless sensor networks using source- based restricted flooding. Chin. J. Comput. **33**(9), 1736–1747 (2010)
10. Fu-zhi, X.: Research and Implementation of Location Privacy Protection Solution in Wireless Sensor Network. University of Electronic Science and Technology of China, Chengdu (2014)
11. Jiangnan, Z., Chun-liang, C.: A Scheme to protect the source location privacy in wireless sensor networks. Chin. J. Sens. Actuators **29**(9), 1405–1409 (2016)
12. Shao, M., Yang, Y., Zhu, S., et al.: Towards statistically strong source anonymity for sensor networks. In: The 27th Conference on Computer Communications, pp. 51–55 (2008)
13. Xiao-yan, H.: Research on the source-location privacy in wireless sensor networks against a global eavesdropper. Central South University, Changsha (2014)
14. Xiaoguang, N., Chuan-bo, W., Ya-lan, Y.: Energy-consumption-balanced efficient source-location privacy preserving protocol in WSN. J. Commun. **37**(4), 23–33 (2016)
15. Ji-xiang, S.: Modern Pattern Recognition, pp. 16–40. Higher Education Press, Beijing (2016)
16. Kumar, K.A, Rangan, C.P.: Privacy preserving DBSCAN algorithm for clustering. In: Proceedings of Advanced Data Mining and Applications, Third International Conference, ADMA 2007, Harbin, China, 6–8 August 2007, DBLP, pp. 57-68 (2007)
17. Xiao-qian, L., Qian-mu, L.: Differentially private data release based on clustering anonymization. J. Commun. **37**(5), 125–129 (2016)
18. Min-rui, C., Hui-hui, F.: Efficient (K,L)-anonymous privacy protection based on clustering. Comput. Eng. **41**(01), 139–142 + 163 (2015)
19. Zhao-yan, M.: Research on anonymous privacy protection algorithm based on clustering, Xi'an University of Technology (2017)
20. Dwork, C.: Differential privacy. In: Proceedings of the 33rd International Colloquium on Automata, Languages and Programming, Venice, Italy, pp. 1–12 (2006)
21. Dwork, C.: A firm foundation for private data analysis. Commun. ACM **54**(1), 86–95 (2011)
22. Ping, X., Tian-qing, Z., Xiao-feng, W.: A survey on differential privacy and applications. Chin. J. Comput. **37**(1), 101–122 (2014)

23. Dwork, C., McSherry, F., Nissim, K., et al.: Calibrating noise to sensitivity in private data analysis. In: Proceedings of the 3rd Conference on Theory of Cryptography, New York, USA, pp. 265–284 (2006)
24. Hardt, M., Rothblum, G.N.: A multiplicative weights mechanism for privacy-preserving data analysis. In: IEEE, Symposium on Foundations of Computer Science, pp. 61–70. IEEE Computer Society (2010)
25. Xue-jun, Z., Xiao-lin, G., Jing-hua, J.: A user-centric location privacy-preserving method with differential perturbation for location-based services. J. Xi'an Jiao-tong Univ. **50**(12), 79–86 (2016)
26. Wei-jun, Z., Qing-huang, Y., Wei-dong, Y., et al.: Trajectory privacy preserving based on statistical differential privacy. J. Comput. Res. Develop. **54**(12), 2825–2832 (2017)
27. Huo-wen, J., Guo-sun, Z., Ke-kun, H.: A graph-clustering anonymity method implemented by genetic algorithm for privacy-preserving. J. Comput. Res. Develop. **53**(10), 2354–2364 (2016)
28. Mcsherry, F.D.: Privacy integrated queries: an extensible platform for privacy-preserving data analysis. In: The 2009 ACM SIGMOD International Conference on Management of Data. Providence, pp. 19–30. ACM, Rhode Island (2009)
29. Yu-hong, W.: General overview on clustering algorithms. Comput. Sci. **42**(S1), 491–499 + 524 (2015)

Distributed Graph Perturbation Algorithm on Social Networks with Reachability Preservation

Xiaolin Zhang[1][⊠] [iD], Jian Li[1] [iD], Xiaoyu He[2] [iD], and Jiao Liu[1] [iD]

[1] School of Information Engineering, Inner Mongolia University of Science and Technology, Baotou 014010, China
zhangxl@imust.cn
[2] School of Computer Engineering and Science, Shanghai University, Shanghai 200000, China
2784899426@qq.com

Abstract. With the rapid development of social networks, the current scale of graph data continues to increase, and the performance of anonymous social network methods is limited. Node reachability query is essential in directed graphs, which can reflect the relationship between nodes and the direction of information dissemination. Aiming at the problem of the reachability of nodes between directed social network privacy technologies, this paper proposes a reachability preserving distribution perturbation (RPDP) algorithm, which is based on the distributed graph processing system GraphX. This algorithm first generates a Random Neighborhood Table (RNT) composed of four tuples for the nodes and then uses the message transmission of GraphX and "probe" mechanism. The proposed algorithm improves the disposal efficiency of the large-scale social network while maintaining the reachability of the nodes. Experiments based on the real social network data show that the proposed algorithm can keep the node reachability and deal with large-scale social network efficiently while protecting the character of the graph structure.

Keywords: Social network · Privacy protection · Distribution · Reachability · Perturbation · GraphX

1 Introduction

With the development and popularization of online social networks, information interaction between individuals has become more convenient. At the same time, the content generated by the individual when using the web service and the personal information registered by the individual constitute the social network data. There are a variety of directed graphs in social networks, such as mutual follow on twitter and the flow of money in online payments. The anonymity problem of studying directed graphs plays an important role in analyzing the hidden information of published graphs. Many researchers wish to obtain these data and use some technologies such as data mining, social network analysis. However, since social network data involves personal information, if it is directly shared or distributed to a third party, it is easy to cause the

© Springer Nature Singapore Pte Ltd. 2019
H. Jin et al. (Eds.): BigData 2019, CCIS 1120, pp. 194–208, 2019.
https://doi.org/10.1007/978-981-15-1899-7_14

disclosure of individual sensitive information. The simple approach is to delete information which directly identifies the individual's identity, such as the ID number when publishing the data. However, paper [1] pointed out that due to the structural background, this simple deletion of the identity attribute does not protect the individual identity. Therefore, researchers have proposed different privacy protection schemes, and the random perturbation technique [2–4] is one of them. The random perturbation refers to reducing the attacker's inferred confidence by randomly modifying the original data. For example, paper [2] proposed to protect the privacy of social network individuals by randomly deleting and adding m edges from the edge sequence. Since the deleted links and added links are chosen from the entire graph, such perturbation incurs considerable structural distortion.

The reachability query [5–8] refers to determining whether there is a path between nodes u and v in the graph. If there is a path between the two nodes, the node u is said to reach the node v. As shown in Fig. 1, in the graph G, the nodes 1 and 7 have paths {<1, 4>, <4, 5>, <5, 7>}. Therefore, node 1 can reach node 7.

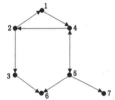

Fig. 1. A directed social network graph G

Reachability query is a common relational query in social network services. It is used by social networking sites to improve the activity and relevance of users. Thus, it is meaningful to maintain the accessibility of nodes. However, the random perturbation technique ignores the protection of node reachability. As shown in Fig. 2, G_1 and G_2 are two anonymous graphs obtained after the random perturbance of the social network graph G. It can be seen that although both G_1 and G_2 hide the link <1, 4>, the effect of protecting the node reachability is not the same. Specifically, if the destination node in <1, 4> is replaced with node 3, node 1 and node 4 become unreachable, but if the destination node in <1, 4> is used with node 5 instead, the two nodes are still reachable, so the former maintains the node's accessibility better than the latter.

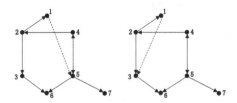

Fig. 2. Two anonymous graphs of the social network G

In addition, with the rapid development of Internet technology, the scale of online social network data has shown a trend of quantitative changes. Faced with such a large scale of social network data, traditional anonymity technology cannot meet the actual needs, the use of parallel algorithms is an effective way to improve the efficiency of execution. At present, the social network anonymous parallel processing technology is mainly divided into two categories, one is based on the Secure Multi Party (SMC) model [9], and the other takes use of the MapReduce which is a data processing framework [10, 11]. However, these two types of parallel processing technologies are aimed at relational data and do not consider the graph properties of individuals such as the degree of nodes, neighbor subgraphs in social networks, which cannot protect private information in social networks. The graph modeling and parallel processing of social networks are effective solutions to the problem.

The research work of this paper is aimed at large-scale social network graph G. Its content is to generate the anonymous graph G^* quickly and efficiently while maintaining the reachability of nodes. The main work and contributions are as follows.

(1) We propose a distributed Random Neighborhood Search (DRNS) algorithm. This algorithm generates a Random Neighbor Table (RNT) for nodes and implements a fast lookup of random neighbor sets based on the message passing mechanism of GraphX.
(2) We propose two different distributed graph perturbation algorithms, Distribution Neighborhood Randomization (DNR) and Reachability Preserving Distribution Perturbation (PRDP). Based on the DRNS and the graph construction operations in GraphX, DNR implements fast edge perturbation of large-scale social networks. RPDP proposes a "probe" mechanism. It is possible to maintain reachability node in the rapid edge perturbance.

2   Related Work

The reachability query is to query whether a node can reach another node in the directed graph [5]. In order to protect the link relationship in the social network, the researchers propose to protect the sensitive link through the random perturbation technology [3]. The technology randomly modifies the social network graph by edge probability, so the attacker cannot accurately guess the real data in the original social network.

An edge perturbation technique [12] based on subgraph structure is proposed which divides the original graph into several subgraphs, and then adds/deletes m edges randomly in the subgraph. However, this increases the degree of some nodes in the anonymous graph and the probability that such nodes will be identified in the subgraph. In order to solve this problem, a random neighbor edge perturbation technique [13] is proposed. The edge <u, v> in the graph is reserved with a certain probability p ($0 \leq p \leq 1$). If <u, v> needs to be deleted, the destination node v is replaced with the r-hop ($r \geq 2$) neighbor node w of the node u. Paper [14] proposed to use secure grouping to protect the link relationship of interactive social networks. The idea is to abstract the network into bipartite graphs, then group the network nodes, and the nodes in the same group are indistinguishable in the link relationship. Paper [15] extend the paper [14] to protect high-latitude data for the high dimensionality of interactive social network data. Paper [16] divided the nodes in the network into the public nodes and the privacy nodes, and propose two link relationship protection methods based on adding virtual nodes and modifying edges. In order to reduce the complexity of algorithm execution and improve the usability of published data, paper [17] proposed to protect private information by adding, deleting and switching edges on the basis of the paper [18]. Paper [19] proposed a method for calculating the link recognition probability based on the link density between two equivalence classes. On this basis, a greedy strategy that minimizes the number of deleted or swapped edges is applied to reduce the possibility of link identification below a given threshold. In order to protect the relevance of the link, paper [20] first used the edge neighbor centrality to calculate the removal probability p of each edge, and removes the edges from the original graph G according to the probability p, meanwhile, and then selects the edges in the anonymous graph. Paper [21] proposed a graph perturbation strategy based on random walk mechanism by introducing Markov chain and establishing transfer matrix P, in order to reduce the impact of perturbance process on data availability while protecting link privacy. Paper [22] designed a LinkMirage model for sensitive link relationships. Under the premise of protecting data availability, this model reduces the social topology in the social network graph and provides untrusted external applications with fuzzy social relationship views to resist the re-identification attack against the link relationships. Paper [23] considered the dynamic evolution characteristics of social networks and propose a dynamic social network privacy protection method based on the LinkMirage model.

It can be seen that the current privacy protection technology based on random perturbation ignores the protection of node reachability, resulting in the low availability of anonymous graphs in node reachability queries and shows the shortcomings of low efficiency and poor processing capacity when dealing with the large-scale anonymous social network. To solve this problem, this paper proposes a distributed graph perturbation algorithm RPDP which can effectively protect node reachability. The RPDP algorithm utilizes the distributed graph processing system GraphX [24–26] programming mode to follow the "node-centric" feature. Through the message transmission between nodes in the anonymous social network, the reachability of nodes is protected while maintaining an anonymous social network.

3   Background Knowledge and Problem Definition

In this paper, the social network is represented as a simple directed graph G = (V, E), where the node set V represents the users in the social network, and E is the set of directed edges, the directed edge <u, v> is represented to a directed social connection between the users u and v in the social network.

Definition 1 (Edge Perturbation). The directed graph G = (V, E) is known, and the nodes u, v are two different nodes in V. <u, v> ∈ E is a directed edge, node w is a node different from nodes u, v in the graph and <u, w> ∉ E. The edge <u, v> is as follows operating:

$$<u, v> \notin E \wedge <u, w> \in E \tag{1}$$

Given a constant β, we use a random function to assign a value to each edge of the graph. If the assignment is equal to p, the edge is preserved. If the assignment is equal to 1-p, the edge is perturbed by the neighbor list. The edge perturbation is deleting the edge $<u, v>$ and adding the edge $<u, w>$ to the graph G, where $<u, w>$ is the perturbed edge, and the node w is called the hypothetical node of u.

As shown in Fig. 1, in the social network graph G, if $<1, 4>$ is deleted and $<1, 5>$ is added to the graph G, an edge perturbation operation is completed, and the node 5 is called the virtual target node of the node 1.

Definition 2 (Reachability). In the directed graph $G = (V, E)$, u, v are two different nodes in the graph. If there is a path L that the node u can reach the node v, the node u to v is reachable. It is recorded as $u \leadsto v$ and (u, v) is called as the reachable node pair.

As shown in Fig. 1, node 2 and node 7 have paths $\{<2, 1>, <1, 4>, <4, 5>, <5, 7>\}$, then node 2 to node 7 is reachable, that is $2 \leadsto 7$.

Definition 3 (Neighbor-r). Given a non-negative integer r, nodes u, v are two different nodes in the graph and $u \leadsto v$. Dist(u, v) is employed to represent the shortest path length from u to v. If node v meets the condition:

$$Dist(u, v) \leq r \tag{2}$$

Then, the node v is the Neighbor-r of the node u. All the nodes that satisfy the condition are denoted by $N_r(u)$, and $N^*(u)$ is used to represent all the reachable nodes of the source node u in the graph G, i.e. $N^*(u) \in \{v| u \leadsto v\}$. Dst (u) represents the neighbors of the source node u, i.e. $Dst(u) \in \{v|Dist(u, v) = 1\}$.

4 Distributed Graph Perturbation Algorithm

GraphX is a processing system for graph and parallel graph calculations on Spark. The entire calculation process consists of several supersteps that are executed sequentially. GraphX follows the "node-centric" mode on the programming model. In the superstep S, the nodes in the graph merge the messages transmitted from other nodes in the superstep (S − 1) and change their state, then send messages to other nodes. After they are synchronized, they will be received and processed by other nodes in SuperStep (S + 1).

4.1 Distributed Random Neighbor Search Algorithm DRNS

To support graph calculations, GraphX provides a basic set of functional operations, such as reverse: Graph[VD, ED]. The reverse operation returns a new graph, and the direction of the edges in the new graph is reversed. Based on the reverse operation, a distributed random neighbor search algorithm called Distribution Random Nestination Search (DRNS) is proposed. First, in the initialization, the reverse operation is used to invert the graph, and a Random Neighborhood Table (RNT) consists of four elements (srcid, dstid, hops, tags) and is called a Random Neighborhood Table Entry (RNTE). It can be seen that each row of RNT is an RNTE. In RNTE, srcid is the node id of the source node, dstid is the node id of the destination node, hops is the shortest path length of the source node and the destination node, tags is the flag bit, and the value of 1 or 0 is used to determine whether to pass the current of the content of RNTE. Initially, the srcid and dstid of the node in the RNTE are their node id, and hops = 0, tags = 1. For example, the node 1, in the initial state, its RNTE = {1, 1, 0, 1}. Then, update its RNT through the message passing between nodes to find a random neighbor set.

The steps of the DRNS algorithm are as follows:

(1) Initialization, we use the reverse operation to invert the graph G, and generate RNT for the node. The nodes with an out-degree are set to the Active state, and other nodes are set to InActive.

(2) When Superstep%2 = 0, the node in the active state detects the value of tags in each RNTE. If tags = 1, the corresponding dstid is sent to the neighbor node and set the tags = 0, otherwise an empty message is sent;

(3) When Superstep%2 = 1, the node receiving the message is set to the Active state, and the other nodes are set to the InActive state. The node in the Active state updates its RNT according to the received distid. If the received value does not appear in the dstid column, then this value is written to RNT along with its own nodeid, where the received value is distid, its own nodeid is srcid, hops = (supertep + 1)/2, tags = 1 and set the node to the InActive state.

(4) Repeat the situation (2), (3) until Superstep = 2r. The program then stops.

The Distribution Random Neighborhood Search algorithm is shown in Algorithm 1.

Algorithm.1 Distribution Random Neighborhood Search
Input: Messages passed between the supersteps
Output: Random neighbor list RNT of node v
```
1)   reverse (G), RNT (v);
2)   long step = getSuperstep();
3)   RNTList = getValue();
4)   if step% 2 = = 0 then
5)     if vertex v is Active then
6)       for each RNTE in RNTList do
7)         if RNTE.tagss=1 then
8)           msgList.add (RNTE.dstid);
9)           sendMessToNeighbors (msgList);
10)       else return;
11)  if step%2==1 then
12)    if messages == null then
13)      voteToHalt(); return;
14)    else then
15)      for each RdstidList in messages do
16)        for each dstid in RdstidList do
17)          if isNotExistInRNTList(dstid)
18)            setValue (RNTList.put (VertexId, dstid, (step+1)/2, 1);
19)          else next dstid in RdstidList
20)  return RNT;
```

For example, the result of the DRNS algorithm for the graph in Fig. 1 is shown in Fig. 3, and only some of the nodes are marked in the graph for convenience. In initialization, the reverse operation inverts the graph G and initializes the RNT of the node, then returns a new graph G, as shown in Fig. 3(a). When superstep = 0, the node detects the value of tags in RNTE. If tag = 1, it sends RNTE.distid to the neighbor and sets RNTE.tags to 0. As shown in Fig. 3(b), the tags are all 1, so send the distid to the neighbor and concatenate tags = 0. In the Fig. 3(c), the node received the message is in the Active state. Taking Node 4 as an example, Node 4 receives messages of nodes 2 and 5, and the RNT of node 4 doesn't contain 2 and 5. Perform step 3 to update RNT, where $N_1(1) = \{1, 4\}$, $N_1(2) = \{2, 1, 3\}$, $N_1(4) = \{4, 2, 5\}$. It should be pointed out that in the next superstep, the node only sends the contents of tags = 1, such as node 5, it only sends 6, 4, 7 to the neighbor when superstep = 2, thus reducing the amount of message transmission.

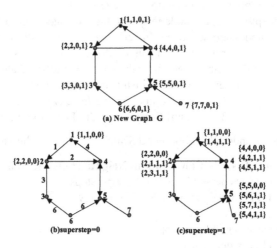

Fig. 3. Random neighborhood search

4.2 Distributed Random Neighbor Perturbation Algorithm DNR

Combined with DRNS algorithm, we parallelizes the random neighbor perturbation method NR (Neighbor Randomization) [13] and proposes a Distributed Neighborhood Randomization (DNR) method. The DNR algorithm requires two steps to complete. Firstly, the node initializes the RNT. If hops = 1, the tag is assigned a probability p or (1-p). Perturb the edge of (1-p) which the perturbation probability is the edge of (1-p) and the result of the perturbance is stored in EdgeRDD. Then use GraphX to build an anonymous graph G^* from EdgeRDD. The Distribution Neighborhood Randomization algorithm is shown in Algorithm 2.

Algorithm.2 Distribution Neighborhood Randomization
Input: Messages passed between the super steps
Output: Perturbed edge list peredgeMapList
1)   peredgeMapList;

```
2)   long step = getSuperstep();
3)   if step=0 then;
4)     if isNotExistInRNT.tags==(1-p)
5)       for every RNTE in RNT
6)         if RNTE.hops=1∧RNTE.tags=p
7)           peredgeMapList←(srcid,distid);
8)         else voteToHalt(); return;
9)   if step=1 then
10)    if isNotExistInRNT.tags==(1-p)
11)      voteToHalt(); return;
12)    else
13)      Perturbation edge;
14)        if RNTE.hops=1∧RNTE.tags=p
15)          peredgeMapList←(srcid,distid);
16)  return peredgeMapList;
```

Take node 1 and node 2 in Fig. 1 as an example. Assume that Vetext1.RNT = {(1, 4, 1, p), (1, 2, 2, 0), (1, 5, 2, 0)}, Vetext2.RNT = {(2, 1, 1, 1-p), (2, 3, 1, 1-p), (2, 6, 2, 0), (2, 4, 2, 0), (2, 5, 3, 0)}. When superstep = 0, it can be known from the line 4 of the algorithm that node 2 is in the InActive state and node 1 is in the active state, then node 1 stores <1, 4> in peredgeMapList. When superstep = 1, node 2 is in the Active state and needs perturbation. The node 2 only needs to be randomly selected from the items of hops ≥ 2, such as Vetext2.RNT = {(2, 6, 1, p), (2, 5, 1, p), (2, 6, 2, 0), (2, 4, 2, 0), (2, 5, 3, 0)}. Then save the edges <2, 6> and <2, 5> into the peredgeMapList. After the algorithm ends, the peredgeMapList is stored in EdgeRDD, and GraphX constructs an anonymous graph G^* from EdgeRDD using the graph construction operation.

4.3 Distributed Graph Perturbation Algorithm to Maintain Node Reachability RPDP

Although the DNR algorithm described in Sect. 4.2 can achieve fast edge perturbations of large-scale graphs, it does not protect the reachability of nodes. Therefore, Reachability Preserving Distribution Perturbation (RPDP) is proposed for protecting the reachability of nodes. The "probing mechanism" means that for the edge <u, v> that needs to be perturbed, the node u considers that there is a node w in the random neighbor set satisfying w ⇝ v; then node u sends the nodeid of node v to the node w; node w receives the message to determine whether it meets w ⇝ v; if it does, return its own nodeid to node u, otherwise return 0. In combination with RNT, the "probing mechanism" is easy to complete. Firstly, the node u judges that whether the hops in (srcid, dstid, hops, tags) is 1. If it is 1, the tags are assigned the probability p or (1-p). If hops = 1∧tags = (1-p), send distid to the random neighbor node w; node w receives message distid and judges whether a RNTE satisfies hops = 1∧tags = p∧w.distid = distid; If the condition is met, node w returns its own nodeid to the node u, otherwise return 0.

The steps of the RPDP algorithm are as follows:

(1) Initialization, if hops = 1, assign the probability p or (1-p) to the tags, store the part of hops = 1∧tags = (1-p) to L1, and store the part of hops ≥ 2 to L2.

2) When Superstep%3 = 0, if L1 is empty, it is in InActive state, otherwise it is in Active state. And do the following operations, where v = L1.first.distid. If v ∈ L2. distid∧L1.first.tags = 1-p, then w = L2.first.next.distid. That is, w points to the next node of L2. Send v to w, and v = w; If v ∈ L2.distid∧L1.first.tags = p, then L1.first = L1.first.next. That is, process the next edge in L1, v = L1.first.distid, w = L2.first.distid. Send v to w, and v = w.

3) When Superstep%3 = 1, the node w which received message is turned into the active state, and it is judged whether w.hops = 1∧w.distid = v∧w.tags = p is satisfied. If it is, the nodeid of the node w i.e., RNT.srcid is returned, otherwise 0 is returned.

4) When Superstep%3 = 2, if the message received by the node is non-zero, change L1.first.tags = 1-p to L1.first.tags = p; if the received message is 0, judge whether L2 is traversed. If the traversal is finished, randomly select L2.distid to replace L1.first.distid, and change L1.first.tags = 1-p to L1.first.tags = p, otherwise there is no operation.

The Reachability Preserving Distribution Perturbation algorithm is shown in Algorithm 3.

Algorithm.3 Reachability Preserving Distribution Perturbation
Input: Messages passed between the super steps
Output: Perturbed edge table edgeMapList
1)   peredgeMapList;
2)   longstep = getSuperstep();
3)   if superstep%3=0 then;
4)   if L1 is Null
5)   voteToHalt(); return;
6)   else
7)   v=L1.first.distid;
8)   if v ∈ L2.distid∧L1.first.tags=1-p
9)   w= L2.first.next.distid and sendMessToNode w (v);
10)   else v ∈ L2.distid∧L1.first.tags=p
11)   v=L1.first.next.distid , w=L2.first.distid and sendMessToNode w (v);
12)   if superstep%3=1 then
13)   v=getValue(messages);
14)   if w.hops=1∧w.distid=v∧w.tags=p;
15)   sendMessToNode u (w.srcid);
16)   else sendMessToNode u (0);
17)   if superstep%3=2 then
18)   w = getValue(messages);
19)   if w == L1.first.distid;
20)   set L1.first.tags=p;
21)   else w == 0
22)   if L2 ergodic complete;
23)   RandomSelect distid from L2.distid, L1.first.distid=distid,

L1.first.tags=p;
24)   else no operation;
25) return peredgeMapList;

Take the nodes 1, 2, 5 in Fig. 1 as an example, and assume that node 1 is the probe node, node 2 and node 5 are the detected nodes. The RPDP algorithm is shown in Fig. 4. For the convenience of node 2 and node 5, we don't mark the contents of hops ≥ 2. When superstep = 0, the node sends 4 to node 2, and (1, 4, 1, 1-p) is modified to (1, 2, 1, 1-p). When superstep = 1, the node 2 receives message 4. Node 2 doesn't satisfy the line 14 of the algorithm, then return 0 to node 1. When superstep = 2, node 1 receives message 0. Since node 5 is not detected, there is no operation. When superstep = 3, since the line 8 of the algorithm is satisfied, node 5 is detected. Node 5 satisfies the line 14 of the algorithm, nodeid = 5 is returned to node 1. When executing to superstep = 5, node 1 receives the message 5, satisfying line 19 of the algorithm, then (1, 5, 1, 1-p) is modified to (1, 5, 1, p). After the perturbance, the items satisfying hops = 1∧tags = p are stored in the form of (srcid, distid) to EdgeRDD.

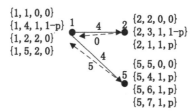

Fig. 4. Reachability edge perturbance

5 Experimental Analysis

5.1 Dataset and Experimental Environment

The experiment uses three real social network datasets to test the algorithm: (1) URV email network. (2) Polblogs political blog dataset. (3) Soc-Pokec dataset. The description of dataset is shown in Table 1. The attributes of the graph include the number of vertives(n), number of edges(m), average indegree(AID), and diameter(d)

Table 1. Dataset in experiment

Dataset	n	m	AID	d
URV email	1133	5451	9.62	8
Polblogs	1224	16715	31.19	8
Soc-Pokec	1632803	30622564	18.75	11

The cluster environment: 10 compute nodes, CPU 1.8 GHz, 16 GB RAM, Hadoop 2.7.2, Spark 2.2.0, programming: Scala 2.11.12.

5.2   Algorithm Performance Analysis

The experiment tests the variation of the execution time of the three algorithms with two different parameters, the neighbor r and the edge perturbance probability δ (δ = 1-p) on Soc-Pokec dataset. The former δ value in Fig. 5(a) is 0.4, the r value in Fig. 5 (b) is 3. It can be seen from the experimental results that as the parameter r and δ increase, the generation of random neighbor sets in the anonymous process requires more time. As δ increases, the number of edges that need to be perturbed is more, so the time increases. It can be seen from the execution time of the algorithm that the execution time of the DNR and RPDP algorithms in the cluster environment is much smaller than that of NR [13]. It shows that the parallel DNR and RPDP methods can process large-scale social networks more efficiently than the traditional methods.

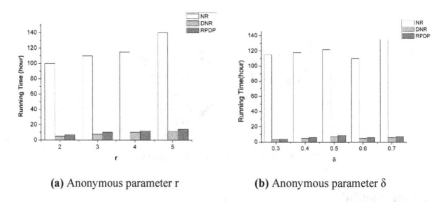

(a) Anonymous parameter r   (b) Anonymous parameter δ

Fig. 5.   Running time

5.3   Loss of Anonymous

Given the original graph G, the perturbation graph G^* and the node reachability variety radio = $|R(G^*)-R(G)|/|R(G^*)|$, where $R(G^*)$ represents the perturbance for all reachable node pairs in graph G^*, $R(G)$ represents all reachable node pairs in the original graph G, the experimental results are shown in Fig. 6. The experimental results show that as the probability of δ increases, the variety of G^* reachability increases. The reason is that the larger the δ is, the more edges need to be perturbed. However, the RPDP algorithm can maintain node reachability more than the DNR algorithm.

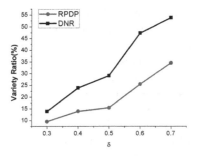

Fig. 6. Variety radio of reachability

5.4 Data Availability

The experiment evaluates the performance of the DNR algorithm and the RPDP algorithm in data availability by testing the structural changes of the graph after anonymity. The experiment mainly considers the influence of the edge perturbance probability δ on the experimental results. The experiment uses the Relative Error = (u-u*)/u to evaluate the performance of the average shortest path in algorithm DNR and RPDP, where u and u* represent the measured values in the original and perturbation graphs respectively. The smaller the Relative Error value is, the better the availability of data. Figure 7 shows the relative error rate change of the average shortest path as the parameter δ changes in the case where the perturbance range r = 3. It can be seen from the experimental results that the larger the value of the parameter δ, the more edges that need to be perturbed, which results in the utility decrease. At the same time, the results show that the RPDP algorithm is more utility than other algorithms. Because in the edge perturbance process, the RPDP algorithm is different from the DNR algorithm. Instead of randomly selecting from the RNT in DNR, RPDP "probes" from small to large according to the random neighbor radius r, to ensure that the shortest path change is relatively minimal.

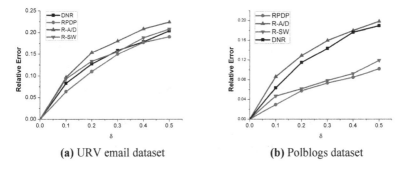

(a) URV email dataset   (b) Polblogs dataset

Fig. 7. Relative error of average shortest path

Betweenness Centrality and Closeness Centrality [27] are indicators of the centrality of social network. Betweenness centrality is often used in social network analysis to describe whether a participant is on the shortest path of two nodes. Closeness centrality is the reciprocal of the sum of the shortest distances from the node to the other nodes. The closeness centrality of a node is high, and the time it takes for information to travel from that node to other nodes is shorter. As can be seen from Figs. 8 and 9, as the perturbance probability δ increases, the relative error of the betweenness centrality and the closeness centrality increase, and the data availability decreases. However, through comparison the proposed distributed perturbation algorithm RPDP has a good effect on protecting data availability and maximally protects data availability. For the R-A/D and R-SW algorithms, the R-SW algorithm is a random exchange edge protection link relationship, and the degree of privacy of the node is unchanged at the same time which is a good protection of the node centrality. The R-A/D algorithm randomly adds and deletes edges, which is greatly destructive to data availability.

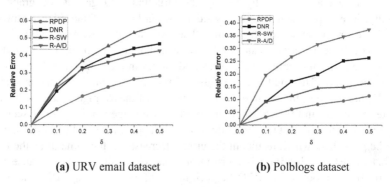

(a) URV email dataset   (b) Polblogs dataset

Fig. 8. Relative error of betweenness centrality

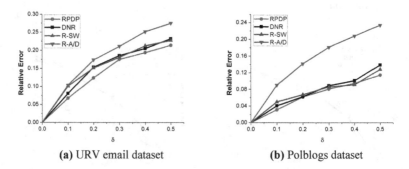

(a) URV email dataset   (b) Polblogs dataset

Fig. 9. Relative error of closeness centrality

6   Conclusion and Future Work

Aiming at the reachability problem of nodes in social network privacy protection, we design the RPDP algorithm based on GraphX to efficiently process execution when dealing with large-scale social network data. The RPDP algorithm first constructs an RNT table for each node in the graph, and then uses the message transmission of GraphX and the proposed "probe" mechanism to perturb the social network graph, improving the efficiency of processing large-scale social networks. Experiments based on real social network data show that the proposed algorithm has excellent protection for other structural properties of the graph, such as betweenness centrality and closeness centrality while maintaining node reachability.

In social networks, the needs of different users for privacy protection are not the same. Adopting fully consistent anonymous protection for users in social networks will overprotect certain users which not only leads to an increase in the cost of anonymity but also reduces the availability of data. Therefore, it is practical to provide users with personalized privacy protection, which will be the next research direction.

Acknowledgments. This work is partially supported by Natural Science Foundation of China (No. 61562065). The authors also gratefully acknowledge the helpful comments and suggestions of the reviewers, which have improved the presentation.

References

1. Backstrom, L, Dwork, C.: Kleinberg, J.: Wherefore art thou r3579x? anonymized social networks, hidden patterns, and structural steganography. In: 16th International Conference on World Wide Web on Proceedings, pp. 181–190. ACM, Canada (2007)
2. Hay, M., Miklau, G., Jensen, D.: Anonymizing social networks. Computer science department faculty publication series, pp. 173–187 (2007)
3. Ying, X.W., Wu, X.T.: Randomizing social networks: a spectrum preserving approach. In: 8th SIAM International Conference on Data Mining on Proceedings, pp. 739–750. Industrial and Applied Mathematics, United States (2008)
4. Ying, X., Wu, X.: On link privacy in randomizing social networks. In: Theeramunkong, T., Kijsirikul, B., Cercone, N., Ho, T.-B. (eds.) PAKDD 2009. LNCS (LNAI), vol. 5476, pp. 28–39. Springer, Heidelberg (2009). https://doi.org/10.1007/978-3-642-01307-2_6
5. Fu, L.Z., Meng, X.F.: Reachability indexing for large-scale graphs: studies and forecasts. J. Comput. Res. Dev. 2(1), 116–129 (2015)
6. Seufert, S., Anand, A., Bedathur, S.: FERRARI: flexible and efficient reachability range assignment for graph indexing. In IEEE 29th International Conference on Data Engineering on Proceedings, pp. 1009–1020. IEEE, Brisbane (2013)
7. Cheng, J., Shang, Z., Cheng, H.: Efficient processing of k-hop reachability queries. VLDB 23(2), 227–252 (2014)
8. Zhou, J.F., Chen, W., Fei, C, P.: BiRch: a bidirectional search algorithm for k-step reachability queries. Communications 36(8), 50–60 (2015)
9. Jurczyk, P., Xiong, L.: Distributed anonymization: achieving privacy for both data subjects and data providers. In: Gudes, E., Vaidya, J. (eds.) DBSec 2009. LNCS, vol. 5645, pp. 191–207. Springer, Heidelberg (2009). https://doi.org/10.1007/978-3-642-03007-9_13

10. Zhang, X., Dou, W., Pei, J.: Proximity-aware local-recoding anonymization with MapReduce for scalable big data privacy preservation in cloud. IEEE Trans. Comput. **64** (8), 2293–2307 (2015)

11. Zhang, X,. Yang, L, T., Liu, C.: A scalable two-phase top-down specialization approach for data anonymization using MapReduce on cloud. IEEE Trans. Parallel Distrib. Syst. **25**(2), 363–373 (2014)

12. Fard, A.M., Wang, K., Yu, P.S.: Limiting link disclosure in social network analysis through subgraph-wise perturbation. In: 15th International Conference on Extending Database Technology on Proceedings, pp. 109–119, ACM, New York (2012)

13. Fard, A.M., Wang, K.: Neighborhood randomization for link privacy in social network analysis. World Wide Web **18**(1), 9–32 (2015)

14. Bhagat, S., Cormode, G., Krishnamurthy, B.: Class-based graph anonymization for social network data. VLDB Endow. **2**(1), 766–777 (2009)

15. Wang, L., Li, X.: A clustering-based bipartite graph privacy preserving approach for sharing high-dimensional data. Int. J. Softw. Eng. Knowl. Eng. **24**(07), 1091–1111 (2017)

16. Wang, Y., Zheng, B.: Preserving privacy in social networks against connection fingerprint attacks. In: IEEE 31st International Conference on Data Engineering on Proceedings, pp. 54–65. IEEE, Seoul (2015)

17. Masoumzadeh, A., Joshi, J.: Preserving structural properties in edge-perturbing anonymization techniques for social networks. IEEE Trans. Dependable Secure Comput. **9**(6), 877–889 (2012)

18. Masoumzadeh, A., Joshi, J.: Preserving structural properties in anonymization of social networks. In: 6th International Conference on Collaborative Computing: Networking Applications and Worksharing on Proceedings, pp. 1–10. IEEE, Chicago (2011)

19. Zhang, L., Zhang, W.: Edge anonymity in social network graphs. In: Proceedings International Conference on Computational Science & Engineering, pp. 1–8. IEEE, Vancouver (2009)

20. Casas-Roma, J.: Privacy-preserving on graphs using randomization and edge-relevance. In: Torra, V., Narukawa, Y., Endo, Y. (eds.) MDAI 2014. LNCS (LNAI), vol. 8825, pp. 204–216. Springer, Cham (2014). https://doi.org/10.1007/978-3-319-12054-6_18

21. Mittal, P., Papamanthou, C., Song, D.: Preserving Link Privacy in Social Network Based Systems. arXiv preprint arXiv:1208.6189 (2012)

22. Liu, C., Mittal, P.: LinkMirage: enabling privacy-preserving analytics on social relationships. In: Network & Distributed System Security Symposium on Proceedings (2016)

23. Liu, C., Mittal, P.: LinkMirage: How to Anonymize Links in Dynamic Social Systems. Eprint arXiv:1501.01361 (2015)

24. Gonzalez, J.E., Xin, R.S., Dave, A.: Graphx: graph processing in a distributed dataflow framework. In: Usenix Conference on Operating Systems Design & Implementation on Proceedings, pp. 599–613 (2014)

25. Xin, R.S., Crankshaw, D., Dave, A.: GraphX: unifying data-parallel and graph-parallel analytics. arXiv preprint arXiv:1402.2394 (2014)

26. Xin, R.S., Gonzalez, J.E., Franklin, M.J.: GraphX: a resilient distributed graph system on Spark. In: First International Workshop on Graph Data Management Experiences and Systems on Proceedings. ACM, 2 (2013)

27. Daly, E., Haahr, M.: Social network analysis for information flow in disconnected delay-tolerant MANETs. IEEE Trans. Mob. Comput. **8**(5), 621 (2009)

Minimum Spanning Tree Clustering Based on Density Filtering

Ke Wang[1], Xia Xie[1(✉)], Jiayu Sun[2], and Wenzhi Cao[3]

[1] National Engineering Research Center for Big Data Technology and System, Services Computing Technology and System Lab, Cluster and Grid Computing Lab, School of Computer Science and Technology, Huazhong University of Science and Technology, Wuhan 430074, China
shelicy@hust.edu.cn
[2] School of Computer Science and Technology, Huazhong University of Science and Technology, Wuhan 430074, China
[3] Hunan University of Technology and Business, Changsha 410205, China

Abstract. Clustering analysis is an important method in data mining. In order to recognize clusters with arbitrary shapes as well as clusters with different density, we propose a new clustering approach: minimum spanning tree clustering based on density filtering. It masks the low-density points in the density filtering step, which reduces the interference of noise and makes the gap between clusters clearer. It uses relative values of adjacent distances to find mutations of density and changes between clusters to divide data sets. It is tested on multiple synthetic data sets and real-world data sets, the results of which show that the algorithm is able to detect clusters with arbitrary shape and it is insensitive to the imbalance of density between clusters. It has achieved great results on multiple data sets.

Keywords: Clustering analysis · Data mining · Minimum spanning tree

1 Introduction

Data analysis and processing help people make better use of varieties of enormous data generated by online activities, industrial production, monitoring record and so on. Many methods and algorithms have been put forward to dealing with multifarious data and mining valuable information. Some researchers proposed a tensor-based cloud-edge computing framework as well as Tree-based and Ring-based Tree algorithms to improve the efficiency in processing the increasing data [1, 2]. A tensor-based big-data-driven routing recommendation framework and a nested anti-collision algorithm for the RFID system were put forward [3, 4], which worked well for heterogeneous networks and RFID systems. Clustering analysis aims to divide data sets according to the similarity of objects. For example, K-means [5] identifies as a cluster that points closest to the cluster center, updating iteratively until the algorithm converges. This algorithm can effectively identify spherical clusters with relatively uniform density, but when the density distribution is imbalanced or the clusters are non-spherical, K-means performs badly. Density-based spatial clustering of applications with noise (DBSCAN) [6] identifies continuous high-density areas as clusters, and the points with density below a certain

© Springer Nature Singapore Pte Ltd. 2019
H. Jin et al. (Eds.): BigData 2019, CCIS 1120, pp. 209–223, 2019.
https://doi.org/10.1007/978-981-15-1899-7_15

threshold are identified as noise. This method is capable of detecting non-spherical clusters, whereas it is impossible to select an appropriate density threshold to identify all the clusters when the densities of clusters are significantly different. Agglomerative hierarchical clustering [7] continuously merges the closest objects to form a hierarchical tree. Hierarchical clustering using single linkage merge strategy can identify contiguity-based clusters, but it is easily affected by noise. Density peak clustering [8] identifies the relatively independent high-density points as the cluster centers, and the remaining points are attached to the nearest points with higher density. It can identify clusters with arbitrary shape under normal conditions, but when there are multiple density peaks in a cluster, this cluster may split.

For the purpose of detecting clusters with arbitrary shapes as well as different densities, we propose a new clustering method: minimum spanning tree clustering based on density filtering. We tested the algorithm on multiple synthetic and real-world data sets, the results of which show that the algorithm meets our expectations. It was not affected by noise and the cluster splitting problem doesn't occur.

2   Related Work

Clustering by fast search and find of density peaks (CFSFDP) [8] was proposed in 2014 by Alex Rodrigue and Alessandro. This algorithm is based on the assumptions that the cluster centers are surrounded by neighbors with lower local densities, and they are further away from the point with higher local density. Each point, denoted by i, has its local density ρ_i as well as the minimum distance to the higher-density point δ_i.

$$\rho_i = \sum_j \chi(d_{ij} - d_c) \tag{1}$$

ρ_i is calculated according to formula (1), in which, $\chi(x) = 1$ if $x < 0$ and $\chi(x) = 0$ otherwise. j stands for another point and d_{ij} is the Euclidean Distance between point i and point j. d_c is the cutoff distance. Formula (1) is the discrete case and we can use formula (2) for higher accuracy.

$$\rho_i = \sum_j e^{-\left(\frac{d_{ij}}{d_c}\right)^2} \tag{2}$$

Every point is placed in a two-dimensional coordinate graph determined by these two quantities (referred to as the decision graph in the original paper). The clustering center is manually selected according to the decision graph. Based on the previous assumptions, the cluster centers should appear at the top right of the decision graph. After choosing the cluster centers, the remaining points are assigned successively according to the dependency relationship.

Density peaks clustering has the advantages of being able to detect clusters with arbitrary shape, but the cutoff distance and the cluster centers are chosen manually, both of which have an impact on the results. What's more, the catastrophic problem is that the cluster with multiple density peaks may split, leading to a poor result.

Many scholars have improved the original algorithm, mainly focusing on improving the robustness of the cutoff distance, density calculation and cluster centers selection. Du and Chen applied the idea of k-nearest neighbor to density peak clustering [9]. The K-nearest neighbor method was applied to calculate the local density of sample points and the self-adaptive d_c' was generated according to the distribution characteristics of different data sets. In the calculation of the second parameter δ_i, the author proposed a new measurement method to improve the identification of the clustering center. Yang and Wang adopted the idea of the weighted K-nearest neighbor and redefined the first parameter ρ_i with the product of gaussian kernel function and inverse function, improving the performance of the algorithm on the data sets with density difference [10]. Gao et al. used the iterative method of splitting and merging clusters to deal with the splitting problem in clusters with multiple density peaks [11]. Lotfi et al. put forward a new voting label propagation strategy to realize the assignment of the remaining points by adopting the idea of K-nearest neighbor [12]. Du et al. adopted the K-nearest neighbor method to optimize the performance of the algorithm in dealing with the datasets with non-uniform density and adopted the principal component analysis method to improve the performance of the algorithm in high-dimensional data at the same time [13].

These above methods improve the clustering capability and stability, but the cluster splitting problem is not eliminated. The MSTCBDF approach presented in this article will solve this problem.

3   Minimum Spanning Tree Clustering

3.1   Basic Idea

Clustering analysis uses the attributes of objects to classify them. Since no label is given, the data set is required to have the following features to complete the classification: objects of the same class are clustered together in the feature space, while objects of different classes are discrete, which is called "cluster feature". Based on the above characteristics, we define a cluster as follows: a cluster is a set of continuous points with relatively higher density, and the density of points within the cluster is higher than that of points between the clusters. That is, the different clusters are separated by regions with relatively lower density. Compared with DBSCAN, the definition is the same in terms of cluster morphology, whereas the DBSCAN algorithm needs to set a global density threshold. If the difference of density between clusters is significant, DBSCAN cannot identify all clusters. MSTCBDF has a lower requirement for clusters, which means that the density of the areas between clusters is relatively low, but not that all clusters have a similar density. When the data set satisfies the characteristics above, we claim that the data set has the characteristics of clusters.

To identify clusters in the above definition, the following capabilities are required: (1) being able to find continuous high-density areas without being affected by the shape of clusters and noise; (2) being able to detect the low-density areas between clusters, namely the gaps between clusters, which are used to distinguish them.

The basic idea of our algorithm is as follows. We hold two sets, the set of classified points C and the set of unclassified points U. At the beginning, all the points are unclassified and they are in U. At this time, C is an empty set. We move the point with the highest density from U to C, then keep looking for the nearest unclassified point to the classified points. We move it to C and record the smallest distance between this point and the classified points, which is called the adjacent distance. We repeat this step until all the points are added to C. The order in that the points are added to C forms a sequence, which is called adjacency chain. The clusters are obtained by cutting the adjacency chain at relatively large intervals. In the above process, we keep expanding set C by adding the nearest point. When all the points of a cluster are added, our algorithm has to cross to another cluster, which leads to a significantly larger adjacent distance. This process is similar to constructing the minimum spanning tree of the data set using the prim algorithm.

Agglomerative hierarchical clustering method using the single linkage merge strategy is easily affected by noise. For instance, if there is a point bridge between two clusters, the single linkage strategy will identify them as one cluster. In order to reduce the influence of noise in our algorithm, we use the density filtering strategy. We mask the points with lower density, then classify the points with higher density. When high-density points are classified, we assign the low-density points. The density filtering step can reduce the interference of low-density points, and it increases the gaps between clusters, making it easier to distinguish different clusters. Flame and aggregation data sets are used to demonstrate the effect of density filtering.

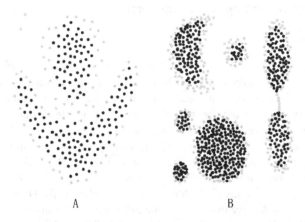

A   B

Fig. 1. Density filtering results

Figure 1A and B show the results of masking 30% low-density points in flame data set and aggregation data set respectively, in which the masked points are marked in yellow. The gap between clusters becomes more obvious and the point bridge in the aggregation data set disappears.

3.2   Main Steps

We demonstrate the basic process of our algorithm with a demo scatter diagram shown in Fig. 2. There are non-spherical clusters in this data set, and the density of the clusters is greatly different. The distance between some points in the left sparse cluster is longer than that between the two dense clusters on the right.

Step 1: density filtering, as shown in Fig. 3. It can be seen that density filtering wipes out the low-density points that may cause interference, making the gaps between clusters clearer.

Fig. 2. Demo scatter diagram   **Fig. 3.** Result of density filtering

Step 2: start from the highest density point, add points to the classified set continuously. As shown in Fig. 4, the point in orange color is the point with the highest density. It is first added to the classified set C, and the order of addition is recorded in the adjacency chain. When all the points in one cluster are added, our algorithm will expand to another cluster, leading to a significantly larger adjacent distance, as shown in Fig. 5.

Point chain: | 0 |

Fig. 4. Iteration a

Similarly, when the first point of the last cluster is added, there will be a significantly larger adjacent distance. When all the high-density points are added, we cut the adjacency chain at the relatively large gaps, and three clusters are obtained, as shown in Fig. 6. We use the ratio of adjacent distance instead of adjacent distance to cut off the chain in order to avoid the interference of the difference of cluster density.

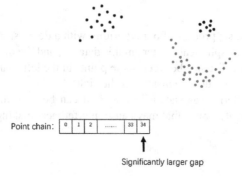

Fig. 5. Iterations b

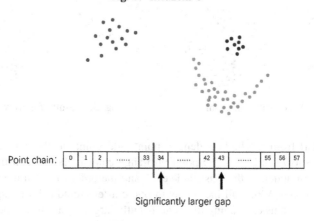

Fig. 6. Iteration c

Fig. 7. Recover low-density points

Fig. 8. Assign low-density points

Step 3: Recover and assign low-density points, as shown in Figs. 7 and 8.

3.3   Algorithms Details

We give the pseudo-code of the algorithm below.

- Define the adjacent distance between set A and set B:

$$AD_{AB} = \min_{a \in A, b \in B} D_{ab} \tag{3}$$

Where D_{ab} is the Euclidean Distance between point a and point b.

- Define the adjacent point pair between set A and set B:

$$a, b = \arg AD_{AB} \tag{4}$$

- Define operation FindAdjacency(A, B): input set A and set B, return the adjacent point pair and the adjacent distance (a, b, AD_{AB})
- Three set: C(classified), U(unclassified), M(masked)
- A list: point_chain, where the element is (idx_i, d_i, gap_i), stand for the index of points, adjacent distance, gap value respectively

---

Algorithm: MSTCBDF

Input: distance matrix, mask rate $m\%$, cutoff pos p, cluster num k
Output: clustering result

S1. initialization: $C = \Phi$, $M = \Phi$, U = data set, point_chain = Φ
S2. calculate the density of every point and sort them, note the point with the highest density as point a, put the $m\%$ low-density points into the set M, $C = C \cup \{a\}$, $U = U - \{a\} - M$, point_chain.append((idx(a), 0, 0))
S3. while $U \neq \Phi$:

   a, b, AD = FindAdjacency(C, U)
   $C = C \cup \{b\}$, $U = U - \{b\}$, point_chain.append((idx(b), AD, 0))

S4. from $i = 2$ to length(point_chain) − 2:

$$gap_i = (\frac{\exp(d_i)}{\exp(max\{d_{i-1}, d_c\})} + \frac{\exp(d_i)}{\exp(max\{d_{i+1}, d_c\})})/2 \tag{5}$$

   in which d_c is the cutoff distance

S5. find the greatest k–1 gap value and cut off the point_chain at these positions, get k parts(gap positions belong to the latter part), each part stands for a cluster, label them
S6. while $M \neq \Phi$:

   a, b, AD = FindAdjacency(C, M)
   $C = C \cup \{b\}$, $M = M - \{b\}$
   label b = label a

S7. sort points according to their index and return the result

---

End

The density calculation in step S2 follows formula (2). The cutoff distance d_c in step S4 comes from the distance at cutoff position p among ascending sorted distances between all point pairs.

3.4 Complexity Analysis

Assuming the data set consists of n points. In step S2 we need to calculate the density of n points, each of which costs $O(n)$, thus the time complexity adds up to $O(n^2)$. Then we sort the n points according to their density and divide them into C, U and M, which costs $O(n \log n)$. In step S3 and step S6, we keep finding the nearest point. For every unclassified point in U and M, we record its adjacent point in C as well as the adjacent distance, and update this information every time a point is added to C. In the beginning, set C contains only the point with the highest density, thus the adjacent point of every unclassified point is that point. To find the nearest point to C, we just traverse all the unclassified points and select the one with the minimum adjacent distance, note it as point i, and then update the adjacency information of other unclassified points by comparing its adjacent distance with its distance to point i. If the former adjacent distance is smaller, the information remains, otherwise the adjacent point changes to point i and the adjacent distance changes to the distance to i. Finding the nearest point and updating the information of other points cost $O(n + n)$, and we have to run $n - 1$ times to add all points to C, leading to $O(n^2)$ time cost. Step S4 costs $O(n)$, step S5 costs $O(kn)$ and step S7 costs $O(n \log n)$. If the cutoff distance d_c is given directly, the total time complexity of our algorithm is $O(n^2)$. If the cutoff position p is given instead, we have to sort all the n^2 distances and choose the distance at p as d_c, in which case the total time cost reaches $O(n^2 \log n)$.

4   Experiments and Analysis

4.1   Experiments Data Sets and Environment

We test MSTCBDF on multiple synthetic data sets and real-world data sets as shown in Table 1. The experiments are implemented using python 3.7 on surface pro 5, Intel Core m3-7y30@1.00 GHz with 4 GB RAM.

4.2   Evaluation

We use accuracy (Acc), rand index (RI) and normalized mutual information (NMI) to evaluate the results. All of them vary from 0 to 1 and a greater value indicates a better result.

Table 1. Experiment data set

	Data set	Size	Cluster	Feature	Source
Synthetic data sets	Circle	500	2	2	[14]
	Moon	100	2	2	[14]
	Flame	240	2	2	[15]
	Spiral	312	3	2	[16]
	Jain	373	2	2	[17]
	R15	600	15	2	[18]
	Aggregation	788	7	2	[19]
Real-world data sets	Iris	150	3	4	[20]
	Wine	178	3	13	[20]
	Digit	300	10	64	[20]
	Har	347	6	561	[20]

4.3 Synthetic Data Sets Analysis

The results of MSTCBDF on synthetic data sets are listed in Fig. 9.

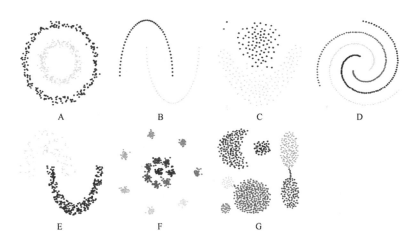

Fig. 9. Clustering results on synthetic data sets

MSTCBDF is able to detect all the synthetic data sets above. K-means identifies clusters in R15, while results on other data sets were far from ideal due to non-spherical clusters or uneven density. CFSFDP correctly identifies moon, flame, spiral, R15 and aggregation. Circle, moon and spiral are identified accurately by hierarchical clustering using single linkage but big errors occur due to the influence of low-density points on other data sets.

The Acc, RI and NMI value of results are listed as follows, in which HC stand for agglomerative hierarchical clustering using single linkage merge strategy (Tables 2, 3 and 4).

Table 2. Acc value of the results on synthetic data sets

Data set	Acc (%)			
	MSTCBDF	K-means	CFSFDP	HC
Circle	100	50.2	53.4	100
Moon	100	74	100	100
Flame	100	83.8	100	64.6
Spiral	100	32.4	100	100
Jain	100	78.6	86.1	81.0
R15	99.3	99.7	99.7	72.3
Aggregation	99.5	78.4	99.8	82.4

Table 3. RI value of the results on synthetic data sets

Data set	RI			
	MSTCBDF	K-means	CFSFDP	HC
Circle	1	0.503	0.505	1
Moon	1	0.630	1	1
Flame	1	0.727	1	0.541
Spiral	1	0.554	1	1
Jain	1	0.662	0.759	0.691
R15	0.998	0.999	0.999	0.910
Aggregation	0.987	0.928	0.999	0.926

Table 4. NMI value of the results on synthetic data sets

Data set	NMI			
	MSTCBDF	K-means	CFSFDP	HC
Circle	1	0	0.004	1
Moon	1	0.173	1	1
Flame	1	0.399	1	0.048
Spiral	1	0	1	1
Jain	1	0.371	0.507	0.267
R15	0.989	0.994	0.994	0.882
Aggregation	0.987	0.882	0.996	0.888

According to the experimental results, we draw a conclusion that MSTCBDF is able to recognize clusters with arbitrary shapes and is insensitive to the density differences and noise. It has achieved good results with multiple artificial data sets.

4.4 Real-World Data Sets Analysis

Iris flower classification needs to divide 150 iris samples into three sub-categories according to petal length, petal width, sepal length and sepal width. Wine classification needs to divide them into three categories according to the content of 13 substances extracted from 178 wine samples. Due to the great difference in the values of different attributes of wine dataset samples, we standardize each attribute value, as shown in Formula (6), in which μ is the mean value of the samples and σ is the standard deviation of the samples.

$$x' = \frac{x-\mu}{\sigma} \tag{6}$$

Table 5. Iris data set result

Evaluation	Iris dataset			
	MSTCBDF	K-means	CFSFDP	HC
Acc (%)	96.0	89.3	90.7	68.0
RI	0.893	0.880	0.892	0.777
NMI	0.893	0.758	0.806	0.736

MSTCBDF achieves the best results on the iris dataset with an accuracy of 96%, which is a great result in unsupervised learning (Table 5).

Table 6. Wine data set result

Evaluation	Wine dataset			
	MSTCBDF	K-means	CFSFDP	HC
Acc (%)	91.0	96.6	89.3	42.7
RI	88.5	95.4	0.865	0.342
NMI	0.753	87.6	0.726	0.032

The K-means method achieves the best results on the wine data set, followed by MSTCBDF. We use principal component analysis (PCA) technology to reduce the data set to three dimensions in order to observe the data distribution. As shown in Fig. 10, we find that the characteristic of cluster of this data set is not very obvious, thus MSTCBDF does not perform well. When the mask rate is set to 0.8, the result of MSTCBDF is comparable to that of CFSFDP (Table 6).

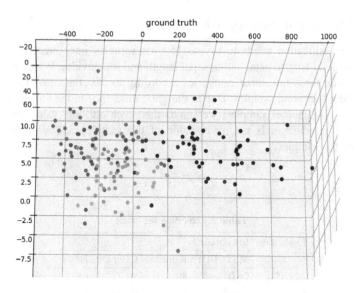

Fig. 10. Wine data set 3d plot

The hand-writing digits data set consists of 1797 8 * 8 hand-writing digits pictures, including 0 to 9 digits. We took the first 300 pieces for experiments. We paint different kinds of numbers with different colors according to the clustering results. The results are shown in Fig. 11, where the category labels are marked above the pictures. Note that the labels here are different from the real numbers represented by the pictures.

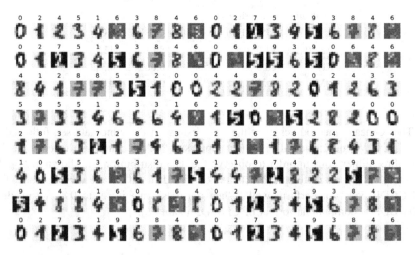

Fig. 11. Hand-writing digits result

Our MSTCBDF gets the best result on hand-writing data set as shown in Table 7.

The human activity recognition data set contains 10299 human activity data. Several volunteers wore mobile phones on their bodies conducting six activities and several parameters were recorded. The six activities are walking, sitting down, standing up, going upstairs, going downstairs and lying down (w, sd, su, gu, gd, ld for short). We select the data of the first two volunteers for the experiments, including 347 samples.

Table 7. Results on digits data set

Evaluation	Digits dataset			
	MSTCBDF	K-means	CFSFDP	HC
Acc (%)	95.7	89.7	95.0	69.3
RI	0.984	0.963	0.981	0.878
NMI	0.950	0.855	0.941	0.855

The confusion matrices of the results of the MSTCBDF algorithm are as follows.

Table 8. MSTCBDF har confusion matrix

Real\prediction	w	gu	gd	sd	su	ld
w	95	0	0	0	0	0
gu	0	53	0	0	0	0
gd	0	0	44	5	0	0
sd	0	0	0	0	47	0
su	0	0	0	0	53	0
ld	0	0	0	0	0	50

As can be seen from Table 8, MSTCBDF can accurately identify three movements: walking, going upstairs and lying down. There is a little deviation in the identification of going downstairs, and it is unable to distinguish sitting from standing. From the results of the other three algorithms we know that: K-means can identify walking, going upstairs and going downstairs, with only minimal deviation, and it can't distinguish the three movements of sitting, standing and lying down. CFSFDP can identify the movements of going upstairs and downstairs. There is a big error in walking recognition. It is completely unable to distinguish sitting, standing up and lying down. Hierarchical clustering can hardly finish clustering because of the interference of noise.

Table 9. Results on har data set

Evaluation	Har dataset			
	MSTCBDF	K-means	CFSFDP	HC
Acc (%)	85.0	81.0	65.0	43.2
RI	0.955	0.942	0.830	0.677
NMI	0.928	0.855	0.737	0.599

The results on har data set are shown in Table 9, from which we can see that MSTCBDF performs best.

5 Conclusion and Future Work

Based on the definition of contiguity and density-based clusters, we propose a new clustering method: minimum spanning tree clustering based on density filtering. It not only has the advantage of hierarchical clustering using the single linkage merge strategy, but also reduces the interference of low-density points by using the density filtering strategy. It uses relative gap values to find the changes between clusters, which greatly reduces the impact of density differences between clusters, and performs better than K-means and CFSFDP in data sets with uneven density distribution. Cluster extension based on the minimum spanning tree overcomes the problem of cluster splitting in CFSFDP. In summary, MSTCBDF is able to recognize clusters with arbitrary shapes and it is insensitive to the difference of density between clusters. It has achieved the best results on many experiment data sets. In addition, the density filtering step in MSTCBDF makes the gap between clusters clearer. This step can be embedded in other clustering algorithms to improve performance. MSTCBDF requires a parameter of mask rate, which has a great influence on the result of the algorithm in some data sets. Although the meaning of this parameter is intuitive and clear, it is still a little hard to select manually. If the parameter can be automatically generated according to the characteristics of data distribution, the robustness of the algorithm will be improved.

Acknowledgements. This work is supported in part by the National Natural Science Foundation of China under Grant No. 61702183.

References

1. Wang, X., Yang, L.T., Xie, X., Jin, J., Deen, M.J.: A cloud-edge computing framework for cyber-physical-social services. IEEE Commun. Mag. **55**(11), 80–85 (2017)
2. Wang, X., Yang, L.T., Chen, X., Deen, M.J., Jin, J.: Improved multi-order distributed HOSVD with its incremental computing for smart city services. IEEE Trans. Sustain. Comput. (2018). https://doi.org/10.1109/TSUSC.2018.2881439:1-1
3. Wang, X., Yang, L.T., Kuang, L., Liu, X., Zhang, Q., Deen, M.J.: A tensor-based big data-driven routing recommendation approach for heterogeneous networks. IEEE Netw. Mag. **33**(1), 64–69 (2019)
4. Wang, X., Yang, L.T., Li, H., Lin, M., Han, J., Apduhan, B.O.: NQA: a nested anti-collision algorithm for RFID systems. ACM Trans. Embed. Comput. Syst. **18**(4), 32 (2019)
5. MacQueen, J.: Some methods for classification and analysis of multivariate observations. In: Proceedings of the Fifth Berkeley Symposium on Mathematical Statistics and Probability, Berkeley, CA, vol. 1, pp. 281–297 (1967)
6. Ester, M., Kriegel, H.P., Xu, X.: A density-based algorithm for discovering clusters in large spatial databases with noise. In: Proceedings of International Conference on Knowledge Discovery and Data Mining, pp 226–231 (1996)
7. Johnson, S.C.: Hierarchical clustering schemes. Psychometrika **32**(3), 241–254 (1967)

8. Rodriguez, A., Laio, A.: Clustering by fast search and find of density peaks. Science **344** (6191), 1492–1496 (2014)
9. Du, P., Cheng, X.R.: Comparative density peaks clustering based on K-nearest neighbors. Comput. Eng. Algorithms **55**(10), 161–168 (2019)
10. Yang, Z., Wang, H.J.: Improved density peak clustering algorithm based on weighted K-nearest neighbor. Appl. Res. Comput. **37**(3), 1–7 (2019)
11. Gao, J., Zhao, L., Chen, Z., Li, P., Xu, H., Hu, Y.: ICFS: an improved fast search and find of density peaks clustering algorithm. In: Proceedings of 2016 IEEE 14th International Conference on Dependable, Autonomic and Secure Computing, 14th International Conference on Pervasive Intelligence and Computing, 2nd I International Conference on Big Data Intelligence and Computing and Cyber Science and Technology Congress (DASC/PiCom/DataCom/CyberSciTech), Auckland, pp. 537–543 (2016)
12. Lotfi, A., Seyedi, S.A., Moradi, P.: An improved density peaks method for data clustering. In: Proceedings of the 6th International Conference on Computer and Knowledge Engineering (ICCKE), Mashhad, pp. 263–268 (2016)
13. Du, M., Ding, S., Jia, H.: Study on density peaks clustering based on K-nearest neighbors and principal component analysis. Knowl.-Based Syst. **99**, 135–145 (2016)
14. Pedregosa, F., Varoquaux, G., Gramfort, A., Michel, V., Thirion, B., Grisel, O., et al.: Scikit-learn: machine learning in python. J. Mach. Learn. Res. **12**(Oct), 2825–2830 (2011)
15. Fu, L., Medico, E.: FLAME, a novel fuzzy clustering method for the analysis of DNA microarray data. BMC Bioinform. **8**(1), 1–15 (2007)
16. Chang, H., Yeung, D.Y.: Robust path-based spectral clustering. Pattern Recogn. **41**(1), 191–203 (2008)
17. Zhu, Y., Dass, S.C., Jain, A.K.: Statistical models for assessing the individuality of fingerprints. IEEE Trans. Inf. Forensics Secur. **2**(3), 391–401 (2007)
18. Veenman, C.J., Reinders, M.J.T., Backer, E.: A maximum variance cluster algorithm. IEEE Trans. Pattern Anal. Mach. Intell. **24**(9), 1273–1280 (2002)
19. Gionis, A., Mannila, H., Tsaparas, P.: Clustering aggregation. ACM Trans. Knowl. Discov. Data (TKDD) **1**(1), 4 (2007)
20. Dua, D., Graff, C.: UCI Machine Learning Repository. University of California, School of Information and Computer Science, Irvine, CA (2019). http://archive.ics.uci.edu/ml

Research of CouchDB Storage Plugin
for Big Data Query Engine Apache Drill

Yulei Liao[1] and Liang Tan[1,2(✉)]

[1] College of Computer Science, Sichuan Normal University,
Chengdu 610101, Sichuan, China
568483222@qq.com
[2] Institute of Computing Technology, Chinese Academy of Sciences,
Beijing 100190, China

Abstract. Currently, the document-oriented database supported by Apache Drill is only MongoDB. However, due to the lack of data model, application interface, security and usability of MongoDB, Apache Drill is limited in querying and processing document data. CouchDB is an emerging document-oriented database. Compared to MongoDB, CouchDB has the advantage of supporting triggers, running in Android and BSD environments, rendering in JSON format, and supporting any language that supports HTTP requests, but CouchDB has low query performance and does not support standard SQL queries. Therefore, the research on the CouchDB storage plugin for Apache Drill makes sense. This paper first researches the basic architecture of CouchDB and Apache Drill and the query flow of Apache Drill, and the ValueVector data structure, then designs and implements CouchDB storage plugin based on Apache Drill's storage plugin specification and CouchDB's application programming interface. With a simple configuration, users can use CouchDB as a data source for the Apache Drill query engine. Experiments show that the CouchDB Storage Plugin not only further enhances Apache Drill's query and management capabilities for document-oriented data, but also enables quick query of CouchDB with SQL and greatly improves CouchDB's query performance.

Keywords: Document-oriented database · Query engine · Apache Drill · CouchDB · Storage plugin

1 Introduction

In recent years, with the rapid increase in the size of big data, more and more real-time big data query engines have emerged. Currently, mainstream real-time big data query engines are Google Dremel [1], Apache Drill [2], Impala [3], Spark [4], and Presto [5]. Compared with other query engines, Apache Drill has the following advantages: (1) Drill supports the standard SQL:2003 syntax. No need to learn a new "SQL-like" language or struggle with a semi-functional BI tool. Drill supports many data types including DATE, INTERVAL, TIMESTAMP, and VARCHAR, as well as complex query constructs such as correlated sub-queries and joins in WHERE clauses. (2) Drill is the world's first and only distributed SQL engine that doesn't require schemas. It

© Springer Nature Singapore Pte Ltd. 2019
H. Jin et al. (Eds.): BigData 2019, CCIS 1120, pp. 224–239, 2019.
https://doi.org/10.1007/978-981-15-1899-7_16

shares the same schema-free JSON model as MongoDB and Elasticsearch. No need to define and maintain schemas or transform data (ETL [6]). Drill automatically understands the structure of the data. (3) Drill is extensible. Users can connect Drill out-of-the-box to file systems (local or distributed, such as S3 and HDFS), HBase [7] and Hive [8]. Users can implement a storage plugin to make Drill work with any other data source. Drill can combine data from multiple data sources on the fly in a single query, with no centralized metadata definitions. Apache Drill not only supports structured data sources, but also supports semi-structured and unstructured data sources such as MongoDB [9] and logfile.

Currently, the document-oriented database supported by Apache Drill is MongoDB. MongoDB is a product between a relational database and a non-relational database. It is the most powerful in non-relational database and most similar to relational database. The data structure supported by MongoDB is very loose and is stored in BSON format which is similar to the JSON, so it can support more complex data types. The powerful query language is the biggest feature of MongoDB. Its syntax is similar to the object-oriented query language. It can realize almost all functions like single-table query of relational database, and also supports indexing data. However, due to the lack of data model, application interface, security and usability of MongoDB, Apache Drill is limited in querying and processing document data. Apache CouchDB is an emerging document-oriented database that provides a REST interface using JSON as a data format and can manipulate the organization and rendering of documents through views. The biggest significance of CouchDB is that it is a new generation storage system for web applications. The comparison between MongoDB and CouchDB is shown in Table 1.

Table 1. Comparison of MongoDB and CouchDB

Features	MongoDB	CouchDB
Data model	BSON	JSON
APIs and other access methods	Proprietary protocol using JSON	RESTful HTTP/JSON API
SQL	Read-only SQL queries via the MongoDB Connector for BI	No
Reliability	Downtime guarantees data consistency	Downtime need to repair data files
Typing	Yes	No
Server operating systems	Linux, OS X, Solaris, Windows	Android, BSD, Linux, OS X, Solaris, Windows

Therefore, adding CouchDB as a data source for Apache Drill is important. On the one hand, this further enriches the data source of Apache Drill, enhances the ability of Apache Drill to query and process document data; on the other hand, querying CouchDB in Apache Drill can not only further improve the query performance of CouchDB, but also allows CouchDB to support standard SQL. This also greatly promotes the application, promotion and popularization of CouchDB.

The rest of this paper is organized as follows. The Sect. 2 is a study of CouchDB and Apache Drill. The Sect. 3 discusses the architectural design of the CouchDB storage plugin based on Apache drill. The Sect. 4 describes the implementation of CouchDB storage plugin. Section 5 discusses the experimental setup and results analysis. Section 6 describes the work related to the Apache Drill storage plugin. Section 7 is summarized.

2 Research on CouchDB and Apache Drill

2.1 The Basic Architecture of CouchDB

CouchDB is a schema-free, document-oriented NoSQL database. It has a built-in web server, and data queries are handled through the HTTP RESTful JSON API, so it provides a common access to any programming language that supports HTTP requests without a custom API. CouchDB natively supports incremental replication databases, so it is very convenient to maintain a database mirror copy with high availability and partition fault tolerance to ensure the ultimate consistency of data.

The client treats CouchDB as a web server, requests via HTTP and gets an HTML or JSON response. Inside CouchDB, the database contains the document, the design, and the view that pre-calculates the results of the query. CouchDB's data is stored on disk as a collection of JSON documents. Each document has a unique identifier (_id) that can be used to construct a unique URL to access the resource. The design document (_design) defines the query or view which can be created on database. The response of each view (_view) is pre-computed and stored on disk. Web applications can use the standard AJAX protocol (such as jQuery) for data interaction. Users can use the GET, POST, PUT, and DELETE methods of HTTP to perform common additions, deletions, and changes to CouchDB.

2.2 The Basic Structure and Core Modules of Apache Drill

Drillbit is a process that runs on each active Drill node to coordinate, schedule, and execute queries. The following are the modules related to the storage plugin.

SQL Parser: Drill uses Calcite, the open source framework, to parse incoming queries. The output of the parser component is a language agnostic, computer-friendly logical plan that represents the query.

Storage Plugin Interfaces: Drill serves as a query layer on top of several data sources. Storage plugins in Drill represent the abstractions that Drill uses to interact with the data sources. Storage plugins provide Drill with the following information:

- Metadata available in the source
- Interfaces for Drill to read from and write to data sources
- Location of data and a set of optimization rules to help with efficient and faster execution of Drill queries on a specific data source.

Distributed Cache: Drill uses a distributed cache to manage metadata (not the data) and configuration information across various nodes. Sample metadata information that is stored in the cache includes query plan fragments, intermediate state of the query execution, and statistics. Drill uses Infinispan as its cache.

2.3 Query Flow in Apache Drill Storage Plugin

When the storage plugin receives the physical plan pushed down by Apache Drill, it begins to execute the query. The Storage Plugin sets the columns to be queried in ScanBatchCreator according to the pushdown physical plan, obtains and sets the database name, table name, database connection and filters in GroupScan, and then passes it to ScanBatchCreator. ScanBatchCreator generates several record readers based on these settings, and calls the API of the target data source to query according to the information provided. Finally, the query results will be returned to Apache Drill.

2.4 ValueVector of Apache Drill

ValueVector is the data structure defined by Apache Drill to pass sequences of columnar data between operators. Reading a random element from a ValueVector must be a constant time operation. To accomodate, elements are identified by their offset from the start of the buffer. Repeated, nullable and variable width ValueVectors utilize in an additional fixed width value vector to index each element. Write access is not supported once the ValueVector has been constructed. The ValueVector is comprised of one or more contiguous buffers; one which stores a sequence of values, and zero or more which store any metadata associated with the ValueVector. A ValueVector stores values in a ByteBuf, which is a contiguous region of memory. Additional levels of indirection are used to support variable value widths, nullable values, repeated values and selection vectors. These levels of indirection are primarily lookup tables which consist of one or more fixed width ValueVectors which may be combined (e.g. for nullable, variable width values). A fixed width ValueVector of non-nullable, non-repeatable values does not require an indirect lookup; elements can be accessed directly by multiplying position by stride.

3 Design of CouchDB Storage for Drill

3.1 Functional Design of CouchDB Storage Plugin

The data source of Apache Drill is extensible. To enable Apache Drill to support CouchDB, we need to design and implement the CouchDB Storage Plugin, which allows users to query the data in CouchDB directly in Apache Drill using SQL.

Designing and programming a new storage plugin for Apache Drill requires six interfaces: AbstractGroupScan, SubScan, RecordReader, BatchCreator, Abstract Storage Plugin, and StoragePluginConfig. Their specific functions are as follows.

AbstractGroupScan: GroupScan contains all the data scanned by the physical plan. It is a superset of all SubScans and is the most important body of the storage plugin, providing basic query information such as which columns to query.

SubScan: SubScan represents data scanned by a specific segment. In contrast to GroupScan, the latter represents all the data that the physical plan scans. It is a subset of GroupScan and the main part of the scan.

AbstractRecordReader: The RecordReader is responsible for actually reading the data and returning it to the drill. RecordReader is generated by BatchCreator according to the corresponding parameters.

BatchCreator: Drill automatically scans the implementation of BatchCreator, so this article doesn't need to care about its source.

AbstractStoragePlugin: StoragePluginConfig is used to configure the plugin. It must be JSON serializable/deserializable. Drill will store the storage configuration in / tmp/drill/sys.storage_plugins. When Drill starts, it will automatically scan the StoragePluginRegistry method in the AbstractStoragePlugin implementation class and establish a mapping from StoragePluginConfig.class to AbstractStoragePlugin constructor.

StoragePluginConfig: Configure for AbstractStoragePlugin, such as plugin switch, check plugin enable status, set data source login authentication, set server, etc.

To implement the above six interfaces to achieve access to CouchDB on the Drill platform, we designed six implementation classes, the relationship between them is shown in Fig. 1.

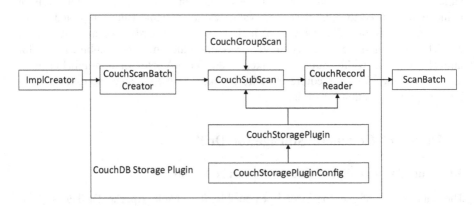

Fig. 1. CouchDB storage plugin architecture

3.2 Performance Design of CouchDB Storage Plugin

Apache Drill is distributed a real-time big data query engine, which is generally only queried. Therefore, when we design the performance of the CouchDB storage plugin, we only consider the query performance, mainly considering two indicators: read time and fetch time. To design the performance of the CouchDB storage plugin, we first analyze the query flow of Apache Drill and CouchDB, as shown in Fig. 2. Figure 2 shows that when a user queries CouchDB through apache drill, Apache Drill first checks whether there is a query plan in the distributed cache infinispan that matches the current query. If it does not exist, it builds the query plan and stores it in the cache; if it exists, it directly reads the query plan and other information in the cache. Then Drill sends this information to the CouchDB storage plugin. The CouchDB storage plugin parses the query plan and sends it to CouchDB. CouchDB uses the information exchange between the client RClient, CouchDB Web Server, and CouchDB Database to find the data. The storage plugin uses 4096 records as a batch, iteratively writes into several ValueVectors and submits them to the user.

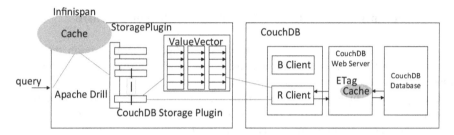

Fig. 2. Fetching process between Apache Drill and CouchDB

This shows that because Apache Drill has operations such as build and get query plan, compared with CouchDB, it will increase the time for fetching, and the storage plugin will write the query results into the ValueVector. Compared with CouchDB, it will greatly improve the throughput and shorten the time for reading. In addition, since CouchDB is based on the HTTP protocol, in order to further improve query performance, we can configure ETag in CouchDB Web Server. We use Tread-CouchDB to represent the time for reading in CouchDB, and Tread-Drill-CouchDB to indicate the time for reading in Apache Drill and CouchDB. Then the quantitative expression of time for reading in Apache Drill and CouchDB we have designed is shown in (1).

$$\mathbf{T}_{read-Drill-CouchDB} = \alpha \mathbf{T}_{read-CouchDB} \tag{1}$$

Where α is a constant and $\alpha \geq 1$. We use $\mathbf{T}_{fetch-Drill-CouchDB}$ to represent the time for fetching in CouchDB, and $\mathbf{T}_{fetch-Drill-CouchDB}$ to represent the time for fetching in Apache Drill and CouchDB. The quantitative expression of time for fetching in Apache Drill and CouchDB we designed is shown in (2).

$$\mathbf{T}_{fetch-Drill-CouchDB} = \beta \mathbf{T}_{fetch-CouchDB} \tag{2}$$

where β is a constant and $1 < \beta < 1.5$.

4   Implementation of CouchDB Storage Plugin for Apache Drill

The implementation of CouchDB Storage Plugin is divided into two parts. One is the implementation of the functionality of the CouchDB Storage Plugin, the other is the implementation of the performance of CouchDB Storage Plugin.

4.1   Functional Implementation of CouchDB Storage Plugin

The implementation of the functionality of the CouchDB Storage Plugin consists of three parts: the first is the implementation of the functional interface, the second is the implementation of the configuration, and the third is the implementation of the registration and loading.

4.1.1   Implementation of the Interface of the Function of CouchDB Storage Plugin

Users can get the existing database and tables in Apache Drill through the "show databases" and "show tables" commands in the Drill console. To implement this function, we need to implement the CouchSchemaFactory class in the CouchDB storage plugin, which inherits the AbstractSchemaFactory class from Drill. In the constructor, the database list in CouchDB and the table in the database are obtained by calling the corresponding inner classes. In the CouchSchemaFactory, two internal classes, DatabaseLoader and TableNameLoader, are implemented. Since CouchDB's records are stored in the database and there is no concept of a table, the same load method is implemented in both classes. When the method is called, the method first gets the URL of CouchDB, then gets a list of all database names by accessing the _all_dbs interface, and returns a list of strings containing all the database names.

1. **The stage of query preparation**

   Upon receiving a query request submitted by the user, Apache Drill will convert it into a physical plan and begin the query initialization process in CouchScanBatch. In CouchScanBatch, we need to implement the getBatch method of the BatchCreator interface. This method can be used to set the query field list, and obtain the database, table, filter conditions, etc. of the query request from the physical plan through CouchGroupScan, CouchSubScan, CouchStoragePlugin, etc. Finally, the above information is combined to initialize the CouchRecordReader and return to Drill.

2. **The stage of query execution**

   In the stage of preparing the query, Drill obtains several record readers with initialization information. When the query is executed, Apache Drill calls the setup and next methods of CouchRecordReader. Since JSON is the storage and read format of CouchDB, it is necessary to set up a JSON record reader provided by Drill in the setup method, and set the text mode, the number format, the problem of invalid and infinite numbers. After the setting is completed, the next step is to execute the step of reading the data. The CouchDB Java driver ektorp tool is used to generate the CouchDB database connection, query according to the existing query information, and write the query result to the JSON record reader. Finally, The results of the query are submitted to Drill for subsequent processing and display operations.

4.1.2 Configuration Implementation of CouchDB Storage Plugin

To use the custom storage plugin, users need to configure and enable the storage plugin. There are two ways to configure a storage plugin, using a RESTful API configuration and configuring it in the Drill Web UI. There are many configuration properties for the Apache Drill storage plugin, and most properties have default settings. But type, connection, and enable are the configuration properties we have to set. Type represents the name of the storage plugin, connection represents the address of the data source, and enable represents the state of the plugin.

1. **Configure the CouchDB storage plugin in the Drill Web UI**
   After launching Drill in the shell, go to the Web UI at http://localhost:8047/, select the Storage tab, enter the storage plugin name "couch" in the New Storage Plugin. The page then automatically jumps to Configuration for detailed setup, and its setup parameters are described using JSON.
   {"type":"couch","enabled":false,"connection":"http://localhost:5984/"}.
2. **Configuring the CouchDB storage plugin using the RESTful API**
   The name and config attributes need to be provided when configuring the storage plugin using the RESTful API. To create a storage plugin for CouchDB, after launching Drill in the shell, you need to enter the following command:
   curl -X POST -H "Content-Type: application/json" -d '{"name": "couch","config": {"type":"couch","enabled":false,"connection":"http://localhost:5984/"}}'http://localhost:8047/storage/couch.json.

4.1.3 Registration and Loading Mechanism Implementation for CouchDB Storage Plugin

The registration of the Apache Drill Storage Plugin is done in the StoragePlugin RegistryImpl, and its registration process is divided into the following steps:

First, Drill will load all the drill-module-conf files in the classpath. This file specifies the path of the package to be scanned. All implementation classes of the StroagePlugin interface will be loaded in the path of this package.

Secondly, Drill verifies the entire storage plugin. First, it is checked whether the configuration method of the StoragePlugin implementation class conforms to the standard. Secondly, it is verified whether the configuration of the plugin is valid. In this process, the Storage Plugin configuration needs to be loaded. If the plugin is started for the first time, Drill will read the bootstrap-storage-plugins.json file in the classpath. Each storage plugin will correspond to one such file. According to the configuration information, the object of the StoragePluginConfig implementation class will be generated. If the plugin is not started for the first time, Drill will read the local configuration file with the path in the /tmp/drill/sys.storage_plugins.

Finally, the storage plugin is initialized and registered according to the configuration information. Only valid plugins that are properly registered will be loaded correctly. The registration steps are shown in Fig. 3.

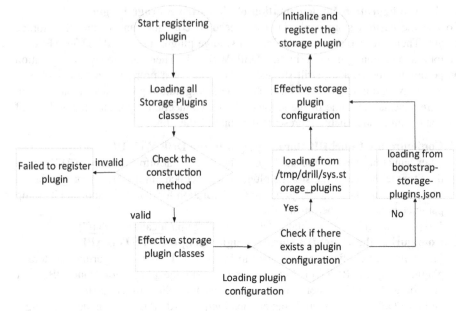

Fig. 3. Registration flow of Apache Drill storage plugin

4.2 Performance Implementation of CouchDB Storage Plugin

In order to achieve the performance goals designed in Sect. 3.2, we did the following two things:

1. **Storing the query results in ValueVectors**
   In the implementation, we need to record the records "owIterator" that have been queried, with the value of "BaseValueVector.NITIAL_VALUE_ALLOCATION" as a batch, and iteratively writes into several ValueVectors. The main code is as Fig. 4.

```
while(docCount        <        BaseValueVector.INITIAL_VALUE_ALLOCATION        &&
rowIterator.hasNext()) {
        JSONObject row = JSONObject.fromObject(rowIterator.next().getDoc());
        jsonReader.setSource(row.toString().getBytes(Charsets.UTF_8));
        writer.setPosition(docCount);
        jsonReader.write(writer);
        docCount ++;
    }
    writer.setValueCount(docCount);
    return docCount;
```

Fig. 4. The main codes for Storing the query results in ValueVectors

Considering the capacity of each ValueVector, the value of BaseValueVector. INITIAL_VALUE_ALLOCATION is set to 4096. writer. setPosition (docCount) sets the starting position for each record to be inserted. Since it is a batch operation, the number of records for each batch will be returned as the return value, in order to count the total number of operation records.

2. **Starting the HTTP cache ETag validation cache response in CouchDB Web sever**

The server passes the authentication token using the ETag HTTP header. The verification token enables efficient resource update checking and does not transfer any data when the resource has not changed. Configuration steps:

(1) Add a file ".htaccess" to the ETag directory and add a line to it: "FileETag MTime Size". If the file ".htacces" already exists, make sure there are no "FileETagNone" param lines in the file. If "FileETagNone" exists, delete the line.

(2) Check if "mod_headers" are used to remove ETag, the following statement should not appear in file "httpd.conf". It is shown as Fig. 5

```
LoadModule headers_module modules/mod_headers.so
Headerunset ETag
```

Fig. 5. Mod_headers configuration

(3) Restart httpd

Through the above work, CouchDB can not only use SQL for data query in Apache Drill, but also effectively reduce the time for reading by improving the throughput.

5   Experiment and Related Analysis

5.1   Environment

The test runs on the Windows 10 1903 x64. The hardware parameters are: AMD Ryzen 5 2600X six-core processor; 16 GB DDR4 3200 MHz; Java virtual machine with version 1.8.0_201. The data we useed is provided by the Yahoo! Cloud Service Benchmark. The experiment is to perform fetching all keys and reading 10 times for 10000–50000 rows in CouchDB, respectively, and get the average running time.

5.2   Experiment Procedure

Case 1: Validation of the functionality of the CouchDB Storage Plugin. The purpose of this experiment is to verify the CouchDB Storage Plugin functionality we developed.

Configure the CouchDB storage plugin according to 4.1.2, and use the "show databases" command in Drill shell to view all the databases. If the result returned contains the databases in CouchDB, it indicates that the CouchDB Storage Plugin can be effectively identified. The query result is shown in Fig. 6.

After Drill recognizes the CouchDB database, the data needs to be queried to confirm that Drill can correctly read the data from the database. In this experiment, we typed a SQL in Drill "select employee_id, department_id, salary_paid from couch. salary.salary limit 10", and observed the results to verify that Drill can successfully read data from CouchDB. The results of the experiment are shown in Fig. 7.

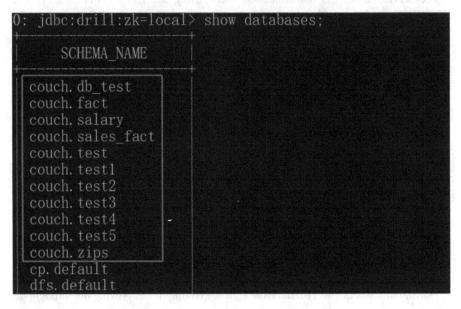

Fig. 6. Verify that CouchDB is recognized by Drill

```
0: jdbc:drill:zk=local> select employee_id, department_id,
salary_paid from couch. salary. salary limit 10;
+-------------+---------------+-------------+
| employee_id | department_id | salary_paid |
+-------------+---------------+-------------+
| 425         | 11            | 7. 38       |
| 496         | 15            | 6. 12       |
| 584         | 16            | 3. 3465     |
| 553         | 15            | 6. 12       |
| 509         | 16            | 3. 3778     |
| 458         | 17            | 4. 8        |
| 1129        | 18            | 3. 456      |
| 605         | 18            | 3. 2673     |
| 637         | 18            | 3. 421      |
| 392         | 16            | 3. 4228     |
+-------------+---------------+-------------+
```

Fig. 7. Verify that Drill can successfully get data from CouchDB

This experiment shows that the development of the CouchDB storage plugin for Apache Drill has been successful, and users can obtain the database and data from CouchDB in Drill.

Thanks to the ValueVector data structure, time for reading in CouchDB in Drill has been greatly reduced compared to CouchDB. Because Drill needs to build/get the query plan before querying, time for fetching all keys in CouchDB in Drill is slightly increased. Experiments show that, except for the first run, which needs to build and cache the query plan, the performance loss is controlled within 50% of the original performance, and gradually decreases as the amount of query data increases.

Case 2: Performance verification of the CouchDB Storage Plugin. The purpose of this experiment is to verify the performance of the CouchDB Storage Plugin we developed.

For this experiment, we used SQL to read and fetch 10000–50000 records in CouchDB. The results of the experiment are shown in Table 2 and Table 3.

Table 2. Time for reading and fetching all keys in CouchDB in Apache Drill.(ms)

Loading records	Read	Fetch all keys
10000	6390	977
20000	12357	1707
30000	18807	2576
40000	24652	3458
50000	30453	4243

Table 3. Time for reading and fetching all keys in CouchDB.(ms)

Loading records	Read	Fetch all keys
10000	20521	683
20000	42729	1290
30000	104399	2009
40000	86485	2646
50000	180205	3276

Case 3: Performance Comparison of Document-Oriented Database Storage Plugins for Apache Drill (CouchDB and MongoDB).

Data in MongoDB is stored on the disk and mapped to memory through "mmap". The additions, deletions, and changes are all done in memory, and then the modified data is written back to disk by flush. If the memory resources are rich enough, the performance of MongoDB will be greatly improved. CouchDB is a disk-based database. All operations are done on disk. Due to the IO limitation of the disk, the performance of CouchDB is much lower. Below is a performance comparison of MongoDB and CouchDB in Drill.

The results of Case 3 show that in the reading operation, MongoDB in Drill runs longer than MongoDB, and CouchDB in Drill runs less than CouchDB. Therefore,

CouchDB performs better than MongoDB for reading in Drill. In the fetching operation, due to the advantages of MongoDB, the performance of CouchDB in Drill is far less than MongoDB in Drill.

Based on the above results, the CouchDB storage plugin in Drill can improve the throughput of CouchDB, thus making time for reading in CouchDB shorter and improving the performance of CouchDB. As the amount of query data increases, time for fetching all keys in CouchDB in Drill will be closer to CouchDB (Tables 4 and 5).

Table 4. Time for reading and fetching all keys in MongoDB in Apache Drill.(ms)

Loading records	Read	Fetch all keys
10000	4882	132
20000	9367	130
30000	14039	135
40000	16701	167
50000	20686	182

Table 5. Time for reading and fetching all keys in MongoDB.(ms)

Loading records	Read	Fetch all keys
10000	2242	132
20000	3834	130
30000	4711	135
40000	5834	167
50000	7614	182

Case 4: Query for multiple data sources in Apache Drill.

Drill can combine data from multiple data sources on the fly in a single query, with no centralized metadata definitions.

In this experiment, we query CouchDB and MongoDB through SQL "select count (*) as total_count from couch.zips.zips a, mongo.city.zips b where a.state=b.state and a. state = 'MA'" and return the correct results. The query results of MongoDB and CouchDB are shown in Fig. 8.

CouchDB can be queried with other data sources of Drill, which improves the user's convenience and adds more usage scenarios to CouchDB, which can improve the usage of CouchDB.

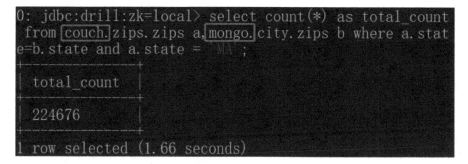

Fig. 8. Query for multiple data sources in Apache Drill.

6 Related Work

At present, the research and development of the Drill memory plugin has achieved some results in the industry.

Apache Drill official and third-party contributed a large number of storage plugins, among which the file system storage plugin has distributed file system plugin, S3 storage plugin, HTTPD storage plugin, system log format plugin, LTSV format plugin, image metadata format plugin, MapR -DB format, etc. In addition, Drill also has storage plugin support for the following database systems:

(1) HBase Storage Plugin: When connecting Drill to an HBase data source using the HBase storage plugin installed with Drill, you need to specify a ZooKeeper quorum. Drill supports HBase version 1.x [14].
(2) Hive Storage Plugin: Prior to Drill 1.13, Drill supported Hive 1.0. Drill 1.13 and later includes version 2.3.2 of the Hive client, which adds support for queries on transactional (ACID) and non-transactional Hive bucketed ORC tables. The updated Hive libraries are backward compatible with earlier versions of the Hive server and metastore [15].
(3) RDBMS Storage Plugin: Apache Drill supports querying a number of RDBMS instances. This allows you to connect your traditional databases to your Drill cluster so you can have a single view of both your relational and NoSQL data sources in a single system [16].
(4) MongoDB Storage Plugin: Drill supports MongoDB 3.0, providing a mongodb storage plugin to connect to MongoDB using MongoDB's latest Java driver. You can run queries to read, but not write, Mongo data using Drill. Attempting to write data back to Mongo results in an error. You do not need any upfront schema definitions [17].

There are no outstanding contributions for the Apache Drill storage plugin in academics.

To sum up, the industry has done a lot of work for the Apache Drill storage plugin, which makes Drill support more data sources and further expand the application. But currently it only supports MongoDB for document-oriented database, which limits Drill's further expansion in big data analysis and processing. The CouchDB storage plugin designed in this article enables Drill to support CouchDB, which not only further promotes the promotion of Drill, but also makes the NoSQL database CouchDB accessible by standard SQL, thus improving the usability of CouchDB.

7 Summary

Compared with other query engine, extensible data source is one of the main advantages of Apache Drill. In support of document-oriented database, the data sources supported by Apache Drill are not rich enough. This paper designs and implements the CouchDB storage plugin based on the storage plugin interface of apache drill, which provides more possibilities for Drill in the support of document-oriented database. it also provides the support of SQL for CouchDB, improves the query performance, and can query CouchDB with other data sources supported by Drill, so that CouchDB can be used in a wider range of scenarios.

During the query process, Drill will query the data and return it to the user according to the where filter. In the future work, the storage plugin can optimize the rules, construct filters before the query, and then query in CouchDB, so that the resources of the CouchDB storage plugin are less occupied and the query efficiency is higher.

References

1. Melnik, S., et al.: Dremel: interactive analysis of web-scale datasets. Proc. VLDB Endow. **3** (1–2), 330–339 (2010)
2. Hausenblas, M., Nadeau, J.: Apache drill: interactive ad-hoc analysis at scale. Big Data **1**(2), 100–104 (2013)
3. Liu, T., Martonosi, M.: Impala: a middleware system for managing autonomic, parallel sensor systems. ACM SIGPLAN Not. **38**(10), 107–118 (2003)
4. Zaharia, M., Chowdhury, M., Franklin, M.J., et al.: Spark: cluster computing with working sets. In: Usenix Conference on Hot Topics in Cloud Computing (2010)
5. Zaharia, M., et al.: Spark: cluster computing with working sets. In: HotCloud 2010, vol. 10, p. 95 (2010)
6. Kimball, R., Caserta, J.: The Data Warehouse ETL Toolkit: Practical Techniques for Extracting, Cleaning, Conforming, and Delivering Data. Wiley, Hoboken (2011)
7. Vora, M.N.: Hadoop-HBase for large-scale data. In: Proceedings of 2011 International Conference on Computer Science and Network Technology, vol. 1, pp. 601–605. IEEE, December 2011
8. Thusoo, A., et al.: Hive: a warehousing solution over a map-reduce framework. Proc. VLDB Endow. **2**(2), 1626–1629 (2009)
9. Chodorow, K.: MongoDB: The Definitive Guide: Powerful and Scalable Data Storage. O'Reilly Media, Inc., Sebastopol (2013)
10. CouchDB vs. MongoDB Comparison. https://db-engines.com/en/system/CouchDB%3BCouchbase%3BMongoDB. Accessed 17 Apr 2019
11. Lamb, J.P., Lew, P.W.: Lotus Notes Network Design. McGraw-Hill, Inc., New York (1996)
12. Androulaki, E., et al.: Hyperledger fabric: a distributed operating system for permissioned blockchains. In: Proceedings of the Thirteenth EuroSys Conference, p. 30. ACM, April 2018

13. Using CouchDB—hyperledger-fabricdocs master documentation. https://hyperledger-fabric.readthedocs.io/en/release-1.4/couchdb_tutorial.html. Accessed 15 July 2019
14. HBase Storage Plugin - Apache Drill. http://drill.apache.org/docs/hbase-storage-plugin/. Accessed 15 July 2019
15. Hive Storage Plugin - Apache Drill. http://drill.apache.org/docs/hive-storage-plugin/. Accessed 15 July 2019
16. RDBMS Storage Plugin - Apache Drill. http://drill.apache.org/docs/rdbms-storage-plugin/. Accessed 15 July 2019
17. MongoDB Storage Plugin - Apache Drill. http://drill.apache.org/docs/mongodb-storage-plugin/. Accessed 15 July 2019

Visual Saliency Based on Two-Dimensional Fractional Fourier Transform

Haibo Xu[1] and Chengshun Jiang[2(✉)]

[1] School of Robot Engineering, Yangtze Normal University, Chongqing,
People's Republic of China
[2] College of Big Data and Intelligent Engineering, Yangtze Normal University,
Chongqing, People's Republic of China
csjiang@yznu.edu.cn

Abstract. Visual saliency is very helpful for image detection and image processing. This paper proposes a novel visual saliency model. First, the proposed model can extract a saliency map with high precision and compound the linear combination of saliency map. Second, based on two-dimensional fractional Fourier transform, the proposed model generates a robust saliency map from the input image with Gaussian or salt-and-pepper noise. In order to reveal the noise influence from the given image, we provide a concept called the noise sensitivity scale (NSS). Third, using the image database from MSRA10K, we analyze the precision-recall and ROC curve and experimentally demonstrate that the proposed model can evaluate human fixation to some extent.

Keywords: Image processing · Computer vision · Fractional Fourier transform

1 Introduction

Visual saliency has become an approach by which a computer imitates the human visual attention system to extract the conspicuous features that can be observed by the introduction of a saliency map in an image. The saliency map of an image supposedly possesses distinctive features compared with other knowledge from background. The research on visual saliency may imply mechanisms of biological visual behavior [1, 2]. The visual attention model is one of the most useful parts of the image processing that can be simulated by a computer.

According to the different processing processes, visual saliency is divided into two parts: bottom-up, top-down. In recent years, visual saliency has been the object of extensive attention from computer researchers and psychologists [3–24]. The research on visual saliency in the frequency domain plays an important role in computer vision and pattern recognition, such as FT, PFT, PQFT, etc., as well [25, 26]. The algorithm is similar to SR, and the model characteristics of PQFT processing are relatively complex; therefore, in order to obtain the image pattern features, however, as in the SR, PFT, PQFT, and AIM approaches, the distinctive saliency map in images is linked to detection accuracy. In [27], the salient regions are classified into several categories. To some extent, one approach can detect a specific category in the theoretical background of the frequency domain. Regarding some images with complex region feature in

H. Jin et al. (Eds.): BigData 2019, CCIS 1120, pp. 240–252, 2019.
https://doi.org/10.1007/978-981-15-1899-7_17

cluttered backgrounds, some visual saliency models show the poor effect, while other studies on visual saliency are demonstrated in terms of Bayes probability or a probabilistic graphical model [28–30]. Several models are shown to compute saliency using global information [23].

2 The Relevant Visual Attention Algorithms

Recently, a visual saliency algorithm used for frequency-domain analysis, called the hypercomplex Fourier transform (HFT), was proposed in [27]. Given a hypercomplex matrix

$$f(m,n) = a + bi + cj + dk \tag{1}$$

the discrete version of the HFT of Eq. (1) is given by

$$\mathcal{F}_{\mathcal{H}}(u,v) = \frac{1}{\sqrt{MN}} \sum_{m=0}^{M-1} \sum_{n=0}^{N-1} e^{-\mu 2\pi\left(\left(\frac{mu}{M}\right) + \left(\frac{nv}{N}\right)\right)} f(m,n) \tag{2}$$

where μ is a unit pure quaternion and $\mu^2 = -1$. The inverse hypercomplex Fourier transform is given as

$$f(m,n) = \frac{1}{\sqrt{MN}} \sum_{m=0}^{M-1} \sum_{n=0}^{N-1} e^{\mu 2\pi\left(\left(\frac{mu}{M}\right) + \left(\frac{nv}{N}\right)\right)} \mathcal{F}_{\mathcal{H}}(u,v) \tag{3}$$

In [2], the Itti model was provided, in which Itti and Koch employed the Gaussian pyramid, Gabor operator and Robert operator to extract the brightness, direction, and color features, respectively. The Itti algorithm can simulate the neural structure and behaviors of the visual system of primates in their early life. It shows a powerful capacity in the real-time processing of complex scenes. In [7], Harel and Koch provided a novel graph-based visual saliency model, in which they define Markov chains over various image maps and treat the equilibrium distribution over map locations as activation and saliency values:

$$d((i,j)\|(p,q)) \triangleq \left| log \frac{M(i,j)}{M(p,q)} \right| \tag{4}$$

where $M(i,j)$ and $M(p,q)$ are dissimilarity matrices.

The directed edge form node (i,j) to node (p,q) will be assigned a weight as follows:

$$w_i((i,j),(p,q)) \triangleq d((i,j)\|(p,q))F(i-p,j-q) \tag{5}$$

where $F(a,b) \triangleq \exp[(a^2 + b^2)/2\sigma^2]$ and σ is a free parameter.

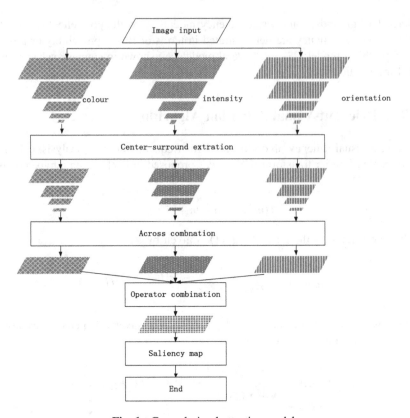

Fig. 1. General visual attention model

In fact, most of visual saliency models have three steps, namely extraction, activation, and normalization or a combination, as can be seen in Fig. 1. Extraction aims at achieving the feature vectors from the background, and the significant step is activation. A saliency map or conspicuous map by activation, last stage normalization, or a combination to normalize the activation map.

In order to facilitate our research, there are some definitions as following: an imaging device is known, whose focal length v is fixed, and the object distance is u, and the images can be classified into three categories:

1. An image contains a large size saliency map if it satisfies: $20v < u \leq 70v$;
2. An image contains a medium size saliency map if it satisfies: $70v < u \leq 500v$;
3. An image contains small size saliency maps if it satisfies: $u \geq 500v$.

In Fig. 2, it can be seen from images 1 to 6 that information entropy is different and the visual saliency models (G&S, ITTI, and HFT) give different results. The first row shows the image with large salient regions and a distinctive background, the second row shows the image with medium salient regions, and the third row shows the image with small salient regions. The next rows show the image with a cluttered background, the image with repeating distracters, and the image with both large and small salient regions. We can observe that the last three images. include complex information compared to the first three images; thus, the detection results are not satisfactory.

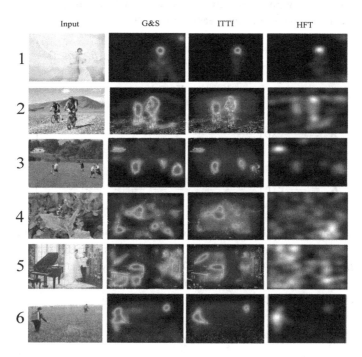

Fig. 2. Results of saliency models based on a distinctive saliency map. First column shows the input images, while the second, third, and fourth columns show the responses of G&S, Itti, and HFT models, respectively.

3   A Novel Visual Saliency Model: Two-Dimensional Fractional Fourier Transform (2D-FRFT)

We believe that a saliency map of an image is vital to the results of experiments by visual saliency models to some extent. Thus, the definition of the saliency map is a basic and principal issue. The Itti model obtains a saliency map by center-surround extraction and across-combination; however, it will give errors when input images are full of high frequency, such as images 3–5 in Fig. 2. In order to detail the intact salient information in images, we employ a novel approach designated two-dimensional fractional Fourier transform (2D FrFT) [31, 32]. A FrFT is a generalized Fourier transform that can transform the image into, in essence, a time-frequency domain, which is transition state between the time domain and frequency domain. Thereby, we can obtain two types of information using 2D FrFT and reveal more and complete features as can be seen in Fig. 3. A parameter p can be defined such that it satisfies $p = 2\xi/\pi$, where $\xi \in \left[0, \frac{\pi}{2}\right]$.

3.1 Two-Dimensional Fractional Fourier Transform

In this paper, we adopt a 2D FrFT to extract feature vectors over the image plane. A discrete fractional Fourier transform (DFrFT) is an analysis tool that has been applied in time-frequency domain in recent years [33–35]. It is defined as

$$X_{(\alpha,\beta)}(m,n) = \sum_{p=0}^{M-1} \sum_{q=0}^{N-1} x(\mu,v) K_{(\alpha,\beta)}(\mu,v,m,n) \tag{6}$$

$$\mathbf{K}_{(\alpha,\beta)} = \mathbf{K}_\alpha \otimes \mathbf{K}_\beta \tag{7}$$

$$K_\alpha = \frac{A_\alpha}{2\Delta x} \exp\left(\frac{i\pi(cot\, cot\alpha)(m^2+n^2) - i2\pi(csc\, csc\alpha)mn}{(2\Delta x)^2} \right) \tag{8}$$

$$K_\beta = \frac{A_\beta}{2\Delta x} \exp\left(\frac{j\pi(cot\, cot\beta)(m^2+n^2) - j2\pi(csc\, csc\beta)mn}{(2\Delta x)^2} \right) \tag{9}$$

where $\mathbf{K}_{(\alpha,\beta)}$ is a transform kernel as defined in Eq. (8). The 2D-FrFT applies a DFrFT on the two parameters of an image signal, $x(p,q)$, successively. $\mathbf{K}_p$ is the kernel with respect to the x and y axes. As is known, a DFT is not reality-preserving, and since a 2D FrFT is a generalized form of a DFT, the transform order can be decided according to different needs, it can be applied in image filtering as well.

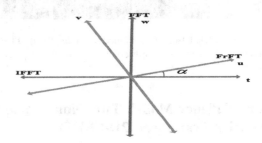

Fig. 3. FFT rotation model: 2D FrFT

Figure 4 shows that the saliency map is related to the parameter p order in the 2D FrFT. We can draw the following conclusion: After applying the 2D FrFT, we reach a time-frequency image, and it reduces some noise of the background when p is equal to specific value. The target feature is conspicuity from the background information, and the object edge is separable.

After applying the 2D FrFT, we adopt Gaussian kernels to smooth the amplitude spectrum. This is regarded as image filter operator to reduce the background noise to obtain a time-domain image. The image – filtering FrFT process is illustrated in Fig. 5, where the IFrFT denotes the inverse FrFT after the image filter operation.

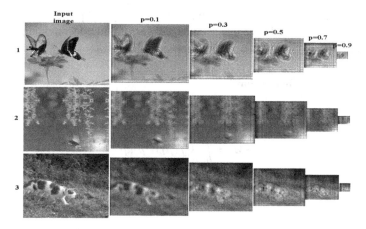

Fig. 4. Results of 2D FrFT in p order. The first row shows the input image with large salient regions, the second row shows the input image with small salient regions, and the last row shows the input image with a cluttered background.

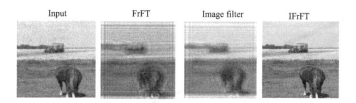

Fig. 5. Image filter process used by 2D FrFT

However, in the real world, the saliency map in an image is too complex for computer to detect, but it can be described by shape, size, color, position, etc. Furthermore, it is also related to density of probability distribution in neighborhood pixels. In this paper, we assume that the saliency map region is related to the area size of the object.

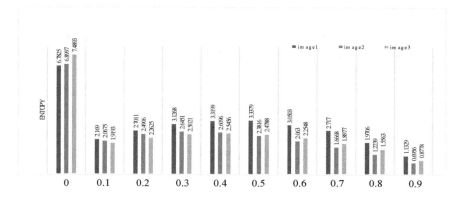

Fig. 6. Parameter p order in 2D FrFT for entropy analysis

In Fig. 6, we analyze the relationship between entropy and p order in the 2D FrFT. The images labeled image1, image2, and image3 were produced from images 1, 2, and 3 in Fig. 4, respectively. p order is a parameter in 2D FrFT.

3.2 Saliency Map Generation

Saliency map (SM) generation is achieved by the following formula:

$$SM = \int_0^1 k_p f(p) M(p) dp \tag{10}$$

$$k_p = \frac{pf(p)}{\int_0^1 f(p) dp} \tag{11}$$

where $M(p)$ denotes the p order 2D IFrFT of the input image, and $f(p)$ is a function of feature probability distribution, as can be seen in Fig. 7. In order to improved computational efficiency, k_p is normalizing parameter of p.

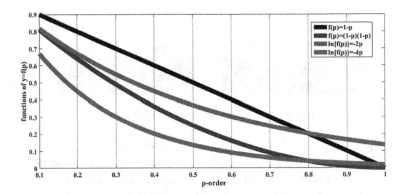

Fig. 7. Selected probability distribution function of $f(p)$

$M(p)$ is continuous matrix-structure function when $p \in [0.1]$, however, it is too complicated and time-consuming for us to achieve a refined calculation and take computational efficiency into account. In order to satisfy the high precision, we adopt Cotes quadrature formulas:

$$\int_a^b F(p) dp \approx (b-a) \sum_{k=0}^n C_k^{(n)} F(p_k) \tag{12}$$

$$F(p) = k_p M(p) f(p) \tag{13}$$

and setting $p = a + th$

$$C_{\mathbf{k}}^{(\mathbf{n})} = \frac{h}{b-a} \int_0^n \prod_{\substack{j=0 \\ j \neq k}}^n \frac{t-j}{k-j} dt = \frac{(-1)^{n-k}}{nk!(n-k)!} \int_0^n \prod_{\substack{j=0 \\ j \neq k}}^n (t-j) dt \qquad (14)$$

we can infer the following formula:

$$SM = (C, F(p)) \qquad (15)$$

where denotes column vector with variable p order, and is a constant column vector with n-dimensional rank vector that denotes the Cotes coefficients as follow:

$$C = \left[c_0^{(n)}, c_1^{(n)}, \cdots, c_n^{(n)} \right] \qquad (16)$$

3.3 SM Optimality Criterion

We assume that the optimal saliency map will appear in the sequence $\{SM_\omega\}$ where ω is included in two variables, n and p. In addition, $\mathcal{H}(x) = -\sum_{i=1}^n p_i \log p_i$ is the definition of the entropy of x, which is a random variable that refers to the amount of information contained in a message. The entropy is determined if the histogram is given, and the amount of information in a message is directly related to its uncertainty:

$$\omega_k = arg \ \min_k \{\mathcal{H}(SM_\omega)\} \qquad (17)$$

4 Experiment

Here, we will show the results of several simulation experiments in different situations using the above-described method. In order to demonstrate the experiment result, we list three variables: noise interference, measurement node in the Cotes quadrature formulas, and the probability distribution of $f(p)$.

4.1 Noise Sensitivity Analysis

First, we take the background noise into account; that is, the input image includes some noise. We then employ the above-mentioned 2D FrFT method to generate the saliency map (SM), and compare and contrast method with other visual models as Fig. 8.

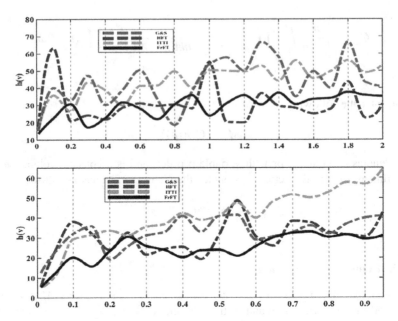

Fig. 8. NSS analysis for saliency models under conditions of Gaussian and salt-and-pepper noise. Top graph shows adding Gaussian white noise of mean v to the image, Bottom graph shows adding salt-and-pepper noise of noise density v to the image

We provide a performance indictor called noise sensitivity scale (NSS) denoted by $h(v)$. Obviously, regarding an algorithm, the stronger its robustness or anti-interference performance exhibited, the lower of $h_{variance}$. We can employ and to depict the noise sensitivity as follows:

$$h(v) = \lim_{\Delta v \to 0} \left(\sum_{(i,j) \in \Omega} \|\mathcal{R}_v[f(i,j)] - \mathcal{R}_{v+\Delta v}[f(i,j)]\|^2 \right)^{1/2} \tag{18}$$

$$h_{width} = h_{max} - h_{min} \tag{19}$$

$$h_{variance} = \frac{1}{n} \sum_{i=1}^{n} \left(h(v_i) - \frac{h_{width}}{2} \right)^2 \tag{20}$$

where Ω denotes a 2D image of f(x, y) and v is noise parameter, where $\mathcal{R}[\cdot]$ denotes an image visual saliency method such as Itti, HFT, or G&B. the image is an input signal.

We show the noise sensitivity for several visual models under condition of Gaussian and salt-and-pepper noise in Fig. 8. Obviously, if the values and computed using the 2D FrFT model are smaller than the values computed using other models, it implies that 2D FrFT possesses a lower NSS value than other models. Thereby, FrFT is superior to other models based on comparisons of above performance indicators.

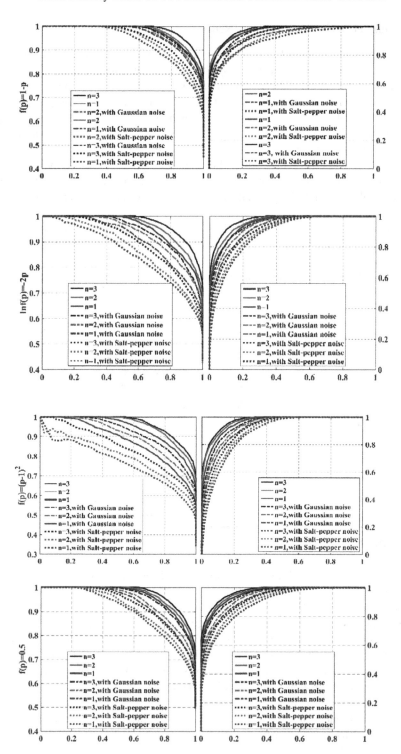

Fig. 9. Results of distinctive probability distribution of f(p) using by MSRA 10K database. First column is the precision-recall curve and the second column is the ROC curve

4.2   Number of Sections in Cotes Nodes

Here, we present the Eqs. 15, 16 and select the number of Cotes nodes for further discussion. We infer that is achieved by the linear combination C and $F(p)$, and the number of the dimension is denoted n. Obviously, it is the same as the number of Cotes coefficient. We employ precision-recall and ROC to show the relationship between the n value (the number of Cotes nodes) and precision in Fig. 9. ROC is a graphical plot that illustrates the performance of a binary classifier system as its discrimination threshold is varied. The curve is created by plotting the true positive rate (TPR) against the false positive rate (FPR) at various threshold settings. The true-positive rate is also known as sensitivity, recall or probability of detection, the false-positive rate is also known as the fall-out or probability of false alarm and can be calculated as (1 − specificity).

In order to analyze performance of the proposed method, we draw the statistical comparison result graph to facilitated comparison with other visual saliency models, such as G&S, Itti, FT, PFT, HFT. The 1000 input images are from the MSRA10K database, and the results of the statistical comparison are shown in Fig. 10.

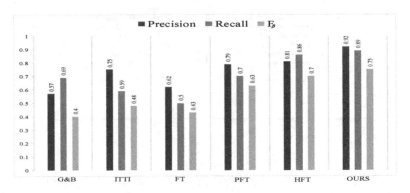

Fig. 10. Statistical comparison results in different saliency detection approaches

5   Conclusion

In this paper, we proposed a new saliency model detection approach based on a two-dimensional fractional Fourier transform (2D FrFT). To achieve high precision, we employed the numerical integration method as the Newton-Cotes formula to show the integral that, using the p-order FrFT, ranges from 0 to 1 as an image feature. In addition, we regard the saliency map (SM) as the linear combination of the Cotes coefficient and, which is a complex matrix obtained by the 2D FrFT operation. Meanwhile, we provided a noise sensitivity scale (NSS) and demonstrated that the proposed approach exhibits stable denoising performance.

To validate the proposed model, we plotted the precision-recall and ROC curves and performed statistical comparisons with other visual saliency models using the MSRA 10K database, the precision, recall, and of proposed model is 0.92, 0.89, and

0.75 respectively, which are higher than in other saliency models. Thereby, the proposed approach can evaluate human fixation to some extent.

Conflicts of Interest. The authors declare that there are no conflicts of interest regarding the publication of this article.

References

1. Alpern, M.: Eye movements, vision and behavior. Am. J. Ophthalmol. **80**(2), 307–308 (1975)
2. Itti, L., Koch, C., Niebur, E.: A model of saliency-based visual attention for rapid scene analysis. IEEE Trans. Pattern Anal. Mach. Intell. **20**(11), 1254–1259 (1998)
3. Kadir, T., Brady, M.: Saliency, scale and image description. Int. J. Comput. Vis. **45**(2), 83–105 (2001)
4. Bruce, N.D.B., Tsotsos, J.K.: Saliency based on information maximization. In: International Conference on Neural Information Processing Systems (2005)
5. Yantis, S.: How visual salience wins the battle for awareness. Nat. Neurosci. **8**(8), 975–977 (2005)
6. De Brecht, M., Saiki, J.: A neural network implementation of a saliency map model. Neural Netw. Off. J. Int. Neural Netw. Soc. **19**(10), 1467 (2006)
7. Harel, J., Koch, C., Perona, P.: Graph-based visual saliency, vol. 19, pp. 545–552 (2006)
8. Meur, O.L., Callet, P.L., Barba, D.: A coherent computational approach to model bottom-up visual attention. IEEE Trans. Pattern Anal. Mach. Intell. **28**(5), 802–817 (2006)
9. Gao, D., Mahadevan, V., Vasconcelos, N.: The discriminant center-surround hypothesis for bottom-up saliency. In: Advances in Neural Information Processing Systems, vol. 20, pp. 497–504 (2007)
10. Gao, D., Vasconcelos, N.: Bottom-up saliency is a discriminant process. In: IEEE International Conference on Computer Vision (2007)
11. Schölkopf, B., Platt, J., Hofmann, T.: A nonparametric approach to bottom-up visual saliency. In: Conference on Advances in Neural Information Processing Systems (2007)
12. Yu, Z., Wong, H.S.: A rule based technique for extraction of visual attention regions based on real-time clustering. IEEE Trans. Multimed. **9**(4), 766–784 (2007)
13. Cerf, M., Harel, J., Einhäuser, W., Koch, C.: Predicting human gaze using low-level saliency combined with face detection. In: Advances in Neural Information Processing Systems, vol. 20, pp. 241–248 (2008)
14. Hou, X., Zhang, L.: Dynamic visual attention: searching for coding length increments. In: Conference on Neural Information Processing Systems, Vancouver, British Columbia, Canada, December 2008
15. Zhang, L., Tong, M.H., Marks, T.K., Shan, H., Cottrell, G.W.: SUN: a Bayesian framework for saliency using natural statistics. Journal of Vision **8**(7), 32.1 (2008)
16. Gao, D., Han, S., Vasconcelos, N.: Discriminant saliency, the detection of suspicious coincidences, and applications to visual recognition. IEEE Trans. Pattern Anal. Mach. Intell. **31**(6), 989 (2009)
17. Itti, L., Baldi, P.: Bayesian surprise attracts human attention. Vis. Res. **49**(10), 1295–1306 (2009)
18. Khan, F.S., Weijer, J.V.D., Vanrell, M.: Top-down color attention for object recognition. In: IEEE International Conference on Computer Vision (2009)

19. Avraham, T., Lindenbaum, M.: Esaliency (extended saliency): meaningful attention using stochastic image modeling. IEEE Trans. Softw. Eng. **32**(4), 693–708 (2010)
20. Chikkerur, S., Serre, T., Tan, C., Poggio, T.: What and where: a Bayesian inference theory of attention. Vis. Res. **50**(50), 2233–2247 (2010)
21. Mahadevan, V., Vasconcelos, N.: Spatiotemporal saliency in dynamic scenes. IEEE Trans. Softw. Eng. **32**(1), 171–177 (2010)
22. Zhang, Q., Liu, H., Shen, J., Gu, G.: An improved computational approach for salient region detection. J. Comput. **5**, 1011–1018 (2010). [cited 5 7]
23. Cheng, M., Mitra, N.J., Huang, X., Torr, P.H.S., Hu, S.: Global contrast based salient region detection. In: Computer Vision and Pattern Recognition (2011)
24. Kim, W., Jung, C., Kim, C.: Spatiotemporal saliency detection and its applications in static and dynamic scenes. IEEE Trans. Circuits Syst. Video Technol. **21**(4), 446–456 (2011)
25. Guo, C., Ma, Q., Zhang, L.: Spatio-temporal Saliency detection using phase spectrum of quaternion fourier transform. In: IEEE Conference on Computer Vision and Pattern Recognition, CVPR 2008 (2008)
26. Achanta, R., Hemami, S., Estrada, F., Susstrunk, S.: Frequency-tuned salient region detection. In: IEEE Conference on Computer Vision and Pattern Recognition, CVPR 2009 (2009)
27. Li, J., Levine, M.D., An, X., Xu, X., He, H.: Visual saliency based on scale-space analysis in the frequency domain. IEEE Trans. Pattern Anal. Mach. Intell. **35**(4), 996–1010 (2013)
28. Jiang, B., Zhang, L., Lu, H., Yang, C., Yang, M.: Saliency detection via absorbing Markov chain. In: IEEE International Conference on Computer Vision (2013)
29. Wang, J., Jiang, H., Yuan, Z., Cheng, M., Hu, X.: Salient object detection: a discriminative regional feature integration approach. In: IEEE Conference on Computer Vision and Pattern Recognition (2013)
30. Li, G., Shi, J., Luo, H., Tang, M.: A computational model of vision attention for inspection of surface quality in production line. Mach. Vis. Appl. **24**(4), 835–844 (2013)
31. Qi, L., Chen, E., Mu, X., Guang, L., Zhang, S.: Recognizing human emotional state based on the 2D-FrFT and FLDA. In: International Congress on Image and Signal Processing (2009)
32. Singh, A.K., Saxena, R.: On convolution and product theorems for FRFT. Wireless Pers. Commun. **65**(1), 189–201 (2012)
33. Gao, L., Qi, L., Wang, Y., Chen, E., Yang, S.: Rotation invariance in 2D-FRFT with application to digital image watermarking. J. Signal Process. Syst. **72**(2), 133–148 (2013)
34. Wang, P., Tian, H., Zheng, W.: A novel image fusion method based on FRFT-NSCT. Math. Probl. Eng. **2013**(11), 1–9 (2013)
35. Wu, J., Luo, X., Zhou, N.: Four-image encryption method based on spectrum truncation, chaos and the MODFrFT. Opt. Laser Technol. **45**(1), 571–577 (2013)

Big Data Analysis

An Information Sensitivity Inference Method for Big Data Aggregation Based on Granular Analysis

Lifeng Cao, Xin Lu$^{(\boxtimes)}$, Zhensheng Gao, and Zhanbing Zhu

He'nan Province Key Laboratory of Information Security,
Zhengzhou 450001, Henan, China
1209774364@qq.com

Abstract. Aiming at solving the problem of deducing sensitive information leakage after big data aggregation, this paper proposes an information sensitivity inference method for big data aggregation based on granular analysis. Firstly, data conceptual objects are generated based on data attributes and attribute values, and data granular sets are formed. By calculating the quality and gravity of data granules, the data granules are analyzed and classified into equivalence classes. Then, subdividing the equivalence classes according to the decision data granules, the data granules sets in positive and critical domains of decision data granules are established respectively, and the data granular characteristic matrix and eigenvalue matrix are constructed. By the approximate data granules dynamic updating algorithm, the dynamic clustering of data granules is realized when the values of objects, attributes and attributes increase dynamically. Finally, a similar granules cloud is established. Based on the attribute fuzzy set probability measure and the contribution of the data granule to the sensitive granules cloud, the possibility of deducing sensitive information by similar data is inferred. What's more, the performance of data granular clustering algorithm and dynamic update algorithm, and the accuracy of the inference algorithm are verified. This method is helpful for the formulation of large data access control strategy and the analysis of similar data, as well as reduce the risk of information leakage.

Keywords: Big data · Granule analysis · Cloud theory · Dynamic clustering · Aggregation inference

1 Introduction

With the development of network bandwidth and terminals, the Internet has become an indispensable part of people's lives, and the data generated every day cannot be estimated. These massive and diverse data sets are called big data [1]. The analysis of big data has achieved relevant research results and has been successfully applied to some typical commercial fields, medical fields, etc., providing people with convenient and easy-to-use service models [2]. However, the analysis of big data mining [3, 4] pays too much attention to business value, user attention, medical information and other information, and rarely has other applications. In a cloud computing environment, a

© Springer Nature Singapore Pte Ltd. 2019
H. Jin et al. (Eds.): BigData 2019, CCIS 1120, pp. 255–270, 2019.
https://doi.org/10.1007/978-981-15-1899-7_18

large amount of tenant and user data is stored in the cloud environment resulting in generating many data "raindrops". And the raindrops cross-integration can form a large amount of different types of data information, namely cloud big data. The convergence of "raindrops" makes it possible for cloud data to be aggregated and deduced, which may lead to the leakage of high-level sensitive information. This will also seriously affect the security of individuals, enterprises and even the country [5, 6].

Data aggregation inference, as its name suggests, is to mine, analyze, and infer derived sensitive information from large amounts of data, which has been proven to be a NP-complete problem. Many similar "raindrops" in the cloud environment seem to be public information, and the information that can be inferred after convergence together has a certain level of sensitivity. If an attacker gains access to these data, using the "raindrops" to aggregate analysis and infer will result in serious loss of confidentiality. Therefore, it is of great significance and research value to analyze complex and heterogeneous cloud big data while making the best use of cloud big data, to find the correlation between the data, to deduce the possibility of leaks caused by aggregation inference of similar "raindrops", and to control user access and data flow.

For this aspect of research, most of the literature focuses on the analysis of data association [7, 8, 13], and there are few studies about aggregate inference. The literature [9] proposed to introduce the relationship between objects in risk access control to prevent privacy leaks caused by the interconnection of objects. The literature [10] summarized the big data correlation analysis methods from four aspects: statistical correlation analysis, mutual information, matrix calculation and distance. The literature [11] solved the inference problem by Bayesian formula, and establishes a multi-mode inference framework. The literature [12] studied the inferential access control of relational databases, but it is not suitable for cloud big data environments with large data volume and many attributes; The literature [13] proposed a level inference method for aggregated information of objects based on associated attributes by in-depth analysis of object relevance, which controlled restricted access of associated objects and reduced the risk on information system. effectively the literature [14] studies the aggregation of similar objects in the hierarchical protection environment and proposes a level inference method for aggregated information of objects based on clustering analysis. However, the previous research is biased towards the same type of attributes, which is relatively simple and cannot be completely satisfied for cloud big data environment. Granular computing can greatly improve computing efficiency by granulating complex data and replacing samples with information granules as the basic unit of computing and granular computing reasoning is a logical method of deduction using known information granules or granular spaces. It is very suitable for data aggregation analysis and reasoning in cloud environment [15–17].

Therefore, based on the previous research of the literature [14], this paper proposes an information sensitivity inference method for big data aggregation based on granular analysis. In this method, the similarity dynamic clustering of cloud data is carried out through the data granular analysis and the construction of composite relationship between data granule attributes, and a similar granular cloud based on attribute combination is formed. And then, based on the contribution of data granules to the concept cloud and the concept-sensitive cloud fuzzy set possibility measure, the possibility of inferring high-level sensitive information by similar data objects is inferred.

2 Dynamic Clustering Based on Data Granules Similarity

In order to infer the possibility of cloud similar "raindrops" convergence to deduce sensitive information, and to control users' access to cloud data and reduce the risk of data leakage, this paper firstly analyzes the similarity of complex data in the cloud and proposes a dynamic clustering method based on data granules. The basic idea is: Extract and segment the cloud data "raindrop" to form a data concept object. For the concept object, data granule analysis is carried out based on the attribute segmentation set, and the similarity of the cloud data is analyzed in the case of the "raindrops" dynamic update. Finally, according to the weight of attribute semantic similarity, the clustering is adjusted reasonably.

2.1 Data Granules and Granular Segmentation

Let U be the global domain of a domain object, A be the attribute space, $cD = \{cd_1, cd_2, \cdots, cd_n\}$ be the data "raindrops" collection in the cloud environment, and $SL = \{sl_1, sl_2, \cdots sl_k\}$ be the sensitive level. A data raindrop in the cloud can be represented as $cd_k = (A(cd_k), SL(cd_k))$, a set of data attributes and sensitivity levels.

Definition 1 (data granules): Let $a \in A$, v be the value of an object of property a on U, $m()$ be the meaning set function on U, (a, v) be the atomic formula on U, denoted as a_v, and the combination of a_v be the well-formed formula on U, denoted as $\varphi = \prod a_v$. Then, $(\varphi, m(\varphi))$ is called a data granule on U. $m(\varphi) = \{x, x \in U | x | \approx \varphi\}$, that is, all the sets of objects satisfying φ on U.

Definition 2 (concept object): Let the attribute set of cd_i be $\omega_i = \langle a_{i1}, a_{i2}, \cdots, a_{ik} \rangle$, the attribute value corresponding to the attribute in ω is represented by $v_i = \langle v_{i1}, v_{i2}, \cdots, v_{ik} \rangle$. The sensitivity level of cd_i is represented by sl_{cd_i}. Then, the "raindrops" concept objects are defined as $\langle cd_i, \langle \omega_i \cdot v_i, sl_{cd_i} \rangle \rangle$.

Definition 3 (granular feature matrix and eigenvalue matrix): Let K be the "raindrops" concept objects collection, $K \subset U$. $K/A = \{dg_1, dg_2, \cdots, dg_l\}$ is a division on K. The feature vector $\overrightarrow{dg_i}$ of the data granule dg_i is $\langle index_i, obj_i, reg_i \rangle$. $index_i$ is the feature index, $index_i \in dg_i$. obj_i is a set of objects in data granule dg_i, $obj_i = \{x_k | x_k \in dg_i\}$. If dg_i is the positive domain of $X = \{x_k | \exists s_{cd_i} \in SL, x_k.s = s_{cd_i}\}$, then reg_i is P. If dg_i is the boundary domain of X, then reg_i is B. If dg_i is the negative domain of X, then reg_i is N. Let $\overrightarrow{dg_{i_a}}$ be the data granule dg_i eigenvalue vector, av_{ik} be the value of attribute a_j ($a_j \in w_h$), and w_h be an attribute set of cd_h ($cd_h \in dg_i$). The granule feature matrix and the eigenvalue matrix are as follows:

$$M_{dg} = \begin{bmatrix} \overrightarrow{dg_1} \\ \overrightarrow{dg_2} \\ \vdots \\ \overrightarrow{dg_l} \end{bmatrix} = \begin{bmatrix} index_1 & obj_1 & reg_1 \\ index_2 & obj_2 & reg_2 \\ \vdots & \vdots & \vdots \\ index_l & obj_l & reg_{l-1} \end{bmatrix} \quad M_{dga} = \begin{bmatrix} \overrightarrow{dg_{1a}} \\ \overrightarrow{dg_{2a}} \\ \vdots \\ \overrightarrow{dg_{la}} \end{bmatrix} = \begin{bmatrix} av_{11} & \cdots & av_{1m} \\ av_{21} & \cdots & av_{2m} \\ \vdots & \vdots & \vdots \\ av_{11} & \cdots & av_{1m} \end{bmatrix}$$

2.2   Cloud Data Clustering Algorithm Based on Granule Gravity

Definition 4 (granule core): Let $dg_i = \{cd_1, \cdots, cd_i, \cdots cd_l, \cdots, cd_k\}$, define the granule core of data granule dg_i as $dg_i^{core} = \langle cd_{core}, \omega_{core} \rangle$, $\omega_{core} = \{a_h | a_h \in \cap cd_i.\omega_i$, $cd_i \in dg_i\}$ represents the similar attribute set of data granule objects. cd_{core} is the data objects corresponding to the granule core, which can be concrete objects or a virtual objects. dg_i^{core} is also called the virtual center of gravity of dg_i.

Definition 5 (granule gravity): For given data granules dg_i, dg_j, the granule cores are dg_i^{core} and dg_j^{core}, the mass of granule cores are $m_{dg_i^{core}}$ and $m_{dg_j^{core}}$, and the distance between granules is $r\left(dg_i, dg_j\right)$. Define the gravity between the data granules as: $g\left(dg_i, dg_j\right) = G\left(m_{dg_i^{core}} m_{dg_j^{core}} / r\left(dg_i, dg_j\right)\right)$. The gravitational constant G is 1, $m_{dg_i^{core}} = \sum_{a_h \in \omega_{core}^{dg_i}} H(a_h)$ represents the sum of attribute information entropy in granule core, $r\left(dg_i, dg_j\right)$ is the distance between the virtual centers of gravity of data granules that is the distance between the granule cores: $r\left(dg_i, dg_j\right) = \sqrt{\sum_{a_p \in \omega_{core}^{dg_i}, a_q \in \omega_{core}^{dg_j}} \left(a_p, a_q\right)^2}$.

Based on the above basis, this paper proposes a cloud data clustering algorithm based on granule gravity (CCDGG), The algorithm is shown in Table 1.

The algorithm is mainly divided into three parts. The first part (lines 1–27) is the classification of cloud data "raindrops". The cloud "raindrops" are abstracted as the concept objects of "raindrops" ($\langle cd_i, \langle \omega_i \cdot v_i, sl_{cd_i} \rangle \rangle$). Each "raindrop" is regarded as a granule. The weight and center of gravity of the granule are calculated. And then, the "raindrops" are clustered according to the gravity of the granule to form a class. The second part (lines 28-57) divides the data granules according to the attribute values in each class, and finally forms the data granules of each class. The third part (lines 58–81) is the division of similar sets of different classes, by calculating the decision data granules, further dividing the data granules of different classes according to the domain values and decision data granules, and forming the positive domain data granules, the critical domain data granules and the negative domain data granules of the decision granules. This paper focuses on the positive domain data granules and constructs granular eigenvalue matrix and granular eigenvalue matrix. The complexity of the algorithm is o(n * (m + k)). n is the number of data "raindrops". m is the number of classifications. k is the number of data grains in each classification. In general, m and k are much less than n.

Table 1. Cloud data clustering algorithm based on granule gravity (CCDGG)

Input: K (Collection of cd_i);sl_c (Sensitive level);α;β (Neighborhood threshold)
Output: POS(Positive domain);BND(boundary domain);NEG(negative domain)

1: p=1;$\mathbb{Z}[p] = \{cd_1\}$;	45: $\mathbb{Z}[i].cd[k] \rightarrow E[i][j]$;
2: for i=2 to |K| do	46: flg=1;
3: begin	47: break;
4: m=1;n=1;	48: }
5: $g_{i1} = g(cd_i, \mathbb{Z}(m))$	49: endfor
6: if p==1 then {	50: if flg==0 then{
7: if $g_{i1} \geq \delta$ then $cd_i \rightarrow \mathbb{Z}(m)$;	51: q=q+1;
8: else {	52: $\mathbb{Z}[i].cd[k] \rightarrow E[i][q]$;
9: p=p+1;	53: $r = r+1$
10: $cd_i \rightarrow \mathbb{Z}(p)$;	54: }
11: }	55: }
12: }else{	56: endfor
13: $g = g_{i1}$;	57: endfor
14: for j=m+1 to p do	58: for i=1 to |$\mathbb{Z}$| do
15: $g_{ij} = g(cd_i, \mathbb{Z}(j))$;	59: n=1;
16: if $g_{ij} > g$ then {	60: for j=1 to |$\mathbb{Z}[i]$| do
17: n=j;	61: if $\mathbb{Z}[i].cd[j].sl = sl_c$ then
18: $g = g_{ij}$;	{$\mathbb{Z}[i].cd[j] \rightarrow E_{sl_c}[i]$;}
19: }	62: endfor
20: endfor	63: for k=1 to |$E[i]$| do
21: if $g \geq \delta$ then {	64: $p = \frac{E[i][k] \cap E_{sl_c}[i]}{E[i][k]}$;
22: $cd_i \rightarrow \mathbb{Z}(n)$;	65: if $p \geq \alpha$ then {
23: }else{	66: $E[i][k] \rightarrow POS[i]$;
24: p=p+1;	67:
25: $cd_i \rightarrow \mathbb{Z}(p)$;	$M_{dg}^{\mathbb{Z}[i]}[n] = \{ E[i][k][1], E[i][k], P\}$;
26: }	68.
27: endfor	$M_{dga}^{\mathbb{Z}[i]}[n] = \{E[i][k].v(\omega(a))\}$;
28: q=1;	69: $n = n+1$;
29: for i=1 to |$\mathbb{Z}$| do	60: } else if $p \geq \beta$ then {
30: $\mathbb{Z}[i].cd[1] \rightarrow E[i][q]$;	71: $E[i][k] \rightarrow BND[i]$;
31: r = 1;s = q;	72:
32: for k=2 to |$\mathbb{Z}[i]$| do	$M_{dg}^{\mathbb{Z}[i]}[n] = \{ E[i][k][1], E[i][k], B\}$;
33: if r == 1then {	73:
34 if $v(\mathbb{Z}[i].cd[1].\omega) = v(\mathbb{Z}[i].cd[k].\omega)$ then{	$M_{dga}^{\mathbb{Z}[i]}[n] = \{E[i][k].v(\omega(a))\}$;
35: $\mathbb{Z}[i].cd[k] \rightarrow E[i][s]$;	74: $n = n + 1$;
36: }else{	75: } else if $p \leq \beta$ then{
37: q=q+1;	76: $E[i][k] \rightarrow NEG[i]$;
38 $\mathbb{Z}[i].cd[k] \rightarrow E[i][q]$;	77:
39: $r = r+1$;	$M_{dg}^{\mathbb{Z}[i]}[n] = \{ E[i][k][1], E[i][k], N\}$;
40: }	78:
41: }else{	$M_{dga}^{\mathbb{Z}[i]}[n] = \{E[i][k].v(\omega(a))\}$;
42 flg=0;	79: $n = n+1$;
43: for j=s to q do	80: }
44: if $v(E[i][j].cd[].\omega) = v(\mathbb{Z}[i].cd[k].\omega)$ then{	81: endfor
	82: endfor

2.3 Dynamic Update of Cloud Data Cluster Set

In the cloud big data environment, data objects, attribute sets, and attribute values may change dynamically, causing more descriptions of cloud data "raindrops" conceptual objects, thereby causing changes in approximate data granules. Cloud big data, whether it is the dynamic change of the object set or the attribute set, will cause changes in the data granules feature matrix and the eigenvalue matrix. Therefore, dynamic maintenance of granular feature matrix can update the approximation set of data granules. Based on this, this paper presents a dynamic updating algorithm of cloud data approximation set based on approximation data granules mining (Table 2).

The algorithm is divided into three parts. The first part (lines 1–11) divides the newly added "raindrops" data into corresponding categories by using the gravity of granules. The second part (lines 12–36) classifies the data granules of newly added data objects in the class under the attributes of granule core. In class $\mathbb{Z}[i]$, according to the attribute core ω of $\mathbb{Z}[i]$, the data granules feature matrix $M_{dg}^{\Delta cd}$, the granules eigenvalue matrix $M_{dga}^{\Delta cd}$ and the granules set under the sensitive level sl_c of the newly added data object in the class $\mathbb{Z}[i]$ are calculated. $M_{dga}^{\mathbb{Z}[i]}$ and $M_{dga}^{\Delta cd}$ are compared as well as $M_{dg}^{\mathbb{Z}[i]}$ and $M_{dg}^{\Delta cd}$ are granulated to form a new grain characteristic matrix E. Matching $M_{dga}^{\mathbb{Z}[i]}$ with $M_{dga}^{\Delta cd}$, $M_{dg}^{\mathbb{Z}[i]}$ and $M_{dg}^{\Delta cd}$ are fused to form a new particle characteristic matrix $M_{dg}^{cd \cup \Delta cd}$. In the third part (lines 37–57), under the condition of new attributes, $M_{dg}^{cd \cup \Delta cd}$ is segmented and fused to form the final data granule characteristic matrix, and the approximate data granules set under the sensitive level sl_c is obtained, which provides the basis for inference about the approximate granules set. The complexity of the algorithm is o(n * m + k * v), n is the number of original data classifications, m is the number of new data "raindrops", K is the total number of data granules after the original data classification, and V is the number of new data granules.

3 Deduction of Aggregate Information Sensitivity Based on Granule Contribution

Based on the mining of cloud large data approximation set, this paper estimates the possibility of deducing a higher sensitive information from similar data by the contribution of similar data granules to sensitive clouds. Its basic ideas are as follows:

A data attribute sensitive level fuzzy set $F_c = \{f_{sl_1}, f_{sl_2}, \cdots, f_{sl_k}\}$ is defined in domain U ($sl_1 \prec sl_2 \prec \cdots \prec sl_k$). Similar granules cloud is constructed based on the intra-class approximation sets (positive and boundary domains). With the dynamic increase of the number of similar data granules and the increase of data granules attributes, the expected values of granule clouds will also migrate, which will lead to the migration of sensitive cloud. As shown in Fig. 1. Therefore, according to the membership degree of the similar data granules attribute sets on the cloud expectation, the contribution of the data granules to the sensitive cloud is calculated, so that the possibility of inferring high sensitive information from similar data granules is fuzzily inferred, and the inference interval of similar data granules is given.

Table 2. Approximate data granules dynamic updating algorithm

---

Input: $\mathbb{Z}; M_{dg}^{Z}; M_{dga}^{Z}; \Delta cd; \Delta A$ (A is for added data raindrops and B is for added attributes.)

Output: N_POS（Approximate data granules）

1: for i=1 to $|\mathbb{Z}|$ do

2: p = 1;

3: for j=1 to $|\Delta cd|$ do

4:    compute g = g($\Delta cd[j]$, $\mathbb{Z}[i]$);

5:    if g ≥ δ then {

6:      $\Delta cd[j] \rightarrow \mathbb{Z}[i]$;

7:      $\mathbb{N}[i][p] = \Delta cd[j]$;

8:      p = p + 1;

9:    }

10: endfor

11:endfor

12:for i=1 to $|\mathbb{Z}|$ do

13: construct $S^{\Delta cd}$ that is the Δcd cluster under $\mathbb{Z}[i].\omega$;

14: compute $E^{\Delta A}$、$M_{dg}^{\Delta cd}$、$M_{dga}^{\Delta cd}$、$E_{slc}^{cd\cup\Delta cd}$、in$\mathbb{Z}[i]$ by CCDGG 28-81;

15: for k=1 to $|M_{dga}^{\Delta cd}|$ do

16:    flg=0;

17:    for j=1 to $|M_{dga}^{Z[i]}|$

18:      if $M_{dga}^{Z[i]}[j] == M_{dga}^{\Delta cd}[k]$ then{

19:        $M_{dg}^{Z[i]}[j].obj \leftarrow M_{dg}^{Z[i]}[j].obj + M_{dg}^{\Delta cd}[k].obj$;

20:        if $M_{dg}^{Z[i]}[j].reg == M_{dg}^{\Delta cd}[k].reg$ then {

21:          if $M_{dg}^{\Delta cd}[k].reg == P$ then

22:            $M_{dg}^{Z[i]}[j].reg == P$;

23:          else if $M_{dg}^{\Delta cd}[k].reg == B$ then

24:            $M_{dg}^{Z[i]}[j].reg == B$;

25:          else $M_{dg}^{Z[i]}[j].reg == N$;

26:        }else{

27:          p = ($M_{dg}^{Z[i]}[j].obj \cap E_{slc}^{cd\cup\Delta cd})/M_{dg}^{Z[i]}[j].obj$;

28:          if p ≥ β ∧ p < α then $M_{dg}^{Z[i]}[j].reg == B$;

29:          if p < β then $M_{dg}^{Z[i]}[j].reg == N$;

30:          }

31:        flg=1;

32   break;

33:      }

34:    endfor

35:    if flg=0 then $M_{dg}^{\Delta cd} \rightarrow M_{dg}^{Z[i]}[|M_{dg}^{Z[i]}| + 1]$;

36: endfor

37: m=1;

38: for i=1 to $|\mathbb{Z}|$ do

39:    for j=1 to $|E_{\mathbb{Z}[i]}^{\Delta A}|$ do

40:      for k=1 to $|M_{dg}^{Z[i]}|$ do

41:        if $E_{\mathbb{Z}[i]}^{\Delta A}[j] \cap M_{dg}^{Z[i]}[k] \neq \phi$ then{

42:          $T = E_{\mathbb{Z}[i]}^{\Delta A}[j] \cap M_{dg}^{Z[i]}[k]$;

43:          N[m].index = T[1];

44:          N[m].obj = T;

45:          p = ($T \cap E_{slc}^{cd\cup\Delta cd})/T$;

46:          if p ≥ α then N[m].reg = P;

47:          else if p ≥ β then N[m].reg = B;

48:          else N[m].reg = N;

49:        }

50:      endfor

51:    endfor

52:    m=m+1;

53: endfor

54: for i=1 to $|N|$ do

55:    if N[i].reg = P then {N[i] → N_POS };

56:    else if N[i].reg = B then {N[i] → N_B };

57:      else if N[i].reg = Nthen {N[i] → N_N }

58: endfor

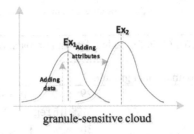

granule-sensitive cloud

Fig. 1. Migration of granule-sensitive cloud

Definition 6 (Attribute Sensitivity-Level Fuzzy Set): Let A be the attribute space on domain U and V be the value range of A. $F_c = \{f_{sl_1}, f_{sl_2}, \cdots, f_{sl_k}\}$ is the fuzzy set on U, f_{sl_k} is the attribute or attribute set on the sensitive level sl_k, and X is the variable taking the value on A. The probability distribution associated with X is denoted as $\prod_x$, and the probability distribution function of $\prod_x$ is denoted by π_X and is defined numerically equal to the membership degree of F_c, that is, $\forall u \in A \; \pi_X(u) = F_c(u)$.

Definition 7 (Granule Cloud): Let U be the quantitative domain, that is, the global domain of a domain object, C be a qualitative concept on U, and $\forall x \in U$ be a random instance of C. If the membership degree $\mu(x)$ of x relative to the qualitative concept C is in the $[0, 1]$ interval, then the distribution of x about the qualitative concept C on U is called cloud. By Definition 2, if $m()$ reflects a certain concept C_1, then all data granules satisfying $m()$ are subordinate to concept C_1 to some extent. The distribution of these data granules about concept C_1 is called granule cloud in this paper. If C_1 represents a sensitive concept on a set of meanings, the granule cloud consisting of related data granules can be called a granule-sensitive cloud.

Definition 8 (granule contribution): In the granule cloud, each data granule contributes to the qualitative concept C. The contribution is called the granule contribution. The contribution of a data granule to the qualitative concept C is as follows:

$$P_{dg_x} = \frac{En - |dg_x - Ex|}{\sqrt{2\pi En}} \mu(dg_x)$$

$\mu(dg_x)$ is the membership degree of data granule dg to qualitative concept C. Ex is the expected value of qualitative concept C, that is, the expected value of data granules set. En is a measure of the fuzziness degree of qualitative concept C. It reflects the range of data granules or attribute sets of data granules accepted by concept C in domain U. The larger the range accepted by Concept C, the more blurred Concept C is. In this paper, C is a sensitive concept on a meaning set, and En is the scale of attribute distance set subordinate to concept C. At the same time, it can be seen in the formula that the greater the membership degree of the data particle, the smaller the distance between its property and the expected property of the qualitative concept C, and the greater its contribution.

Definition 9 (Deduction Threshold): Let $\rho = \sum_{x=1}^{\psi} P_{dg_x}$. If $\rho \geq \tau$, it means that more sensitive information can be inferred from ψ data granules, wherein, the τ is the probability threshold for inferring high-sensitivity information from similar approximation data granules and the ψ is the threshold of the number of data granules.

Based on this, this paper presents a deduction algorithm for approximate data granules to infer the possibility of higher sensitive information based on cloud theory. As shown in Table 3.

Table 3. Possibility inference algorithm to infer highly sensitive information based on approximate data granules

Input: N_POS;C	
Output: ρ; ψ	
1: k=0;	8: $\psi = j$;
2: for i=1 to |N_POS| do	9: $\Sigma = \Sigma + G[k][j]$;
4:	10: $\rho = \Sigma$;
$G[k][i] = \dfrac{C.En - \|N\_POS[i] - C.Ex\|}{\sqrt{2\pi}C.En}\mu(N\_POS[i])$;	11: if $\rho \geq \tau$ then break;
	12: else $p = p + 1$;
5: endfor	13: endfor
6: $\Sigma = 0; p = 1$;	14: if $p > \psi$ then output(ρ, ∞);
7: for j=1 to |G[k]| do	15: else output(ρ, ψ);

Firstly, the algorithm calculates the contribution of the approximate data granules set to the sensitive qualitative concept C under the sensitivity level sl_c. Then, the contribution degree of each approximate data granule is accumulated in turn, and whether it exceeds the deduction threshold is judged in the process of accumulation. If the contribution exceeds the deduction threshold in the cumulative process, the maximum number of accesses to the approximate data granules at the sensitive level sl_c is finally obtained. The complexity of the algorithm is o(n), and n is the number of approximate data granules under sensitive level sl_c.

4 Experimental Simulation Analysis

In order to verify the execution effect of an information sensitivity inference method for big data aggregation based on granular analysis proposed in this paper, we carried out simulation experiments on this method. The simulation environment is as follows:

Python3.6.1, Intel(R) Core(TM) i5-4200 CPU @1.60 GHz 2.30 GHz, 4 GB RAM, the system environment is Windows 10.

The experimental data mainly come from laboratory and Internet scientific data with sensitive level and easy to classify attributes as well as artificial synthetic data. The purpose of synthetic data is to: One is to divide the real sensitive data into different approximate data files and generate the approximate data files of the real sensitive data. The other is to satisfy the requirement of experiment for multiple attribute relationships among data. The attribute relationships between experimental data objects mainly include: no association, intersection and inclusion relationship of attributes.

In the experiment, we extract data attributes according to the characteristics of experimental data, quantify data attributes and attribute values, and construct a prior knowledge base of data, the probability distribution of attributes and attribute sets as well as the fuzzy sets at various sensitive levels. The experimental evaluation indexes are divided into clustering effect and inference accuracy. We divide the experimental data into two parts, one is to verify the clustering effect based on the cloud data clustering algorithm based on granule gravity, and the other is the dynamically added data, which is used to verify the clustering effect of approximate data granules dynamic updating algorithm and the accuracy of the possibility inference algorithm for inferring highly sensitive information for approximate data granules.

4.1 Clustering Effect Analysis of the Algorithms

We use the first part of the experimental data to evaluate the performance of the cloud data clustering algorithm and perform clustering experiments on 600, 1400, 2000, 3000, and 4000 data, respectively. Firstly, in order to determine the appropriate gravitational threshold δ, this paper introduces evaluation factors, which are divided into effect factor (EF) and noise factor (NF):

① **Effect Factor (EF):** Let g_n be the average gravitation between the same kind of data and g_o be the average gravitation between different kinds, then $g_{n\_\mathbb{Z}_k} = \sum_{cd_i,cd_j \in \mathbb{Z}_k} g(cd_i, cd_j)/|\mathbb{Z}_k|$, $g_o = g(\mathbb{Z}_m, \mathbb{Z}_n)$ where $\mathbb{Z}_k$ represents the kth class. The gravitation between classes is the gravitation between core of all kinds of granules. The calculation formula of EF is as follows:

$$EF = \overline{g_n}/\overline{g_n} + \overline{g_o}$$

Wherein, $\overline{g_n} = \sum_{k=1}^{P} g_{n\_\mathbb{Z}_k}/P$, $\overline{g_o} = 2\sum g_o/P \times (P-1)$, P is the number of classifications.

② **Noise Factor (NF):** There are t noise data (called noise data that are not classified), then $g_{noise\_cd_i} = \sum_{k=1}^{P} g(cd_i, \mathbb{Z}_k)/P$, $g(cd_i, \mathbb{Z}_k)$ represents the gravitation between data and granule core of class k. The calculation formula of EF is as follows:

$$NF = \overline{g_n}/\overline{g_n} + \overline{g_{noise}}, \overline{g_{noise}} = \sum_{i=1}^{t} g_{noise\_cd_i}/t$$

According to the above two formulas, the more EF and NF approaches to 1, the better the clustering effect is. Therefore, in the experiments, we constantly change the gravitational threshold δ, and try our best to select the value of delta when NF and EF are large. (Selecting experimental data with a sample size of 3000). As shown in Fig. 2:

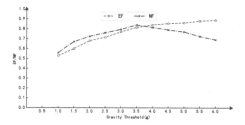

Fig. 2. Schematic diagram of the relationship between gravity threshold and EF/NF

The initial value of δ is 1, and the experiment is performed by incrementing 0.5 each time, and the values of EF and NF are calculated. It can be seen from Fig. 6 that EF gradually increases with the increase of gravity. Formula analysis shows that EF approaches 1 when gravity approaches infinity and 0.5 when gravity approaches zero. In order to ensure the accuracy of the cluster, we require that the EF value be greater than 0.8; The extreme point of NF occurs when δ = 3.5, which indicates that when δ is large, the unclassified data will gradually increase, thus affecting the accuracy of clustering. Therefore, in order to find the approximate optimal NF value, we conducted several experiments in the range of δ [3, 4]. As shown in Fig. 3:

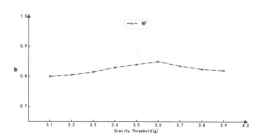

Fig. 3. Schematic diagram of NF approximate maximum

Figure 3 shows that the approximate maximum value of NF is 0.85 when δ is 3.6. Meanwhile, EF = 0.82, which satisfies the experimental accuracy.

In order to verify the clustering effect of the clustering algorithm, we use the error rate σ to reflect it. The formula for calculating σ is as follows:

$$\sigma = \frac{\sum_{i=1}^{n} \Delta sum + t}{SUM\_Data} * 100\%.$$

Wherein, Δsum is the gap between each experimental classification and the standard classification. n is the number of experimental classifications, SUM_Data is the total number of experimental data samples, and t is the number of noise data. The experimental results are shown in Fig. 4.

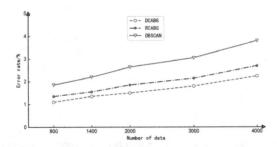

Fig. 4. Schematic diagram comparison of clustering effect between the proposed algorithm and other algorithms

In the experiment, we used four different sets of data to compare the DCABG algorithm proposed in this paper with the recursive object clustering algorithm based on recursive clustering algorithm based on similar concept (RCABC) [1] and the density-based clustering algorithm (DBSCAN) [2]. As can be seen from Fig. 4, DCABG clustering algorithm is better than the other two clustering algorithms, and the clustering accuracy is high. In the final clustering result, the DCABG algorithm further subdivides the similarity data in each classification according to the attribute and the attribute value to form a data granularity feature matrix. To ensure the accuracy of approximate sensitive information deduction, the characteristic matrices of different data granules are divided into positive domain, boundary domain and negative domain according to the sensitivity level.

In addition, we also give approximate data granules dynamic updating algorithm to ensure the accuracy of clustering when the data changes dynamically. In this paper, we use the second part of the experimental data to observe the effect of dynamic clustering by adding 500 data to the experimental data of 800, 2000 and 3000, respectively. As shown in Fig. 5:

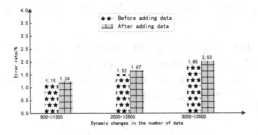

Fig. 5. Schematic diagram of approximate data granules dynamic updating algorithm

From Fig. 5, we can see that the clustering effect and accuracy can reach the expected goal after the data are updated dynamically by the algorithm in this paper. The approximate data granular dynamic update algorithm provides a reasonable basis for the possibility of deducing high-level sensitive information after the dynamic change of data by dynamically maintaining the granularity of the feature matrix.

In the experiment, we also evaluated the efficiency of the three clustering algorithms and analyzed their running time for different amounts of data. As shown in Fig. 6:

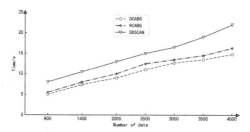

Fig. 6. Schematic diagram of the efficiency comparison of between the clustering algorithm and other algorithms

It can be seen from Fig. 6 that the DCABG clustering algorithm is much more efficient than the traditional DBSCAN algorithm. The time complexity of DBSCAN algorithm is $o(n^2)$, that of DCABG algorithm is $o(n * m)$, and that of DCABG algorithm is relatively low. Because by classification, m is generally much smaller than n. For DCABG clustering algorithm and RCABC clustering algorithm, the execution efficiency of both algorithms is related to the number of experimental data, attribute space and gravitational threshold δ. From the graph, there is little difference in the efficiency between them. DCABG clustering algorithm is slightly better than RCABC clustering algorithm. Because the algorithm of this paper further optimizes the selection of the gravitational threshold, the setting of the gravitational threshold is more reasonable, and the number of iterations of the algorithm in the clustering is reduced accordingly.

From the above experimental results, the closer the EF and NF are to 1, the better the effect of data clustering. In this paper, by designing and adjusting the thresholds of effect factors and noise factors, we can find the gravitational threshold which is more suitable for data clustering under different conditions, which can better guarantee the effect of clustering. In addition, from the DBSCAN algorithm, we can see that the algorithm can automatically classify according to gravity, does not need to specify the number of classifications, is insensitive to noise data, has low time complexity, can dynamically update the clustering of data, and has strong adaptability.

4.2   Algorithm Deduction Accuracy Analysis

In order to verify the accuracy of clustering information sensitivity inference algorithm based on approximate data granular contribution, we use the accuracy (AR), error rate (WR) and error rate (ER) of the deduction algorithm as the measurement indicators. Let T be a set of experimental data with clustering problems inferred from the deduction algorithm in this paper, N be an experimental data set with clustering problems under standard conditions, and S a set of experimental data with data clustering problem which is not deduced by the algorithm. The indicators are calculated as follows:

① $AR = \frac{|T \cap N|}{|N|}$, That is, the ratio of the number of data intersected by T and N to the total number of data in the standard case;

② $WR = \frac{|T - (T \cap N)|}{|N|}$, that is, the proportion of deduction errors;

③ $ER = \frac{|S|}{|N|} = \frac{|N - T \cap N|}{|N|} = 1 - AR$, that is, 1-AR, the ratio of the number of experimental data with clustering problems that are not inferred to the total number of data in the standard case.

Under $\tau = 0.9, 0.8, 0.7$ and 0.6, we use the proposed algorithm to infer the probability of high-level sensitive information from the approximate granule feature matrix and the approximate data granule set under sensitive level A. According to the results of the deduction, the accuracy and error rate of the algorithm are obtained statistically. As is shown in Figs. 7 and 8.

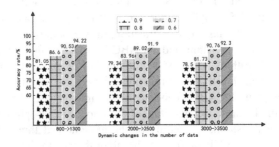

Fig. 7. Schematic diagram of the accuracy of the algorithm under different inference thresholds

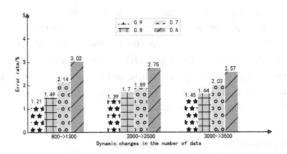

Fig. 8. Schematic diagram of the error rate of the algorithm under different inference thresholds

It can be seen from Figs. 7 and 8 that the accuracy and error rate of the inference algorithm are different under different values of the inference threshold τ. The higher τ is, the more data with low contribution are released in the process of inference, so the accuracy rate is low. It is easy to release data with sensitive information, leading to leaks.

The lower τ, the more data with approximate data clustering problem can be inferred, and the accuracy is higher, but a small number of positive criteria of misjudgments are prone to occur, and the error rate is increased. The reasonable setting of τ is crucial for the accuracy of the inference. By analyzing the accuracy and error rate of the inferred results, when the threshold $\tau = 0.7$, the accuracy can basically reach 90% and control the error rate at about 2%, which can better prevent leaks and misjudgments.

5 Conclusion

In this paper, the similarity dynamic mining of the same business big data set is carried out by data granules analysis. Based on the attribute-sensitive fuzzy set, granules cloud and granules contribution degree, the possibility of inferring sensitive information is deduced by aggregating similar data. This research is of great significance for the formulation of big data access control strategies and the prevention of leakage risk caused by aggregation analysis of big data. In the next step, we will conduct in-depth research on the data with the association rule attributes, and further solve the problem of leaking caused by the aggregation and deduction of association data.

Acknowledgements. This work is supported by the National Natural Science Foundations of China (grant No. 61502531 and No. 61702550) and the National Key Research and Development Plan (grant No. 2018YFB0803603 and No. 2016YFB0501901).

References

1. Yao, Z.: Research overview of big data. J. Ningbo Polytech. **21**(5), 36–40 (2017)
2. Wang, J.M.: Key technologies in big data applications development and runtime support platform. J. Softw. **28**(6), 1516–1528 (2017)
3. Li, T.R.: The Principle and Method of Big Data Mining. Science Press, Beijing (2016)
4. Cheng, C.: Analysis of data mining. Comput. Eng. Softw. (4), 130–131 (2014)
5. Cao, J.F., Dong, X.L., Zhou, J., Shen, J.C.: Research advances on big data security and privacy preserving. J. Comput. Res. Dev. **53**(10), 2137–2151 (2016)
6. Emilin, C., Swamynathan, S.: Reason based access control for privacy protection in object relational database systems. Int. J. Comput. Theory Eng. **3**(1), 1793–8201 (2011)
7. Jiang, J.B.: Research on cloud theory and its application. Master, Guangxi University (2008)
8. Wu, J., Wang, C.Z.: Multiple correlation analysis and application of granular matrix based on big data. Comput. Sci. (B11), 407–410 (2017)
9. Xie, W.C.: Research on risk-based access control framework and correlation technologies. Master, Information Engineering University (2013)
10. Liang, J.Y., Feng, C.J., Song, P.: A survey on correlation analysis of big data. Chin. J. Comput. **39**(1), 1–18 (2016)
11. Parno, M.D.: A multiscale framework for Bayesian inference in elliptic problems. Massachusetts Institute of Technology (2011)
12. Vasilios, K., Dimitrios, V.: A framework for access control with inference constraints. In: 35th IEEE Annual Computer Software and Applications Conference, pp. 289–297. IEEE Xplore (2011)

13. Cao, L.F., Chen, X.Y., Du, X.H.: A level inference method for aggregated information of objects based on associated attributes. Acta Electron. Sin. (7), 1442–1447 (2013)
14. Cao, L.F., Chen, X.Y., Du, X.H.: A level inference method for aggregated information of objects based on clustering analysis. J. Electron. Inf. Technol. 1432–1437 (2012)
15. Liang, J.Y., Qian, Y., Li, D.Y., Hu, Q.H.: Large data processing based on granular. Computing **45**(11), 1355–1369 (2015)
16. Xu, J., Wang, G.Y., Yu, H.: Review of big data processing based on granular computer. Chin. J. Comput. **38**(8), 1497–1517 (2015)
17. Wan, M.X., Ye, A.S.: Analysis on the big data processing technology based on granular computing. Wirel. Internet Technol. (1), 75–76 (2018)

Attentional Transformer Networks for Target-Oriented Sentiment Classification

Jianing Tong[1](✉), Wei Chen[2](✉), and Zhihua Wei[1]

[1] Tongji University, Shanghai, China
{tongjianing,zhihua_wei}@tongji.edu.cn
[2] Shanghai Institute of Criminal Science and Technology, Shanghai, China
weichen_82@163.com

Abstract. Text classification task includes total sentence sentimental classification as well as target-based sentimental classification. Target-based sentimental analysis and classification is aiming at locating sentimental classes of given sentences over different opinion aspects. Recurrent neural network is perfectly suitable for this kind of assignment, and it does achieve the state-of-the-art (SOTA) performance by now. Most of the previous works model target and context words with Recurrent Neural Network (RNN) with attention mechanism. However, RNN can hardly parallelize to train and cause too much memory occupancy. What's more, for this task, long-term memory may cause confusion. For example, *the food is delicious but the service is frustrating*, where the model may think the *food* is good while the *service* is bad. Convolutional neural network (CNN) seems vital in this situation as it can learn the local n-grams information while RNN cannot make it. To address these issues, this paper comes up with an **A**ttention **T**ransformer **Net**work (**ATNet**) which can perfectly address issues above. Our model employs attention mechanism and transformer component to generate target-orient representation, along with CNN layers to extract N-grams information. On open benchmark datasets, our proposed models achieve state-of-art results, namely, 70.3%, 72.1% and 83.4% in three benchmarks. Also, this paper applies pretrained BERT in the encoder part and acquires SOTA achievement. We performed many contrast experiments to elaborate effectiveness of our method.

Keywords: Target-based sentimental analysis · Attention mechanism · Transformer component

1 Introduction

Target-based Sentimental Classification [1] is quite a fine-grained work of sentiment classification. Many web sentences in reviews, debates, etc., often comprise of multiple target words with varied sentiment polarities. The difference between target-based sentiment classification and aspect-based sentiment classification is

© Springer Nature Singapore Pte Ltd. 2019
H. Jin et al. (Eds.): BigData 2019, CCIS 1120, pp. 271–284, 2019.
https://doi.org/10.1007/978-981-15-1899-7_19

that, the aspects are determined outside such as "convenience, service or taste", while the target is shown within the sentences. Fined-grained target-based sentiment analysis still hasn't been studied sufficiently, although there have been many works talking about it. Researchers often design neural network models to automatically extract helpful information to represent both context and target sentences end to end. There are also some obstacles needing to been solved effectively. The first problem with previous works is that the encoder for embeddings, which are supposed to encoder the input data, are Recurrent Neural Network [2] (RNN). Even Long-short Term Memory [3] (LSTM) can solve the gradient exploding problem and prevent gradient disappearing problem, they are still quite expensive to train. RNNs are difficult to be run by parallelizing. What's more, backpropagation through time steps (BPTT) consumes plethoric memory and computation resources. In fact, the BPTT training step for RNN is truncated, which has largely influence on model performance to capture long dependencies over longer time span. Although,the LSTM network could ease gradient disappearing situation in some degree, there has to be a quantity of training epoch.

Another question is that, we should also pay more attention to short scale information, because long scale information can bring sentimental confusion. For example, *I like the food in this restaurant quite a lot, however, this environment is too annoying, which makes me want to leave as soon as possible!* In this example, the positive sentiment *"like"* is corresponding to *"food"*, while the negative sentiment *"annoying"* is corresponding to *"environment"*. As previous works have proved, CNN is quite effective in capturing local information. So far as we can learn, there isn't any work before to capture local information in target-based sentiment analysis assignment.

In this paper, we propose an CNN-based architect neural network with attention, to solve the sentimental classification problem. Concretely, the model in this paper eschews recurrence and employs attention mechanism, which has been proved quite effective much times, and the encoder parts are utilized to capture introspective and interactive semantics between context sentences and targets. Along with the Target-Specific Transformation (TST) component as [4] has mentioned in their work, we proposed a more effective model. It achieves the SOTA in the open source benchmark dataset SemEval-2014 task4 [5], and 2014 Twitter dataset [6] (2014).

2   The Related Works

Target-based sentimental classification is quite hot research focus, as many previous works have been done to explore better methodology. As we can learn from previous works, modeling performance significantly depends on the quality of dependency parsing. Tang et al. [9] come up with a method to connect target embeddings and raw sentence embeddings individually, and feed them into their network. Wang et al. [10] proposed ATAE-LSTM to calculate the attention weights between sentence and target. Ma et al. [12] proposed IAN which learns the representations of the targets and context with two attention networks interactively. Then, Fan et al. [15] proposed multi-grained attention network to aggregates context2target weights and target2context weights to capture information. Chen et al. [13] proposed RAM in the same year. It combines attention result to gated recurrent units.

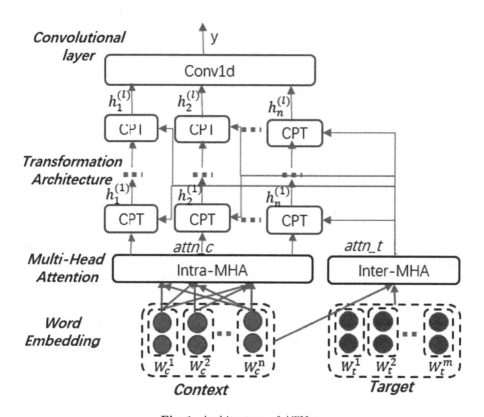

Fig. 1. Architecture of ATNet

3 Methodology

The input of our model is a context sequence $w^c = \{w_1^c, w_2^c, \ldots, w_n^c\}$ and a target sequence $w^t = \{w_1^t, w_2^t, \ldots, w_n^t\}$, where w^t is a subsequence of w^c. The target of our work is to model the input and make prediction for sentimental polarity $y \in \{Positive, Negative, Neutral\}$ of given target w^t towards the context sequence w^c.

Figure 1 shows our architecture of Attention Transformer Network (ATNet), including an embedding layer, an intra-attention, an inter-attention, Point-wise Convolutional Transformer (PCT) and an output layer. We propose two kind of embeddings: GloVe embedding and BERT embedding.

3.1   Embedding

GloVe. Assuming $E \in R^{d_{emb} \times |V|}$ is the GloVe Pennigton et al. [17] word vector matrix. In the equation, d_{emb} is dimension of embeddings. Word number is represented as $|V|$ as we can see in the equation.

BERT. BERT has been quite a hot top since now (2018) [18]. We use pretrained BERT embedding to turn sentences into word embedding.

3.2   Encoder Layer

We propose inter/inner attention architect to take the place of bidirectional LSTM, because it can be parallelizable and thus a superior interactive alternative of LSTM. We use it to calculate input embeddings and interactive weight between context and target. In this paper, multi-head attention has been proved to be better than ordinary attention layer. Multi-Head Attention (**MHA**) [19] firstly been proposed by [20], which can perform multiple attention calculation in parallel between token to token regardless the distances between any of them. In this paper, we model context embeddings with (**Intra-MHA**) for introspective attention weight, in order to find overall attention within the context embeddings. In the meanwhile, we model the target and context embeddings with (**Inter-MHA**) for context-perceptive target embeddings. Attention mechanism builds mapping function on a key sequential embedding $k = k_1, k_2, \ldots, k_n$ and a query sequential embedding $q = q_1, q_2, \ldots, q_n$. The output is also a sequence:

$$Attention(k, q) = softmax(f_s(k, q)) * k \tag{1}$$

where f_s denotes the attention function, like dot product, and alignment function which capture the semantic correlationship between q_j and k_i:

$$f_s(k_i, q_j) = tanh([k_i; q_j] \cdot W_{attn}) \tag{2}$$

where $W_{attn} \in R^{(2*d_{hid})}$ are derived from training. MHA is multi-head and thus, can learn multiple different attention score weight. It is very suitable for parallelization and aligning. We concat the multi-head outcome and map them into an individualized hidden feature space d_{hid}. Specifically,

$$MHA(k, q) = concat[o^1; o^2 \ldots; o^{(n_{head})}] \cdot W_{mh} \tag{3}$$

where o^i denotes the output of attention layer and $W_{mh} \in R^{(d_{hid} \times d_{hid})}$, n_{head} represents the number of heads. Outcome of head-attention is o^h, and h is between 1 and n_{head}.

Intra-MHA. This is because the input matrix are the same as ordinary attention function where $k = q$. We could infer Intra-MHA output representation c^{intra} by semantic layer e^c,

$$c^{intra} = MHA(e^c, e^c) \tag{4}$$

The outputs of encoded context representation are $c^{intra} = c_1^{intra}$, $c_2^{intra}, \ldots, c_n^{intra}$.

Inter-MHA is an ordinary used attention modality which k matrix varies q matrix. Inter-MHA output representation t^{intra} could be derived by,

$$t^{intra} = MHA(e^c, e^t) \tag{5}$$

Through this interactive training process, each given target word e^t would compose a selected representation from context embedding e^c. And thus, we acquire the context-perceptive target words $t^{inter} = t_1^{inter}, t_2^{inter}, \ldots, t_n^{inter}$.

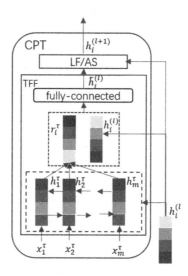

Fig. 2. The architecture of CPT module

3.3 Context-Perserving Transformation

Previous multi-head attention layers extract necessary feature weight matrix, but havn't use the original input context for sentiment polarity classification. Thus, in this layer, inspired by Xin Li et al. work, we incorporate the Context-Preserving Transformation (CPT) [4] into the model for learning target-specific word representations in the deep neural network architecture. The details of CPT are shown in Fig. 2.

Target-Specific Feature Fusion Layer. Target-specific Feature Fusion layer (**TFF**) component is shown in Fig. 2, in which the details of TST are. However, as Fig. 2 has shown, we proposed the Target-specific Feature Fusion layer component to better consolidating word representation.

Former researches [21,22] just make average of the embeddings of target sentence sequence as target extracted representation. However, it may not be suitable for all situations due to inequity contribution of every words in sentences. For instance, *"Burning yellow chicken"*, the word *"chicken"* is obviously much more important than *"Iburning"* and *"yellow"* because sentimental words will modify the polarity towards "chicken"., but merely conveyed over modifiers. Some researches [12] tried to modify this issue by calculating similarity scores between targets words and context words. However, it may be inefficient for sentences which has over one sentimental polarity, like *"I like burning yellow chicken here a lot, however the service makes my crazy, I'd been waiting for more than one hour for one chicken."*

In this paper, we use the Inter-MHA to calculate the importance or each word in target sequence, and thus, we achieve better experiment results. We also proposed to compute the context sequence dynamically by using bidirectional lstm to require context word representation $r_i^\tau \in [\overrightarrow{LSTM}(x_j^\tau); \overleftarrow{LSTM}(x_j^\tau)], j \in [1, m]$.

We relate the word x_j^τ in context sentence to tailor-make target representation r_i^τ at the time step i:

$$r_i^\tau = \sum_{j=1}^{m} h_j^\tau * F(h_i^l, h_j^\tau) \tag{6}$$

where function $F(\cdot)$ of Eq. 6 measures the correlation between the j-th target token embedding h_j^τ and the i-th token embedding h_i^τ:

$$F(h_j^{(l)}, h_j^\tau) = \frac{exp(h_j^{(l)\top} h_j^\tau)}{\sum_{k=1}^{m} exp(h_i^{(l)\top} h_k^\tau)} \tag{7}$$

At last, we introduced the combination of r_i^τ and $h_i^{(l)}$ into a total connection network.

$$\tilde{h}_i^{(l)} = g(W^T[h_i^{(l)} : r_i^\tau] + b^\tau) \tag{8}$$

where the g($\cdot$) function is the nonlinear activation function. $h_i^{(l)}$ and r_i^τ are concatenated matrix. W^T and b^τ are weights that should be trained in this layer.

Context-Preserving Layer. Previous layer is a non-linear layer that context information is captured. Along with the computation of layers and layers, the contextualized representations from the bidirectional LSTM layer would be lost since the variance and mean of the original feature space will be altered. For sake of reminding more previous layer information, we as well investigate two kinds of strategies: Lossless **F**orwarding (**LF**) and **A**daptive **S**caling (**AS**), to convey lower layer semantics, as shown in Fig. 2. Therefore, the model followed are divided into ATNet-LF and ATNet-AS.

Lossless Forwarding. This method retains all the context semantics in lower layers and feed context information into the features directly.

$$h^{(l+1)} = h_t^{(l)} + \tilde{h}_i^{(l)}, i \in [1, n], l \in [0, L] \tag{9}$$

where $h_t^{(l)}$ is l-th layer's input vector and $\tilde{h}_i^{(l)}$ is target Inter-MHA outcome.

Adaptive Scaling. Lossless Forwarding introduces the context information without loss, which may cause context information redundancy and thus loss in training process becomes hard to decrease. This strategy can help us overcome this issue by dynamically update the weight number of each layer. It utilized the gate mechanism, which similar to RNN's gate [26]. The gate helps us control the proportion o input transformed feature.

$$t_i^{(l)} = \sigma(W_{trans} \cdot h_i^{(l)} + b_{trans}) \tag{10}$$

where $t_i^{(l)}$ is the gate for layer l, σ is activation function and conduct connection of $h_t^{(l)}$ and $\tilde{h}_i^{(l)}$ based on the gate product:

$$h_i^{(l+1)} = t_i^{(l)} \odot h_i^{(l)} + (1 - t_i^{(l)}) \odot h_t^{(l)} \tag{11}$$

And thus, the lower layer context information and target information can be integrated into each upper layer and the semantics representation proportions are controlled by the gate function in each layer.

3.4   Convolutional Feature Extractor

As we all can see, the closer of position between sentiment words and target words, the higher probability between them. For instance, *"I like burning yellow chicken here a lot, however the service makes my crazy, I'd been waiting for more than one hour for one chicken."* The word *"like"* is only associated with "burning yellow chicken", but don't have any connection with *"service"*. In this situation, we proposed to use Convolutional neural network (CNN) to solve the problem. CNN has been seen quite effective in many previous works. It can help model the extract n-gram information.

$$c_i = RELU(w_{conv}^T h_{i:i+s-1}^{(L)} + b_{conv}) \tag{12}$$

Where $h_{i:i+s-1}^{(L)} \in R^{sdim_h}$ is the concatenated vector of $(h_i^{(}L)$ and s is the kernel size. w_{conv} and b_{conv} are trainable parameters of convolutional neural network during training process. For sake of most useful information, we use max pooling [27] strategy after the dot production. We use CNN in one dimension as the data is sequence data. The sentence representation $z \in^{n_k}$ and n_k kernels:

$$z = [max(c_1), \cdots, max(c_{n_k})]^T \tag{13}$$

3.5   Sentiment Polarity Classifying Layer

Finally, we feed the z vector in to the last layer to predict sentiment polarity, we achieve this layer by the fully connect neural network:

$$p(y \mid w^\tau, w) = Softmax(W_f \cdot z + b_f) \tag{14}$$

where W_f and b_f are trainable weights.

4 Experiment

4.1 Experiment Set and Dataset

There are three dataset: ACL dataset collected and SemEval dataset for task 4, which composing *Restaurant* and *Laptop* comments (Pontiki et al. 2014). The sentimental polarities in these datasets are the same: *positive*, *neural*, and *negative*. Data statistics of each sentimental polarity of them both train and test are shown in Table 1.

We set two kinds of word embedding as we have mentioned in previous chapter. The GloVe embedding will not be trained in the training process. The word embedding dimension d_dim for GloVe is 300, while 768 for pretrained BERT. Our model is initialized by the Glorot initialization [23]. In the process, we set the coefficient ϕ of L_2 regularization to 10^{-5} and dropout rate is 0.0003. The label parameter ε is set to 0.1. We use Adam as our model optimizer. Accuracy and Macro-F1 are adopted for model performance evaluation.

Table 1. Datasets statistic.

Data		Positive	Negative	Neural
Twitter	Train	1567	1563	454
	Test	33	118	161
Rest.	Train	2148	790	628
	Test	725	185	186
Laptop	Train	974	839	450
	Test	340	125	169

4.2 Model Comparisons

In our experiments, we conduct 9 kinds of different baseline models for comparisons. All of them are deep neural network. Also, we design three kinds of model variants for our model.

- **AOA** [16]: This paper proposed Attention-over-Attention Neural Networks, taking attention layer twice to extract semantics information.
- **AdaRNN** (Dong et al. 2014): By semantic composition over dependency tree, the proposed model extract sequence representation toward target for sentimental polarity prediction.
- **IAN** [12]: This paper takes advantage of two LSTM to model input sequence representation including context and target interactively.
- **CABASC** [21]: Using a memory mechanism and GRUs to learn representation between context and target interactively, the model also composed a weighted memory component to fuse feature.

- **AEN** [19]: It proposed an attention encoder network and label smoothing regularization, aiming to solve the parallelization problem and label unreliability issue.
- **RAM** [13]: The model in this paper is a stacked layer architecture in which every part composed of attention-based concatenation of token representation along with GRUs to learning sentence semantics.
- **Td-LSTM** [28]: The model takes advantage of RNNs by extracting both parts sentence of contexts respectively. After that, the model combines the representation from two sides to make final prediction.
- **MemNet** [9]: The same author as Td-LSTM. The model applies attention mechanism on token embeddings many times, and make prediction on the top-most representation.

In the experiments, we set the pretrained GloVe word embeddings and the dimension is set 300 (i.e., $dim_w = 300$). We set 80% of total words for the out-of-bag words number and we uses uniform distribution initialization for them. We set CNN with 3 kernel size because it has been proved effective for text sequence feature extracting. For sake of overcoming the overfitting issue, we apply dropout layer to input embedding layers and CPT components part. The dropout rate is 0.2 and all matrix weights are set uniform distribution. Our loss objective is cross-entropy, and Adam is utilized as our optimizer. Learning rate η is set as 0.001.

Table 2. Experiments results.

	Models	Restaurants		Laptop		Twitter	
		Accu.	F1	Accu.	F1	Accu.	F1
Baseline	AOA	0.753	0.616	0.683	0.619	0.665	0.637
	MGAN	0.778	0.683	0.702	0.635	0.708	0.690
	TNet_LF	0.778	**0.696**	0.701	0.632	0.612	0.561
	CABASC	0.779	0.661	0.691	0.626	0.672	0.654
	AEN	**0.789**	0.684	0.703	**0.651**	0.682	0.667
	MemNet	0.765	0.644	**0.707**	0.650	**0.716**	**0.702**
	Td_LSTM	0.771	0.656	0.694	0.630	0.714	0.692
	ATAE_LSTM	0.772	0.652	0.683	0.629	0.701	0.681
	RAM	0.769	0.653	0.651	0.590	0.697	0.670
ATNet variants	ATNet_LF	0.790	0.689	0.713	0.641	0.677	0.598
	ATNet_AL	0.811	**0.711**	0.701	0.636	0.678	0.600
	ATNet_BERT_LF	**0.834**	0.697	0.720	0.642	**0.703**	**0.610**
	ATNet_BERT_AL	0.821	0.694	**0.721**	**0.644**	0.701	0.607

4.3   Main Results

As shown in Table 2, ATNet with BERT pretrained word embedding achieves best results in most times. In the restaurant and laptop dataset, we can see that ATNet variants can acquire higher results than the baselines methods. However, in the twitter dataset ATNet variants can't beat some of the baseline methods. In the ATNet variants ATNet with BERT pretrained model can always achieve better results. In restaurant dataset ATNet_BERT_LF is 0.05% higher than the best result in baseline methods. In laptop result, ATNet_BERT_AL is 0.2% higher than the best baseline method. Also, in Fig. 3 we can have a look at the models' training time comparison, where epoch is set 20 and learning rate is set 3e−5. We can figure out that ATNet_LF can achieve better result with less time consuming. Even ATNet_BERT_LF need more time, it can always achieve best results. As shown in Table 3, intra-MHA makes the biggest influence on ATNet model. Conv1d part and inner-MHA almost equally affect the performance of our proposed model. CPT component has least influence on the model.

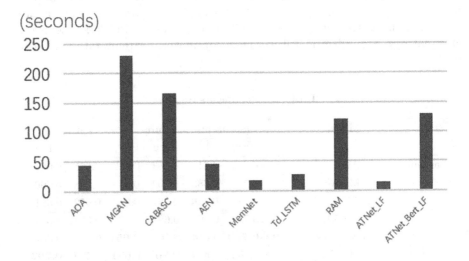

Fig. 3. Training time of different models

4.4   Case Study

For sake of elaborating the effectiveness of ATNet variants, we select some of the sentences in the sentiment analysis datasetTable. Table 4 shows some cases.

We add brackets around the input goal words and label them in the columns with different models' prediction. For each sentence, before feeding them to the embedding layer, we pad them with "PAD" to keep the consistency of sentence length. In the Table 4, P means positive attitude, N means negative attitude, and O means neural attitude.

As we can figure out in the table, for the first case, there three target to figure, which near closely. ATNet-AS is forced to pay attention to nearest sentimental words, which leading to the mistake in the second sentiment analysis. When target number is big, our model are more likely to make mistakes. What's more, from the last case, we can see that ATNet-LF and ATNet-AS tend to focus the negative words, such as "no", "only", which also may increase the error rate.

4.5   Ablated Test

In this part, ablated test experiments are executed to analyze the contribution of every components in the neural network. Contrast experiments are conducted on ATNet_LF model only due to passage capacity limitation.

ATNet_LF w/o conv1d ablates conv1d layer.
ATNet_LF w/o intra-MHA ablates intra-MHA layer.
ATNet_LF w/o inner-MHA ablates inner-MHA layer.
ATNet_LF w/o CPT ablates CPT layer.

As shown in Table 3, when eliminating the conv1d part, we get 0.117 loss for the precision and 0.101 loss for the F1-score. Intra_MHA achieves the biggest contribution, reaching 0.231 loss for precision and 0.190 loss for F1-score. However, CPT take the least loss in the whole four experiments, only 0.019 and 0.006 for precision and F1-score correspondingly. For inner_MHA part, it also contributes decent devotes, which causes 0.126 and 0.084 loss on precision and F1-score separately.

Table 3. Ablated test of ATNet

Model	Precision	F1-score
ATNet_LF - conv1d	−0.117	−0.101
ATNet_LF - intra_MHA	−0.231	−0.190
ATNet_LF - inner_MHA	−0.126	−0.084
ATNet_LF - CPT	−0.019	−0.006

Table 4. Example predictions. X in the right header indicates incorrect predictions

Sentence	ATNet-LF	ATNet-AS	ATNet-Bert-LF	ATNet-Bert-AS
They did not have [mayonnaise], forgot our [toast] and the [bacon] was so over cooked	(N, N, N)	(N, O^X, N)	(N, N, N)	(N, N, N)
The [service] was typical short-order, dinner type	O	O	P^X	O
From the [appetizers] we ate, the dim sum and other variety of foods, it was impossible to criticize the food	P	O^X	P	P
The [Bagels] have an outstanding taste with a terrific texture, both chewy yet not gummy	O^X	N	N	N
Not only was the [food] outstanding, but the little '[perks]' cannot be better	(P, N^X)	(P, N^X)	(P, P)	(P, P)

5 Conclusion

We proposed an Attention Transformer Network for target sentimental classification in this paper. We use the multi-head attention mechanism to take the place of Bi-LSTM feature extractor in order to speed up training, meanwhile achieving better results. What's more, we also apply the pretrained BERT embeddings to obtain newer state-of-art result in the benchmark datasets. We found that Transformer self-attention can largely improve model performance and CNN is also a superior n-gram feature extractor. Experiments elaborate effectiveness by the proposed models. In the Restaurant dataset, our model improve about two percentage of both the accuracy and f1-score. Comparing to other former proposed models, ANTet variants eliminate training time and capture the n-gram information between words. For target-based sentiment analysis, ATNet variants reach SOTA performance.

Acknowledgement. This work is sponsored by National Key Research and Development Project (No. 213), the National Nature Science Foundation of China (No. 61573259, No. 61673299, No. 61673301, No. 61573255) and the Special Project of the Ministry of Public Safety (No. 20170004). Supported by Key Laboratory of Information Network Safety, Ministry of Public Safety No. C18608. It is also supported by Shanghai Health and Family Planning Commission Chinese Medicine Science and Technology Innovation Project (ZYKC201702005).

References

1. Jiang, L., Yu, M., Zhou, M., et al.: Target-dependent twitter sentiment classification. In: Proceedings of the 49th Annual Meeting of the Association for Computational Linguistics: Human Language Technologies-Volume 1, pp. 151–160. Association for Computational Linguistics (2011)
2. Mikolov, T., Karafiát, M., Burget, L., et al.: Recurrent neural network based language model. In: Eleventh Annual Conference of the International Speech Communication Association (2010)
3. Hochreiter, S., Schmidhuber, J.: Long short-term memory. Neural Comput. **9**(8), 1735–1780 (1997)
4. Li, X., Bing, L., Lam, W., et al.: Transformation networks for target-oriented sentiment classification. arXiv preprint arXiv:1805.01086 (2018)
5. http://alt.qcri.org/semeval2014/task4/
6. Mohammad, S., Kiritchenko, S., Sobhani, P., et al.: A dataset for detecting stance in tweets. In: LREC (2016)
7. Suykens, J.A.K., Vandewalle, J.: Least squares support vector machine classifiers. Neural Process. Lett. **9**(3), 293–300 (1999)
8. Scholkopf, B., Smola, A.J.: Learning with Kernels: Support Vector Machines, Regularization, Optimization, and Beyond. MIT Press, Cambridge (2001)
9. Tang, D., Qin, B., Liu, T.: Aspect level sentiment classification with deep memory network. arXiv preprint arXiv:1605.08900 (2016)
10. Wang, Y., Huang, M., Zhao, L.: Attention-based LSTM for aspect-level sentiment classification. In: Proceedings of the 2016 Conference on Empirical Methods in Natural Language Processing, pp. 606–615 (2016)
11. Yang, M., Tu, W., Wang, J., et al.: Attention based LSTM for target dependent sentiment classification. In: Thirty-First AAAI Conference on Artificial Intelligence (2017)
12. Ma, D., Li, S., Zhang, X., et al.: Interactive attention networks for aspect-level sentiment classification. arXiv preprint arXiv:1709.00893 (2017)
13. Chen, P., Sun, Z., Bing, L., et al.: Recurrent attention network on memory for aspect sentiment analysis. In: Proceedings of the 2017 Conference on Empirical Methods in Natural Language Processing, pp. 452–461 (2017)
14. Dey, R., Salemt, F.M.: Gate-variants of gated recurrent unit (GRU) neural networks. In: 2017 IEEE 60th International Midwest Symposium on Circuits and Systems (MWSCAS), pp. 1597–1600. IEEE (2017)
15. Fan, F., Feng, Y., Zhao, D.: Multi-grained attention network for aspect-level sentiment classification. In: Proceedings of the Conference on Empirical Methods in Natural Language Processing, pp. 3433–3442 (2018)
16. Huang, B., Ou, Y., Carley, K.M.: Aspect level sentiment classification with attention-over-attention neural networks. In: Thomson, R., Dancy, C., Hyder, A., Bisgin, H. (eds.) SBP-BRiMS 2018. LNCS, vol. 10899, pp. 197–206. Springer, Cham (2018). https://doi.org/10.1007/978-3-319-93372-6_22
17. Pennington, J., Socher, R., Manning, C.: GloVe: global vectors for word representation. In: Proceedings of the 2014 Conference on Empirical Methods in Natural Language Processing (EMNLP), pp. 1532–1543 (2014)
18. Devlin, J., Chang, M.W., Lee, K., et al.: BERT: pre-training of deep bidirectional transformers for language understanding. arXiv preprint arXiv:1810.04805 (2018)
19. Song, Y., Wang, J., Jiang, T., et al.: Attentional encoder network for targeted sentiment classification. arXiv preprint arXiv:1902.09314 (2019)

20. Vaswani, A., Shazeer, N., Parmar, N., et al.: Attention is all you need. In: Advances in Neural Information Processing Systems, pp. 5998–6008 (2017)
21. Liu, Q., Zhang, H., Zeng, Y., et al.: Content attention model for aspect based sentiment analysis. In: Proceedings of the 2018 World Wide Web Conference on World Wide Web, pp. 1023–1032. International World Wide Web Conferences Steering Committee (2018)
22. Zhang, L., Wang, S., Liu, B.: Deep learning for sentiment analysis: a survey. Wiley Interdisc. Rev.: Data Min. Knowl. Discovery $8(4)$, e1253 (2018)
23. Glorot, X., Bengio, Y.: Understanding the difficulty of training deep feedforward neural networks. In: Proceedings of the Thirteenth International Conference on Artificial Intelligence and Statistics, pp. 249–256 (2010)
24. Zhou, Z., Zhang, W., Wang, J.: Inception score, label smoothing, gradient vanishing and-log $(d(x))$ alternative. arXiv preprint arXiv:1708.01729 (2017)
25. Kingma, D.P., Ba, J.: Adam: a method for stochastic optimization. arXiv preprint arXiv:1412.6980 (2014)
26. Jozefowicz, R., Zaremba, W., Sutskever, I.: An empirical exploration of recurrent network architectures. Int. Conf. Mach. Learn., 2342–2350 (2015)
27. Kim, Y.: Convolutional neural networks for sentence classification. arXiv preprint arXiv:1408.5882 (2014)
28. Tang, D., Qin, B., Feng, X., et al.: Effective LSTMs for target-dependent sentiment classification. arXiv preprint arXiv:1512.01100 (2015)

Distributed Logistic Regression for Separated Massive Data

Peishen Shi[1], Puyu Wang[1], and Hai Zhang[1,2(✉)]

[1] School of Mathematics, Northwest University, Xi'an 710127, Shaanxi, China
zhanghai@nwu.edu.cn
[2] Faculty of Information Technology
and State Key Laboratory of Quality Research in Chinese Medicines,
Macau University of Science and Technology, Macau, China

Abstract. In this paper, we study the distributed logistic regression to process the separated large scale data which is stored in different linked computers. Based on the Alternating Direction Method of Multipliers (ADMM) algorithm, we transform the solving of logistic problem into the multistep iteration process, and propose the distributed logistic algorithm which has controllable communication cost. Specifically, in each iteration of the distributed algorithm, each computer updates the local estimators and interacts the local estimators with the neighbors simultaneously. Then we prove the convergence of distributed logistic algorithm. Due to the decentralized property of computer network, the proposed distributed logistic algorithm is robust. The classification results of our distributed logistic method are same as the non-distributed approach. Numerical studies have shown that our approach are both effective and efficient which perform well in distributed massive data analysis.

Keywords: Distributed · Logistic regression · ADMM algorithm

1 Introduction

With the rapid development of the science and information technology, it is possible for us to acquire massive data in many areas such as medical science, economics and biological information science. Due to the large scale of these observations, analyzing this type of data is a new challenge for us. Parallel computing and distributed computing are two common methods for processing large amounts data. In parallel computing, complex tasks are split into multiple subtasks, and multiple processors processing these subtasks simultaneously. Every processors in a parallel system can exchange information, because they all can access to the shared memory. While in distributed computing, the whole data stored in different networked computers, and all computers collaborate to achieve a common goal by exchanging messages with neighbors. Each computer in the distributed system has its own private memory, and is more independent of other computers. Therefore, distributed algorithms are more suitable for

© Springer Nature Singapore Pte Ltd. 2019
H. Jin et al. (Eds.): BigData 2019, CCIS 1120, pp. 285–296, 2019.
https://doi.org/10.1007/978-981-15-1899-7_20

processing the separated stored massive data than parallel algorithms, and are more robust, especially when the network is decentralized. Moreover, distributed storage is a common form of storing massive data in modern society, which has well scalability and high availability. Due to the above advantages of distributed computing, we study the distributed approach suitable for processing large scale data in distributed storage.

The simplest distributed statistical estimation algorithm is the average mixture (Avgm) algorithm. Specifically, each computer computes estimator on its own data set, and finally output the average of all estimators across the computers. McDonald et al. [1,2] and Zhang et al. [3,4] have studied the Avgm algorithm for some classification and estimation problems. Another distributed framework is based on a connected computer network which can exchange information in the iteration of the algorithm. Mateos et al. [5] proposed the distributed sparse linear regression via Lasso, Wang et al. [6] generalized the distributed nonconvex regularization methods, as well as a Newton-type method for distributed optimization by Zhang et al. [7].

In this paper, we focus on classification issue with a wide range of applications, and it is the major part of machine learning [8–11]. Currently there are many classification methods, such as logistic regression, naive bayes, decision trees, support vector machines (SVM) and so on. Now the above approaches are already the most popular tools in data mining. As a typical classification model, logistic regression has been widely used and studied extensively.

Let us consider the logistic model. Suppose we have the data set $S = \{x_i, y_i\}_{i=1}^n$, $x_i \in R^{1 \times p}$ is an input vector, $y_i \in \{0, 1\}$ is a response variable. The conditional probability of $y_i = 1$ given x_i and $y_i = 0$ given x_i given by

$$p(y_i = 1 | x_i) = h_\theta(x_i) = \frac{1}{1 + \exp(-x_i\theta)},$$

$$p(y_i = 0 | x_i) = 1 - h_\theta(x_i) = \frac{1}{1 + \exp(x_i\theta)},$$

where $\theta \in R^{p \times 1}$ is the vector of coefficients, and $h_\theta(\cdot)$ is sigmoid function. Maximum likelihood estimation (MLE) is the common method to estimate θ. The log-likelihood of dataset S is

$$l = \frac{1}{n} \sum_{i=1}^n \{y_i \log(h_\theta(x_i)) + (1 - y_i) \log(1 - h_\theta(x_i))\}.$$

By minimizing the negative log-likelihood function, the ML estimator is as follow

$$\min_\theta \frac{1}{n} \sum_{i=1}^n \{(y_i - 1) \log(1 - h_\theta(x_i)) - y_i \log(h_\theta(x_i))\}. \tag{1.1}$$

(1.1) is a convex optimization and it is easy to obtain its solution.

In this paper, we focus on distributed logistic regression suitable for processing large amounts data. The existing works of large scale logistic regression include the distributed ADMM based algorithm [12], and parallel logistic algorithms [13–16]. Boyd et al. [12] studied distributed ADMM logistic method with global variables. There exists a master computer in their distributed computer network who can exchange information with all computers, and the master updates the global variables per iteration. The above algorithm is vulnerable due to the centralization of the computer network, that is, the algorithm completely breaks down when the master is attacked. While we focus on the distributed robust algorithm without global variables. Computers in the decentralized network are physically connected with others, and could only exchange information with neighbors. To reduce the communication cost, computers in our algorithm only transmits the local estimators which are $p \times 1$ vectors with neighbors.

The remainder of this paper is shown as follows. In Sect. 2, based on ADMM algorithm, we presented the distributed logistic algorithm which is robust and has controllable communication cost. In each iteration of the distributed algorithm, each computer updates the local estimators, and interacts the local estimators with the neighbors simultaneously. We then prove the convergence of distributed algorithm. In Sect. 3, several experiments show the effectiveness of our distributed logistic method in dealing with distributed storage data. And Sect. 4 we conclude the paper.

2   Distributed Logistic Regression

In this section, we present the distributed logistic model, and propose distributed logistic algorithm suitable for processing distributed storage data. Then we prove its convergence.

2.1   Model Set−Up

Suppose data set $S = \{x_i, y_i\}_{i=1}^n$ is stored in a distributed computer system. Computers in the distributed system are physically connected which only exchange messages with neighbors, and all of them form a network. Let N_j represents the neighborhood set of computer j, and $|N_j|$ is the number of the neighbors of computer j. The network can automatically modeled as an undirected graph $G = (C, E)$, where $C := \{1, 2, 3, \cdots, J\}$ denotes computer sets, and $E_{J \times J}$ is the adjacency matrix, if computer j can exchange information with computer i, $E_{ij} = 1$, otherwise, $E_{ij} = 0$. Then, we assume that any pair of computers can exchange information through a path, that is, $G = (C, E)$ is connected.

Computer j in the network stores the local data $\{x_{ji}, y_{ji}\}_{i=1}^{n_j}$, where n_j denotes the number of samples in computer j. Here, we assume that the domain $\mathcal{X}$ is bounded. According to the data belongs to different linked computers, model (1.1) can be reformulated as

$$\min_{\theta} \frac{1}{n} \sum_{j=1}^{J} \sum_{i=1}^{n_j} \{(y_{ji} - 1) \log(1 - h_\theta(x_{ji})) - y_{ji} \log(h_\theta(x_{ji}))\}. \tag{2.1}$$

(2.1) divides the entire data into J parts, it's not difficult to find that (2.1) is equal to (1.1).

To distribute the (2.1), we consider using the local variables $\{\theta_j\}_{j=1}^J$ instead of the global variable θ, where θ_j denotes the computer j's local estimator. Now, we can rewrite model (2.1) as the following distributed convex minimization problem

$$\min_{\{\theta_j\}_{j=1}^J} \frac{1}{n} \sum_{j=1}^J \sum_{i=1}^{n_j} \{(y_{ji} - 1)\log(1 - h_{\theta_j}(x_{ji})) - y_{ji}\log(h_{\theta_j}(x_{ji}))\} \tag{2.2}$$
$$s.t. \ \theta_j = \theta_{j'}, j \in \mathcal{C}, j' \in N_j.$$

where N_j is the neighborhood set of computer j. Until now, we have rewritten the original problem as a distributed constrained optimization problem.

Theorem 1. *If the undirected graph $G = (\mathcal{C}, E)$ is connected, then the optimization problem (2.1) is same as (2.2).*

Proof. Based on the assumption $G = (\mathcal{C}, E)$ is connected, computer $j \in \mathcal{C}$ can transmits information with others, that is, for any $j, j' \in \mathcal{C}$, there exist $j_1, \cdots, j_n$, such that $\hat{\theta}_j = \hat{\theta}_{j_1} = \cdots = \hat{\theta}_{j_n} = \hat{\theta}_{j'}$. Thus, $\exists \ \hat{\theta}$, s.t. $\hat{\theta} = \hat{\theta}_1 = \cdots = \hat{\theta}_J$, so the constraints in (2.2) can be eliminated by replacing all the $\{\theta_j\}_{j=1}^J$ with a global variable θ, say, in which case the cost in (2.2) reduces to the one in (2.1). Vice versa, then (2.2) is same with (2.1).

2.2 Distributed Logistic Algorithm

To solve (2.2) in distributed fashion, we resort to alternative direction method of multiplier (ADMM), which has been proposed in early 1970s [17,18] and has been widely studied [19,20]. ADMM algorithm is an alternating iterative process consisting of three steps per iteration, and more details are in the Appendix.

First of all, we consider adding the auxiliary variables $\{R_j\}_{j=1}^J$, model (2.2) can be rewritten as

$$\min_{\{\theta_j\}_{j=1}^J, \{R_j\}_{j=1}^J} \frac{1}{n} \sum_{j=1}^J \sum_{i=1}^{n_j} \{(y_{ji} - 1)\log(1 - h_{\theta_j}(x_{ji})) - y_{ji}\log(h_{R_j}(x_{ji}))\} \tag{2.3}$$
$$s.t. \ R_j - \theta_i = 0, for \ any \ j, i = 1, \cdots, J.$$

Because we assume that graph $G = (\mathcal{C}, E)$ is connected, then constrained optimization problem (2.3) is same as (2.1).

For notational convenience, we rewrite model (2.3) to the form of vectors

$$\min_{R, \theta} f(\theta) + h(R) \tag{2.4}$$
$$s.t. \ MR + M'\theta = 0.$$

where $R = [R_1^T, \ldots, R_J^T]^T$, $\theta = [\theta_1^T, \ldots, \theta_J^T]^T$. And $M = [M_1^T, \ldots, M_J^T]^T$, $0 \in R^{(J \times p) \times p}$ is zero matrix, where $M^* = [I_{p \times p}, \ldots, I_{p \times p}]^T \in R^{(J \times p) \times p}$,

$M_1 = [M^*, 0, \ldots, 0]$, $M_2 = [0, M^*, \ldots, 0]$, $\ldots, M_J = [0, \ldots, 0, M^*]$. $M' = -[I_{(J \times p) \times (J \times p)}, \ldots, I_{(J \times p) \times (J \times p)}]^T$. $h(R) = \frac{1}{n} \sum_{j=1}^{J} \sum_{i=1}^{n_j} -y_{ji} \log h_{R_j}(x_{ji})$, $f(\theta) = \frac{1}{n} \sum_{j=1}^{J} \sum_{i=1}^{n_j} (y_{ji} - 1) \log(1 - h_{\theta_j}(x_{ji}))$.

Then we can write the quadratically augmented Lagrangian function

$$L(R, \theta, V) = f(\theta) + h(R) + \langle V, MR + M'\theta \rangle + \frac{c}{2} \| MR + M'\theta \|_2^2, \quad (2.5)$$

where $V := \{V_j\}_{j \in \mathcal{C}}$ is Lagrange multipliers, and $c > 0$ is a pre-selected constant. We now give some notations. Let A and B represents the smallest strictly-positive eigenvalue of $M^T M$ and $M'^T M'$. If a function is differentiable and its gradient is Lipschitz continuous, then we call this function Lipschitz differentiable. Obviously, $f(\theta)$ in (2.4) is Lipschitz differentiable, and the Lipschitz constant is L_f. For any $j \in \mathcal{C}$, let

$$g(R_j) = \frac{1}{n} \sum_{i=1}^{n_j} (-y_{ji}) \log(h_{R_j}(x_{ji})),$$

then there exist $\omega \geq 0$, for any R_j, R'_j,

$$g(R_j) + \frac{\omega}{2} \| R_j - R'_j \|_2^2 \geq g(R'_j) + \langle \nabla g(R'_j), R_j - R'_j \rangle. \quad (2.6)$$

Based on ADMM, we update variables R, θ, and Lagrange multipliers V in turn until algorithm converges. Now, we present the main convergence theorem of ADMM algorithm.

Theorem 2. *Suppose that $c > \max \left\{ \frac{2L_f}{B}, L_f B, \frac{\omega}{A} \right\}$, then the sequence generated by the (A.1),(A.2),(A.3) converges to the minimum point of $L(R, \theta, V)$.*

For the convenience of the reader, we put the proofs in the Appendix.

Remark 1. Compared to distributed algorithm that contains global variable, our method increases the communication cost, but it is more robust. This is because the algorithm contains the global variable which is stored on the network master computer, and the global variable is updated at the master. When master is attacked, the algorithm fails. And our approach without master computer avoids this situation.

2.3   Estimation of Parameters

To solve (A.1), (A.2), (A.3), we rewrite $L(R, \theta, V)$ as follows

$$L(\{R_j\}, \{\theta_j\}, \{V_j\}) = \sum_{j=1}^{J} \{ \frac{1}{n} \sum_{i=1}^{n_j} (y_{ji} - 1) \log(1 - h_{\theta_j}(x_{ji}))$$

$$- \frac{1}{n} \sum_{i=1}^{n_j} y_{ji} \log(h_{R_j}(x_{ji})) + \frac{c}{2} \sum_{i=1}^{J} E_{ij}(\| R_i - \theta_j + \frac{V_{ij}}{c} \|_2^2 - \| \frac{V_{ij}}{c} \|_2^2) \}. \quad (2.7)$$

where E_{ij} is the i-th row and the j-th column of adjacency matrix E, which represents the connection condition between computer i and computer j.

Consider step (A.1)

$$R^{k+1} = \arg\min_{R} L(R, \theta^{(k)}, V^{(k)}).$$

From the decomposable structure of (2.7), (A.1) decouples into J sub-problems

$$R_j^{k+1} = \arg\min_{R_j} L_{1j}$$

$$= \arg\min_{R_j} \frac{c}{2} \sum_{i=1}^{J} E_{ij} \| R_j - \theta_i^k + \frac{V_i^k}{c} \|_2^2 - \frac{1}{n} \sum_{i=1}^{n_j} y_{ji} \log(h_{R_j}(x_{ji})). \quad (2.8)$$

We used gradient descent method to solve (2.8), the iteration formula is

$$R_j^{k+1} = R_j^k - \alpha \frac{\partial L_{1j}}{\partial R_j}, \quad (2.9)$$

where α is learning rate, and $\frac{\partial L_{1j}}{\partial R_j} = c \sum_{i=1}^{J} E_{ij}(R_j^k - \theta_i^k + \frac{V_i^k}{c}) - \frac{1}{n} \sum_{i=1}^{n_j} y_{ji}(1 - h_{R_j^k}(x_{ji}))x_{ji}^T$. Using the gradient descent method until convergence is overly accurate, pursuit of the relaxation here, we update R_j^{k+1} by (2.9) without iteration.

Next, consider the step (A.2)

$$\theta^{k+1} = \arg\min_{\theta} L(R^{k+1}, \theta, V^k),$$

similarly, it can also be resolved into J sub-problems

$$\theta_j^{(k+1)} = \arg\min_{\theta_j} L_{2j}$$

$$= \arg\min_{\theta_j} \frac{1}{n} \sum_{i=1}^{n_j} (y_{ji} - 1) \log(1 - h_{\theta_j}(x_{ji})) + \frac{c}{2} \sum_{i=1}^{J} E_{ij} \| R_i^{k+1} - \theta_j + \frac{V_i^k}{c} \|_2^2.$$

$$(2.10)$$

By using the gradient descent method, we have

$$\theta_j^{k+1} = \theta_j^k - \alpha \frac{\partial L_{2j}}{\partial \theta_j}, \quad (2.11)$$

where $\frac{\partial L_{2j}}{\partial \theta_j} = \frac{1}{n} \sum_{i=1}^{n_j} (1 - y_{ji}) h_{\theta_j^k}(x_{ji})x_{ji}^T - c \sum_{i=1}^{J} E_{ij}(R_i^{k+1} - \theta_j^k + \frac{V_i^k}{c})$. Same as (2.9), we update θ_j^{k+1} via (2.11) directly.

Now, After replacing (A.1) and (A.2) with (2.9) and (2.11), we present the distributed logistic algorithm (DLA), which is tabulated as Algorithm 1.

The operation of the algorithm is described as follows. In the $k + 1$ iteration, each computer j first updates $R_j(k + 1)$ via (2.9), then transmits local estimator $R_j(k + 1)$ to neighbors. Next, each computer j updates $\theta_j(k + 1)$ via (2.11) and transmits it to neighbors. Finally, computer j updates $V_j(k + 1)$ via (2.12) locally and $V_{ji}(k + 1)$ to neighbors.

Algorithm 1. Distributed Logistic Algorithm(DLA)

Input: $S_{all} = \{S_j\}_{j=1}^{J}$, preselect parameter c, step size α, maximum number of iterations K.

For all $j \in \mathcal{C}$, let $\theta_j(0) = (0,\ldots,0)^T$, $R_j(0) = (0,\ldots,0)^T$, $V_j(0) = (0,\ldots,0)^T$.

Locally run

for $k = 1$ to $K - 1$ do

  Update R_j^{k+1} via (2.9), then transmits $R_j(k+1)$ to neighbors.

  Update θ_j^{k+1} via (2.11), then transmits $\theta_j(k+1)$ to neighbors.

  Update V_{ji}^{k+1} via $V_{ji}^{k+1} = V_{ji}^k + c(R_j^{k+1} - \theta_i^{k+1})$, for all $i = 1,...,J$. (2.12)

  Computer j transmits $V_{ji}(k+1)$ to neighbors.

end for

Output: θ^*.

3   Experiments

In this section, we simulate distributed computer systems though a single computer and then apply our method to two application experiments.

3.1   Example 1 (Simulation Experiment)

For the classification model

$$p(y_i = 1|x_i) = \frac{1}{1 + \exp(-x_i\theta)}, \quad i = 1,\ldots,n.$$

where $x_i \in \mathbf{R}^{1\times p}$ is the input vector, $\theta \in \mathbf{R}^p$ is the coefficient vector, and n is sample number, $x_i \sim N(0,\Sigma)$, covariance matrix $\Sigma = \rho^{|i-j|}$, $1 \leq i,j \leq n$. If $p(y_i = 1|x_i) \geq 0.5$, let $y_i = 1$, else, $y_i = 0$. We simulated two cases, in order to show the effectiveness of our distributed logistic approach. In case 1, we separated the entire data into $J = 5$ parts, and chose matrix E as Fig. 1 and set preselected parameter c as 0.4.

$$E_1 = \begin{pmatrix} 1 & 0 & 1 & 0 & 0 \\ 0 & 1 & 1 & 0 & 1 \\ 1 & 1 & 1 & 1 & 0 \\ 0 & 0 & 1 & 1 & 0 \\ 1 & 1 & 0 & 0 & 1 \end{pmatrix} \qquad E_2 = \begin{pmatrix} 1 & 1 & 0 & 0 & 1 \\ 1 & 1 & 1 & 0 & 0 \\ 0 & 1 & 1 & 1 & 0 \\ 0 & 0 & 1 & 1 & 1 \\ 1 & 0 & 0 & 1 & 1 \end{pmatrix}$$

Fig. 1. Adjacency matrix E_1.   **Fig. 2.** Adjacency matrix E_2.

Case 1. In this case, we compare the classification results between non-distributed logistic method and DLA. We set $p = 8$, specifically, $\theta = (3, 1.5, 2, 1, 5, 1, 0.5, 2)$, $x_i = (x_{i1}, ..., x_{i8})$, and set covariance parameter ρ equal 0, 0.25, 0.5, 0.6 respectively. For different covariance parameter ρ, we simulated 100 datasets, each containing 150 observations. Each dataset was divided into two parts: a training set which contains 100 observations and a test set which contains 50 observations. Next, we run the non-distributed logistic method and distributed logistic method to the 100 sets. We set $J = 1$ to solve non-distributed logistic problem. The classification error (CE) and mean squared error (MSE) on test set of the 100 datasets were recorded. The results can be seen in Table 1.

From Table 1, we can see that when logistic method and DLA are applied, the CE of them are equal, and MSE of them have small difference, this is because the algorithms are truncated when the gradient is close to 0. And as ρ increase, the MSE and CE increase. This case shows that classification results of non-distributed logistic algorithm is same to distributed logistic algorithm.

Table 1. Results of the non-distributed logistic method and DLA in case 1.

Method	Logistic		DLA	
	MSE	CE	MSE	CE
$\rho = 0.00$	5.744×10^{-3}	0.35	5.740×10^{-3}	0.35
$\rho = 0.25$	8.322×10^{-3}	0.50	8.322×10^{-3}	0.50
$\rho = 0.50$	1.372×10^{-2}	0.92	1.372×10^{-2}	0.92
$\rho = 0.60$	1.418×10^{-2}	0.96	1.415×10^{-2}	0.96

Case 2. In this experiment, we show the computational efficiency of DLA. We set $n = 10^4, p = 10^3$, $n = 10^4, p = 10^4$, $n = 10^5, p = 10^3$ and $n = 10^4$, $p = 10^5$ respectively, and simulated 4 datasets. For each data set, we ran the distributed logistic algorithm with $J = 5$, $J = 10$ and $J = 100$. Specifically, we split the entire data into 5 groups, 10 groups and 100 groups. When $J = 5$, we chose adjacency matrix as Fig. 1. When $J = 10$ and $J = 100$, we set adjacency matrix as $1^{J \times J}$, which means each computer is connected with each other. To compare the computing time, we also run the glmfit function (in Matlab). The results of computing time on each computer of DLA and glmfit is shown in Table 2.

From Table 2, we can see that when $n = 10^4, p = 10^3$, $n = 10^4, p = 10^4$, and $n = 10^5, p = 10^3$, both glmfit and DLA can analyze data effectively. When $n = 10^4, p = 10^5$, glmfit failed, but DLA still works, and its computation costs was acceptable. It is noteworthy that as the number of computers increases, the computing time of each computer decreases, but it often incurs inevitable communication cost. This case shows that DLA is efficient when non-distributed algorithm fails.

Table 2. Computing time on each computer of DLA and glmfit on 4 datasets.

Data size	DLA ($J = 5$) time(s)	DLA ($J = 10$) time(s)	DLA ($J = 100$) time(s)	glmfit time(s)
$n = 10^4$, $p = 10^3$	91.32	45.01	4.98	114.02
$n = 10^4$, $p = 10^4$	969.87	471.24	44.35	6327.79
$n = 10^4$, $p = 10^5$	41932.65	20641.20	2196.63	More than 24 h
$n = 10^5$, $p = 10^3$	683.17	348.25	37.34	1268.64

3.2   Example 2 (Real Data)

To show our distributed method is suitable for real data analysis, in this experiment, we applied DLA to two publicly available datasets: the Colon dataset and the Adult dataset. When using distributed algorithm, we split the training data into $J = 5$ parts, let adjacency matrix E as Fig. 2. Preselected parameter c are chosen as 0.1 for DLA.

The Colon Dataset. The Colon dataset in [21] contains 62 samples, which including 24 normal tissues and 38 tumor tissues. Each sample contains 2000 gene expression values. We split the whole data into two parts: a training set with 40 samples and a test set with 22 samples. After preprocessing the data, we applied non-distributed logistic method and DLA to this problem.

The CE on test set of non-distributed logistic method and DLA both are 8. From this case, we can conclude that our distributed method is skilled in gene data analysis.

The Adult Dataset. The second dataset is the Adult dataset from [22]. The dataset contains survey information about 45222 individuals, including 15 variables such as age, gender, education level, nationality, job information and so on, and our task is to predict whether a person's income is below or exceeding 50k. In order to manipulate Adult dataset into a form suitable for classification, we convert each classification attribute to a binary vector. For example, we convert gender into dummy variables, namely there are two variables here, $(1, 0)$ represent male, and $(0, 1)$ represent female. After preprocessing, the dimension p of each sample is 105.

Now, we split the entire data into two parts: the training set with 30162 samples and the test set with 15060 observations. We applied DLA to this problem. The CE on test set of non-distributed logistic method and DLA both are 3276, error rates are 21.75%.

This case shows that our distributed approach is suitable for processing large-scale distributed storage data analysis.

4   Conclusion

In this paper, we studied the distributed logistic regression to process the separated large scale data which is stored in different linked computers. Based on the ADMM algorithm, we proposed the distributed logistic algorithm with well scalability and high availability. In each iteration of the distributed logistic algorithm, the computer updates the local estimators and interacts the local estimators with the neighbors simultaneously. Due to the decentralized property of the computer network, the proposed distributed logistic algorithm is robust. Moreover, the algorithm has the controllable communication cost. Then we prove the convergence of the method. The classification results of the our approach is same with the results of the non-distributed approaches. Several experiments prove the effectiveness of the distributed logistic algorithm in dealing with distributed storage data.

Appendix

ADMM algorithm

Consider the following optimization problem

$$\min_{R,\theta} \ f(\theta) + h(R) \tag{2.4}$$

$$s.t. \ MR + M'\theta = 0,$$

First, we form the quadratically augmented Lagrangian function

$$L(R,\theta,V) = f(\theta) + h(R) + \langle V, MR + M'\theta \rangle + \frac{c}{2}\|MR + M'\theta\|_2^2, \tag{2.5}$$

where $V := \{V_j\}_{j \in \mathcal{C}}$ is Lagrange multipliers, and $c > 0$ is a preselected penalty coefficient. Then we use ADMM algorithm to solve this problem.

ADMM algorithm entails three steps per iteration

Step 1: R updates,

$$R^{k+1} = \arg\min_{R} L(R,\theta^k,V^k). \tag{A.1}$$

Step 2: θ updates,

$$\theta^{k+1} = \arg\min_{\theta} L(R^{k+1},\theta,V^k). \tag{A.2}$$

Step 3: V updates,

$$V^{k+1} = V^k + c(MR^{k+1} + M'\theta^{k+1}). \tag{A.3}$$

For details, in iteration $k + 1$, we update R^{k+1} via minimum $L(R,\theta^k,V^k)$ with respect to R, update θ via minimum $L(R^{k+1},\theta,V^k)$ with respect to θ, and update Lagrange multiplier via $V^{k+1} = V^{(k)} + c(MR^{(k+1)} + M'\theta^{(k+1)})$, until the algorithm converges.

Proof of Theorem 2

Proof. Theorem 2 can be considered as a special case of [20]. This paper analyzed the convergence of ADMM for minimizing possible nonconvex objective problem. It's enough to show that our distributed optimization problem satisfies the convergence conditions A1-A5 in [20]. It's obvious that A1 holds. Since both M and M' are full rank, then A2 and A5 are established. And $f(\theta)$ is Lipschitz differentiable with constant L_f, A4 holds. For any $j \in J$, and any R_j, R'_j, (2.6) holds, thus A3 established. Then Theorem 2 has been proved.

References

1. Mcdonald, R., Mohri, M., Silberman, N., Walker, D., Mann, G.: Efficient large-scale distributed training of conditional maximum entropy models. Advances in Neural Information Processing Systems, vol. 1, pp. 1231–1239. NIPS, La Jolla (2009)
2. McDonald, R., Hall, K., Mann, G.: Distributed training strategies for the structured perceptron. In: Human Language Technologies: The 2010 Annual Conference of the North American Chapter of the Association for Computational Linguistics, pp. 456–464. ACL, Los Angeles (2010)
3. Zhang, Y., Duchi, J., Wainwright, M.: Communication-efficient algorithms for statistical optimization. J. Mach. Learn. Res. **14**(1), 3321–3363 (2013)
4. Zhang, Y., Duchi, J., Wainwright, M.: Divide and conquer Kernel ridge regression: a distributed algorithm with minimax optimal rates. J. Mach. Learn. Res. **30**(1), 592–617 (2013)
5. Mateos, G., Bazerque, J., Giannakis, G.: Distributed sparse linear regression. IEEE Trans. Signal Process. **58**(10), 5262–5276 (2010)
6. Wang, P., Zhang, H., Liang, Y.: Model selection with distributed SCAD penalty. J. Appl. Stat. **45**(11), 1938–1955 (2017)
7. Wang J., Kolar M., Srebro N., Zhang T.: Efficient distributed learning with sparsity. In: International Conference on Machine Learning, vol. 70, pp. 3636–3645. PMLR, Sydney (2017)
8. Menendez, M.L., Pardo, L., Pardo, M.C.: Preliminary *phi*-divergence test estimators for linear restrictions in a logistic regression model. Stat. Pap. **50**(2), 277–300 (2009)
9. Pardo, J.A., Pardo, L., Pardo, M.C.: Minimum ϕ−divergence estimator in logistic regression models. Stat. Pap. **47**(1), 91–108 (2006)
10. Revan, O.M.: Iterative algorithms of biased estimation methods in binary logistic regression. Stat. Pap. **57**(4), 991–1016 (2016)
11. Lange, T., Mosler, K., Mozharovskyi, P.: Fast nonparametric classification based on data depth. Stat. Pap. **55**(1), 49–69 (2015)
12. Boyd, S., Parikh, N., Chu, E.: Distributed optimization and statistical learning via the alternating direction method of multipliers. Found. Trends Mach. Learn. **3**(1), 1–122 (2011)
13. Xie, P., Jin, K., Xing, E.: Distributed machine learning via sufficient factor broadcasting. Arxiv, http://arxiv.org/abs/1409.5705. Accessed 7 Sep 2015
14. Gopal, S., Yang, Y.: Distributed training of large-scale logistic models. In: Proceedings of the 30th International Conference on Machine Learning, vol. 28, pp. 289–297. PMLR, Atlanta (2013)

15. Peng, H., Liang, D., Choi, C.: Evaluating parallel logistic regression models. In: IEEE International Conference on Big Data, pp. 119–126. IEEE, Silicon Valley (2013)
16. Kang, D., Lim, W., Shin, K.: Data/feature distributed stochastic coordinate descent for logistic regression. In: Proceedings of the 23rd ACM International Conference on Conference on Information and Knowledge Management, pp. 1269–1278. ACM, Shanghai (2014)
17. Gabay, D., Mercier, B.: A dual algorithm for the solution of nonlinear variational problems via finite element approximation. Comput. Math. Appl. **2**(1), 17–40 (1976)
18. Glowinski, R., Marroco, A.: On the solution of a class of non linear Dirichlet problems by a penalty-duality method and finite elements of order one. In: Marchuk, G.I. (ed.) Optimization Techniques IFIP Technical Conference. LNCS, pp. 327–333. Springer, Berlin (1974). https://doi.org/10.1007/978-3-662-38527-2_45
19. Bertsekas, D., Tsitsiklis, J.: Parallel and Distributed Computation: Numerical Methods, 2nd edn. Athena Scientific, Belmont (1997)
20. Wang, Y., Yin, W., Zeng, J.: Global convergence of ADMM in nonconvex nonsmooth optimization. J. Sci. Comput. **78**(1), 29–63 (2019)
21. Alon, U., et al.: Broad patterns of gene expression revealed by clustering analysis of tumor and normal colon tissues probed by oligonucleotide arrays. Proc. Natl. Acad. Sci. U.S.A. **96**(12), 6745–6750 (1999)
22. Blake, C., Merz, C.: UCI repository of machine learning databases (1998). http://www.ics.uci.edu/~mlearn/MLRepository.html

Similarity Evaluation on Labeled Graphs via Hierarchical Core Decomposition

Deming Chu[1], Fan Zhang[2(✉)], and Jingjing Lin[1]

[1] East China Normal University, Shanghai, China
ned.deming.chu@gmail.com, jingle_1984@foxmail.com
[2] Guangzhou University, Guangzhou, China
fanzhang.cs@gmail.com

Abstract. Graph similarity is essential in network analysis and has been applied to various fields. In this paper, we study the graph similarity between labeled graphs, i.e., every vertex is assigned to a label. Since few methods take account of the structure of a graph and most existing methods cannot extend to massive graphs, we develop a novel graph similarity measure that overcomes the above limitations. Given two labeled graphs, our proposed method first utilizes the concept of k-core to organize the connected cohesive subgraphs of each graph in a tree-like hierarchy. Then, the graph similarity between them is computed from their tree-hierarchies. An efficient algorithm is also developed for the proposed measure. Extensive experiments are conducted on 6 public datasets, where our proposed algorithm successfully identifies similar graphs and extends to large-scale graphs.

Keywords: Labeled graph · Graph similarity · Core decomposition

1 Introduction

Graph structures model entities and their relationships in the real world. In recent years, scientific communities have been increasingly interested in graph structures and their analysis. The issue of evaluating the similarity between graphs, knowns as the graph similarity problem, lies in the core of various applications in graph analysis [6,17]. For example, we predict the functionality of an unknown chemical compound by a well-studied compound if they share a similar graph structure [6].

In a chemical compound, we take the atomic number of an atom as its label in the graph. The graph of this type is known as a labeled graph, in which every vertex is assigned to a label. Another scene of labeled graphs, which is more implicit, is the graphs with vertex correspondence. Specifically, graphs with the same vertex set and different structures. For instance, in a social network, by regarding each vertex itself as a label and computing the graph similarity for labeled graphs, we can analyze the evolution of social relationships throughout time. This application becomes more and more important due to the rapid advance of both data acquisition and data analysis.

H. Jin et al. (Eds.): BigData 2019, CCIS 1120, pp. 297–311, 2019.
https://doi.org/10.1007/978-981-15-1899-7_21

Maximum Common Subgraph and Graph Edit Distance are typical metrics that evaluate graph similarity on labeled graphs. Unfortunately, both of them are NP-Complete problems. Recently, many graph kernel based algorithms have been developed to tackle the measure of graph similarity [8,16]. However, the mentioned algorithms suffer from performance issues, therefore, cannot be applied to a large graph.

To develop a measure of graph similarity that is both scalable and effective, we utilize the model of k-core. The k-core of a graph is a maximal subgraph where every vertex has at least degree of k. Moreover, subgraphs given by the model of k-core, namely core decomposition, form a nested chain such that each subgraph is contained in the previous one: 0-core, 1-core, $\cdots$, k_{max}-core (k_{max} is the maximal possible k for k-core). Thus, k-core can decompose a graph into a hierarchy of subgraphs. k-core is also able to build a hierarchy for the cohesive connected components of a graph.

Our proposed method captures the similarity between labeled graphs in the hierarchies of connected cohesive subgraphs. Intuitively, for two labeled graphs, we first transform each graph into a tree where each node of the tree is a cohesive connected component detected by the model of k-core. Then, the graph similarity is computed from bottom to up according to the similarity of labels in the same layer of both trees.

The concept of k-core has following advantages. *(1) Efficiency:* the computation of core decomposition costs $O(m + n)$ time; *(2) Structural Infomation:* the existence of a vertex in a k-core ensures its participation in cohesive subgraphs; *(3) Extensive Application:* k-core related algorithms are applicable in extensive domains, e.g., graph visualization [1], hierarchical structure analysis [2], and influential spreader identification [13]. k-core is also effective for the network analysis in neuroscience [9], ecology [15], and biology [6];

Contribution. Our paper studies the classical graph similarity problem and develops a novel method for the graph similarity between labeled graphs. Our principal contributions are concluded as follows.

Novel Definition. Upon the hierarchy in core decomposition, we develop a novel measure of graph similarity on labeled graphs. To the best of our knowledge, this is the first attempt to apply the hierarchy of core decomposition to evaluate graph similarity. An existing measure of graph similarity [16] also adopts the model of k-core, which merely optimizes the traditional graph kernels for graph similarity via core decomposition.

Practical Algorithm. We develop a practical algorithm for our proposed definition of graph similarity. The algorithm runs in $O(n^2)$ time, and the experimental performance is very efficient.

Extensive Experiments. We perform experiments on 6 real-world networks, one of which has over ten million edges. The results are detailedly analyzed to present the real-world performance of our algorithm.

Organization. Section 2 formally defines the problem statement of our paper. Section 3 explains the basic idea of core decomposition. Section 4 discusses the hierarchy of connected components in a core decomposition. Based on the description in Sects. 3 and 4. Section 5 proposes a novel definition to the graph similarity measure between labeled graphs, along with a practical algorithm for the measure. Section 6 evaluates our proposed method on real-world networks. Section 7 concludes the related works, while Sect. 8 summarizes this paper.

2   Problem Statement

Table 1 lists the frequently used notations. Consider an undirected labeled graph $G = (V, E, L)$, with the vertex set V, the edge set E, and a labeling function $L : V \to \mathcal{L}$. The function L assigns to each vertex v a label $L(v)$ from a finite label set $\mathcal{L}$. Two vertices are considered the same entity if they share the same label.

Table 1. The summary of notations

Symbol	Description
$G = (V, E, L)$	An undirected and unweighted simple graph, where V is the vertex set, E is the edge set, and L is a label function
$n; m$	Number of vertices/edges in G, $n = |V|$, $m = |E|$
S	A subgraph of G
C	A connected subgraph of G
$N(v); N(v, S)$	The set of neighbors of v in G/S
$d(v); d(v, S)$	Degree of v in G/S
$P(S)$	Connected component decomposition of S
G_k	The k-core of G
$c(v)$	Coreness of v in G, $\max\{k \mid v \in G_k\}$
k_{max}	Maximal coreness of any vertex in G
H_k	The k-shell of G, $\{v \mid c(v) = k, v \in V\}$
P_k	The k-core component set of G, $P(G_k)$
$\mathcal{V}$	The core component set of G, $\mathcal{V} = \bigcup_{i=0}^{k_{max}} P(G_k) = \{C_1, C_2, \cdots, C_{|\mathcal{V}|}\}$
$\mathcal{T}$	The core tree of G

Given two labeled graphs G_1 and G_2, with the aid of the structural information provided by core decomposition, we aim at evaluating the extent to which G_1 is similar to G_2.

3   Core Decomposition

A labeled graph $S = (V_S, E_S, L_S)$ is the subgraph of $G = (V, E, L)$ if all of the following conditions are true: (i) $V_S \subseteq V$; (ii) $E_S \subseteq E$; (iii) $L_S = L$; Given two subgraphs S and S' of G, S is the subgraph of S', denoted by $S \subseteq S'$, if $V_S \subseteq V_{S'}$ and $E_S \subseteq E_{S'}$. Given a subgraph S and a vertex $v \in S$, the neighbors of v in S are denoted by $N(v, S)$, namely $\{u \mid (u, v) \in E_S\}$. The degree of v in S, that is $|N(v, S)|$, is represented by $d(v, S)$. We abbreviate $N(v, G)$ to $N(v)$ and $d(v, G)$ to $d(v)$ when the context is clear. A subgraph S of G is called a subgraph induced by V_S if $E_S = \{(u,v) \in E \mid u, v \in V_S\}$. A subgraph induced by any vertex set is also called an induced subgraph.

Definition 1 (k-core). *Given an integer k and a graph G, the k-core of G, represented by G_k, is a maximal subgraph of G such that each vertex $v \in G_k$ satisfies $d(v, G_k) \geq k$.*

The degeneracy of G, denoted by k_{max}, is the largest k such that there is a k-core in G. Moreover, for any positive k, the k-core is always the subgraph of the $(k-1)$-core, that is, subgraphs given by k-core model form a nested chain:

$$G_{k_{max}} \subseteq \cdots \subseteq G_1 \subseteq G_0 = G$$

Definition 2 (Coreness). *Given a graph G, for each vertex $v \in V$, the coreness of v, denoted by $c(v)$, is the largest k such that v is in a k-core, i.e., $c(v) = \max\{k \mid v \in G_k\}$.*

Algorithm 1. the core decomposition algorithm in [3]

    Input: $G = (V, E, L)$
    Output: core, the coreness of each vertex in G
1 **for** $i \leftarrow 1$ **to** n **do**
2      $\deg[i] \leftarrow d(i)$;

3 $k \leftarrow 1$;
4 Remain $\leftarrow V$;
5 **while** $|$Remain$| > 0$ **do**
6   **foreach** $v \in$ Remain **such that** $\deg[v] < k$ **do**
7   core$[v] \leftarrow k - 1$;
8   **for** $u \in N(v)$ **do**
9             $\deg[i] \leftarrow \deg[i] - 1$;
10   Remain $\leftarrow$ Remain $\setminus v$;
11      $k \leftarrow k + 1$;
12 **return** core;

Core decomposition is the problem of calculating coreness for every vertex in a given graph. Algorithm 1 presents a naïve $O(m)$ algorithm for core decomposition [3]. The algorithm repeatedly removes those vertices whose degree is less than k and updates the degree of their neighbors accordingly. If no vertex can be removed, then the remaining graph is a k-core.

Definition 3 (k-shell). *Given an integer k and a graph G, the k-shell of G, namely H_k, is the set of vertices with coreness of k, i.e., $H_k = \{v \mid c(v) = k, v \in V\}$.*

The k-shell of a graph is exactly the difference between the k-core and the $(k+1)$-core, i.e., $H_k = G_k \setminus G_{k+1}$. More importantly, a k-core equals the union of all vertices whose coreness is no less than k, that is, $G_k = \bigcup_{i=k}^{k_{max}} H_i$. Given any k, the k-core can be constructed from the coreness array.

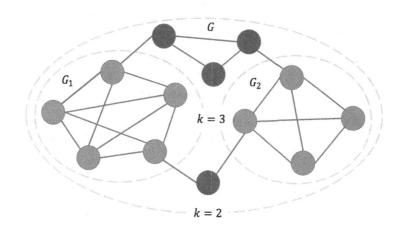

Fig. 1. $G_1 \cup G_2$ is 3-core; G is 2-core;

Example 1. Fig. 1 illustrates the model of k-core. The whole graph is a 2-core (and also a 0-core and a 1-core), while $G_1 \cup G_2$ forms a 3-core. The 3-shell of the graph is $H_3 = V(G_1) \cup V(G_2)$ and the 2-shell is $H_2 = V(G) \setminus H_3$. The coreness of each vertex can be learned from the k-shell of the graph, e.g., vertices in H_2 have coreness of 2.

4 Hierarchy of Connected Components in a Core Decomposition

Core decomposition models the hierarchy of a graph. Furthermore, the connected components detected by core decomposition also form a tree-like hierarchical structure, namely **core tree**. In the following paragraphs, we define a series of concepts so as to explain the idea of core tree.

Definition 4 (k-core Component Set). *Given an integer k and a graph G, the k-core component set of G is the set of connected components of the k-core, i.e., $P_k = P(G_k)$, where $P(S)$ is the connected components of a graph S.*

Definition 5 (k-core Component). *The k-core component of a graph is the connected component in the k-core component set of the graph.*

Property 1 (Hierarchy of k-core Component). The hierarchy of k-core component builds on:

- (NESTING) Each k-core component is the subgraph of one and only one $(k-1)$-core component.
- (DISJOINTNESS) The k-core components are disjoint from each other.

Property 1 is an extension to the nested chain of k-core, that is, there is also a hierarchy in the connected components given by a core decomposition. The core component set of a graph is defined as follows.

Definition 6 (Core Component Set). *Given a graph G, the core component set of G, namely $\mathcal{V} = \bigcup_{i=0}^{k_{max}} P(P_k) = \{C_1, C_2, \cdots, C_{|\mathcal{V}|}\}$, is the set of distinct connected components detected in a core decomposition.*

Definition 7 (Core Component). *Each element in the core component set of a graph is a core component of the graph. The coreness of a core component S is the largest k such that S is in a k-core of the graph.*

Example 2. In Fig. 1, G_1 and G_2 are two 3-core components, and G is a 2-core (also 0-core and 1-core) component. The core component set of the graph is $\mathcal{V} = \{G, G_1, G_2\}$. The coreness of G_1 and G_2 are 3, while G has coreness of 2.

Consider each core component as a hyper-vertex in a new graph, then the nesting property in Property 1 suggests that each hyper-vertex has exactly one father, while the disjointness in Property 1 indicates that each hyper-vertex has disjoint children. To conclude, the core component set of a graph is organized in a tree structure.

Definition 8 (Core Tree). *Consider a rooted tree, the vertices of which are core components of a graph, the root of which is the only 0-core component (the whole graph), the edges of which derived from the subgraph relationship in the nesting property of Property 1.*

Implementation of Core Tree. The definition of core tree in Fig. 2 is inefficient in terms of space usage. To compress redundant space in the implementation of core tree, we adopt the following strategies:

- Each core component appears only once in the core tree, even if it satisfies the condition of k-core component for multiple ks.
- In the core tree, a non-leaf hyper-vertex only records vertices that are not recorded in the children.

In the above implementation, a non-leaf core component C is compressed and can be reconstructed from the vertices of its offsprings in $O(|C|)$ time by DFS algorithm. The space of this implementation is bounded by $O(n)$.

Algorithm 2. construction of core tree

```
1  Function build_tree(G)
      Input: G = (V, E, L)
      Output: T, the core tree of G
2     T ← [ ];
3     V ← ∅;
4     for k ← k_max to 0 do
5        foreach C ∈ P_k do
6           if C ∉ V then
7              V ← V ∪ C;
8              tv ←a new vertex in core tree;
9              foreach C' ∈ P_{k+1} and C' ⊆ C do
10                tv.son.append(element of T whose get_vertex() is C');
11             tv.delta ← C \ ⋃_{s∈tv.son} T[s].get_vertex();

12    return T;
```

Specification of Core Tree. The core tree of a graph, denoted by T, is an array of core components. The elements in T are sorted in descending order of coreness, that is, in a bottom-up order of a core tree. For the $(i+1)$-th core component in T, namely $T[i]$: $T[i]$.son stores the position of its children in T; $T[i]$.delta stores vertices that are not contained in its children; $T[i]$.size() returns the number of vertices in $T[i]$, i.e., $|T[i]$.get_vertex()$|$; $T[i]$.get_vertex() returns the full set of vertices reconstructed from $T[i]$'s children;

Algorithm 2 displays a naïve algorithm for the construction of a core tree. The algorithm builds each core component from bottom to up. For each core component, we compress a core component based on the strategies mentioned above and build the link to children.

Example 3. The core tree of the graph in Fig. 1 is shown in Fig. 2, where the core component set $(G, G_1,$ and $G_2)$ form a tree structure. Note that G is also a 1-core component and a 0-core component, but only one hyper-vertex is constructed in the core tree. Moreover, a core component is not fully recorded in the core tree, instead, substructures that can be obtained from children is omitted (see the light color part in G).

5 The Core Hierarchy Based Graph Similarity

We first discuss the Modified Jaccard Similarity Coefficient in this section, then a novel measure of graph similarity on labeled graphs, and finally the algorithm to compute the proposed measure.

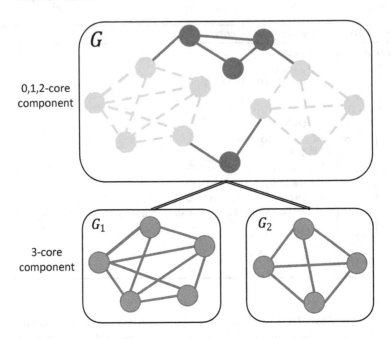

Fig. 2. Core Tree of Fig. 1 Tip: the light color part is not recorded;

5.1 Modified Jaccard Similarity Coefficient (MJSC)

The Jaccard Similarity Coefficient (JSC) of set x and y measures the extent to which set x is similar to set y. This concept is extensively adopted in information retrieval and data miningFlo.

$$Jaccard(x, y) = \frac{|x \cap y|}{|x \cup y|} = \frac{|x \cap y|}{|x| + |y| - |x \cap y|}$$

Our proposed method adopts a modified definition of JSC that computes the similarity of labels among two core components. In the similarity evaluation, as the number of occurrences of a label matters, we cannot directly utilize the original JSC that neglects the number of occurrences of an element.

In Algorithm 3, we compute the Modified Jaccard Similarity Coefficient (MJSC) of two core components. Algorithm 3 first collects the sorted sequence of labels for $T_1[i]$ and $T_2[j]$ by sorted_labels. Then, the algorithm keeps track of indexes ix, iy and counts the number of matches. For each label, we compute its number of occurrences in x and y, and regard the lower one as a contribution to the total matches. Finally, the algorithm computes the MJSC of $T_1[i]$ and $T_2[j]$ in similar way to JSC.

Algorithm 3. compute the MJSC of two core components

Input: core components $T_1[i]$ and $T_2[j]$, graph G_1 and G_2, where T_1 is the core
tree of G_1 and T_2 is the core tree of G_2
Output: MJSC of $T_1[i]$ and $T_2[j]$

1 **Function** sorted_labels($G, T[i]$)
2      $res \leftarrow [\ ]$;
3   **foreach** $v \in T[i]$.get_vertex() **do**
4         $\lfloor$ res.append($G.L(v)$);
5   **return** sorted(res);

6 **Function** MJSC($G_1, T_1[i], G_2, T_2[j]$)
7      $x \leftarrow$ sorted_labels($G_1, T_1[i]$);
8      $y \leftarrow$ sorted_labels($G_2, T_2[j]$);
9   **if** $|x| = 0$ **and** $|y| = 0$ **then return** 1.0;
10    $ix \leftarrow 0, iy \leftarrow 0$;
11    $match \leftarrow 0$;
12   **while** $ix < |x|$ **and** $iy < |y|$ **do**
13   **if** $x[ix] < y[iy]$ **then** $ix \leftarrow ix + 1$;
14   **else if** $x[ix] > y[iy]$ **then** $iy \leftarrow iy + 1$;
15   **else** $match \leftarrow match + 1, ix \leftarrow ix + 1, iy \leftarrow iy + 1$;

16   **return** $match/(|x| + |y| - match)$;

5.2 Graph Similarity Definition and Computation

Given two labeled graphs G_1 and G_2, and their respective core tree T_1 and T_2.
We define an array of similarity Sim whose length is equal to T_1.

$$
Sim[i] = \begin{cases} \max\limits_{c(T_1[i])=c(T_2[j])} MJSC(G_1, T_1[i], G_2, T_2[j]), & \text{if } T_1[i] \text{ is leaf} \\[2mm] \sum\limits_{s \in T_1[i].son} Sim[s] \times \dfrac{T_1[s].\text{size}()}{\sum_{c \in T_1[i].son} T_1[c].\text{size}()}, & \text{if } T_1[i] \text{ is non-leaf} \end{cases}
$$

In case that $T_1[i]$ is leaf and has coreness of k, we compute the MJSC between
$T_1[i]$ and each k-core component in T_2, then select the largest one as $Sim[i]$. In
case that $T_1[i]$ is non-leaf, we gather similarity from the children of $T_1[i]$ in
proportion to the number of vertices.

Given that the length of T_1 and T_2 equals n_1 and n_2 respectively. The root
of T_1 is $T_1[n_1 - 1]$, and the graph similar of G_1 and G_2 is the similarity stored
at the root:

$$
GS(T_1, T_2) = Sim[n_1 - 1]
$$

Note that our proposed measure of graph similarity is not symmetric, that
is, $GS(T_1, T_2)$ is not necessarily equal to $GS(T_2, T_1)$.

Algorithm 4. compute our proposed graph similarity

Input: graph G_1 and G_2, core tree T_1 and T_2, where T_1 is the core tree of G_1 and T_2 is the core tree of G_2

Output: graph similarity of G_1 and G_2

1 remove core components in T_1 whose coreness is greater than the k_{max} of G_2;

2 $n_1 \leftarrow |T_1|, n_2 \leftarrow |T_2|$;

3 $Sim \leftarrow [0.0, \dots, 0.0]$;

4 **for** $i \leftarrow 0$ **to** $n_1 - 1$ **do**

5   **if** $T_1[i]$ *is leaf* **then**

6   **foreach** $T_2[j]$ **such that** $c(T_1[i]) = c(T_2[j])$ **do**

7             $Sim[i] \leftarrow \max\{Sim[i], \texttt{MJSC}(G_1, T_1[i], G_2, T_2[j])\}$;

8   **else**

9         $sum \leftarrow \sum_{s \in T_1[i].son} T_1[s].\texttt{size}()$;

10   **foreach** $s \in T_1[i].son$ **do**

11           $Sim[i] \leftarrow Sim[i] + Sim[s] \times \frac{T_1[s].\texttt{size}()}{sum}$;

12 **return** $Sim[n_1 - 1]$;

Algorithm 4 presents the algorithm for the mentioned graph similarity measure. Given two labeled graphs, first, we construct core tree for each. Then, on the basis of Algorithm 2, Algorithm 3, and the definition above, Algorithm 4 computes the graph similarity of two graphs. Algorithm 1 is intended to prohibit the cases that no $T_2[j]$ has the same coreness as $T_1[i]$. The algorithm costs $O(n^2)$ time.

6   Experiment

In this section, we report the real-world performance of our proposed method. Our experiments are performed on a CentOS Linux server (Release 7.5.1804) with ten Quad-Core Intel Xeon CPU (E5-2630 v4 @ 2.20GHz), and 128G memory. We implement the algorithms in C++ and compile the code by GCC (7.3.0) under O3 optimization.

Table 2. Datasets

Dataset	n	m	$\bar{d}$	k_{max}	Description
PubChem	≤ 120	≤ 120	NA	NA	Chemical compounds, No. 1–No. 100
Astro-Ph	18,772	198,110	21.11	43	Collaboration network
DBLP	317,080	1,049,866	6.62	113	Collaboration network
As-Skitter	1,696,415	11,095,298	13.08	111	Internet topology graph
LiveJournal	3,997,962	34,681,189	17.35	360	Social network
Orkut	3,072,441	117,185,083	76.28	253	Social network

Datasets. We utilize 6 public datasets in the experiments, all of which are from SNAP [1] except PubChem[2]. The details of the networks are presented in Table 2 in ascending order of the number of edges, in which $\bar{d}$ is the average degree and k_{max} is the maximal coreness. As we collect 100 graphs from PubChem, we only report the range of n and m here.

6.1   Case Study: DBLP Multi-layered Graph

DBLP[3] provides public bibliographical data on major publications and proceedings in computer science. The details of all publications are also made public in the form of XML[4], thus, we can easily analyze and process the collaboration relationship in DBLP via XML.

From the XML provided by DBLP, we extract a 10-layer graph whose layers have vertex correspondence to each other. Each layer in the graph corresponds to the publications in one of the ten consecutive years, and reflects the cooperation (edge) of the authors (vertex and label) in that year. Note that a common author list is shared by these layers, therefore, the author list can be regarded as a label list. Each layer in this 10-layer graph is assigned to an ID in ascending order of the corresponding year.

Table 3. Pairwise graph similarity on DBLP multi-layered graph

ID1	ID2									
	1	2	3	4	5	6	7	8	9	10
1	1	0.092	0.0391	0.0308	0.0225	0.0177	0.0286	0.0149	0.0127	0.0116
2	0.0872	1	0.0599	0.035	0.0417	0.0238	0.0284	0.0165	0.0137	0.0121
3	0.0474	0.0393	1	0.0577	0.0322	0.0256	0.0205	0.0193	0.0139	0.0127
4	0.0207	0.0352	0.0449	1	0.0451	0.0282	0.0218	0.0195	0.0148	0.0127
5	0.0229	0.0605	0.0274	0.0531	1	0.0482	0.0289	0.0213	0.0174	0.0152
6	0.0123	0.0156	0.0244	0.0307	0.0436	1	0.0433	0.0308	0.0171	0.0165
7	0.0103	0.0129	0.0142	0.0194	0.0269	0.0331	1	0.0441	0.0286	0.0166
8	0.0082	0.0113	0.0105	0.0136	0.0195	0.0253	0.0365	1	0.0403	0.0231
9	0.0083	0.0081	0.0088	0.0107	0.0125	0.0144	0.0255	0.0304	1	0.0285
10	0.0079	0.0062	0.0066	0.0077	0.009	0.0118	0.0126	0.0198	0.0303	1

Table 3 reports the experimental results on DBLP multi-layered graph, where each element of Table 3 is $GS(G_{ID1}, G_{ID2})$. As shown in Table 3, two graphs have higher similarity if they have a closer ID (corresponding year). The patterns of graph similarity in Table 3 exactly matches our expectation on the evolution of

---

[1] http://snap.stanford.edu.
[2] https://pubchem.ncbi.nlm.nih.gov/.
[3] https://dblp.org/.
[4] https://dblp.org/xml/release/.

scientific collaboration. Specifically, the greater the time interval, the greater the change of collaboration.

6.2   Case Study: PubChem

PubChem[5] is a freely accessible database on chemical information. We have downloaded the detailed structure of No. 1–No. 100 compounds in PubChem. Each chemical compound here is a labeled graph, because each atom (vertex) in the compound is assigned to an atomic number (label), and is linked to each other by chemical bonds (edge).

We compute the graph similarity of every pair in No. 1–No. 100 compounds of PubChem, and sort pairs in descending order of similarity. In Figs. 3 and 4, we visualize two groups of compounds. The members of each group are identified as highly similar to each other by our proposed method.

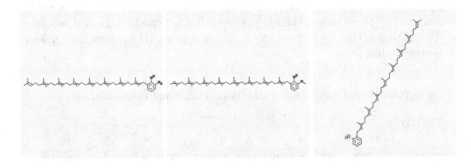

Fig. 3. Group-A: No. 55, No. 56, No. 57 in PubChem

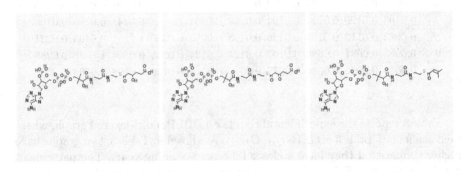

Fig. 4. Group-B: No. 90, No. 94, No. 99 in PubChem

---

[5] https://pubchem.ncbi.nlm.nih.gov/.

6.3   Running Time

Table 4 shows the efficiency of our proposed algorithm. Even though these datasets are not labeled graphs originally, we regard the label of each vertex as the same so as to measure the efficiency of our method on large-scale networks. In Table 4, the evaluation time means the time cost of computing our proposed graph similarity, and the total time considers both the construction of core tree and the evaluation of graph similarity.

In our proposed method, the cost of construction dominates the overall cost. The evaluation time relates to the structure instead of the size of the graph, while the cost of core tree construction increases with respect to graph size.

Table 4. Performance of running time

Dataset	Astro-Ph	DBLP	As-Skitter	LiveJournal	Orkut
Evaluation time(s)	0.01	0.07	2.26	1.00	0.01
Total time(s)	0.11	0.52	8.09	34.82	254.12

7   Related Work

Works related to this paper are grouped into:

Algorithms for k-core Analysis. [18] introduced the idea of k-core to network analysis. [3] proposes an algorithm for core decomposition that runs in $O(m)$ time. [7,20] explore the core decomposition in which graphs cannot fit into memory. [14] presents distributed core decomposition. [23] presents an algorithm for efficient maintaining of k-core in a dynamic graph. [5] proposes the anchored k-core to avoid the unraveling of a graph, while [21] develops an practical solution to this problem. (k,r)-core [22] considers both the engagement and similarity of a community. [12] extends k-core to temporal graphs. [11] takes into consideration the influence of a subgraph, and [10] extends the influential k-core to multi dimensions.

Applications of k-core Analysis. [1] visualizes a graph via the hierarchy of k-core. [19] discovers three meaningful patterns of k-core and identify their applications in anomaly detection. [2] analyzes the hierarchical structure of Internet graphs by core decomposition. [13] utilizes k-core to identifie influential spreader in a network. Moreover, k-core is identified as both efficient and effective in other fields of science, such as neuroscience [9], ecology [15], and biology [6].

Graph Similarity Measurement. Traditional measures for graph similarity include Maximum Common Subgraph and Graph Edit Distance. Recent works on graph similarity are naturally organized into two classes: *(1) Known Vertex Correspondence:* [8] proposes a similarity approach based on graph features. [17] describes five graph similarity functions for web graphs. *(2) Unknown Vertex Correspondence:* [16] proposes a core-based graph kernel for graph similarity. [4] identifies the discontinuity of social networks by a graph similarity method.

8 Conclusion

Our paper studies the graph similarity for labeled graphs. We develop a novel graph similarity measure which takes advantage of both the efficiency and the structural information provided by core decomposition. Moreover, we propose a practical algorithm for the proposed measure. Given the core trees of two labeled graphs, our algorithm runs in $O(n^2)$ time. We perform experiments on 6 public datasets, one of which has over ten million edges. The experiments on real-world datasets illustrate that our proposed method can efficiently identify similar graphs.

References

1. Alvarez-Hamelin, J.I., Dall'Asta, L., Barrat, A., Vespignani, A.: Large scale networks fingerprinting and visualization using the k-core decomposition. In: Advances in Neural Information Processing Systems, pp. 41–50 (2006)
2. Alvarez-Hamelin, J.I., Dall'Asta, L., Barrat, A., Vespignani, A.: K-core decomposition of internet graphs: hierarchies, self-similarity and measurement biases. Netw. Heterogen. Media 3(2), 371–393 (2008)
3. Batagelj, V., Zaversnik, M.: An O(m) algorithm for cores decomposition of networks. arXiv preprint cs/0310049 (2003)
4. Berlingerio, M., Koutra, D., Eliassi-Rad, T., Faloutsos, C.: NetSimile: a scalable approach to size-independent network similarity. arXiv preprint arXiv:1209.2684 (2012)
5. Bhawalkar, K., Kleinberg, J., Lewi, K., Roughgarden, T., Sharma, A.: Preventing unraveling in social networks: the anchored k-core problem. SIAM J. Discrete Math. 29(3), 1452–1475 (2015)
6. Borgwardt, K.M., Ong, C.S., Schönauer, S., Vishwanathan, S., Smola, A.J., Kriegel, H.P.: Protein function prediction via graph kernels. Bioinformatics 21(suppl-1), i47–i56 (2005)
7. Cheng, J., Ke, Y., Chu, S., Özsu, M.T.: Efficient core decomposition in massive networks, pp. 51–62, May 2011. https://doi.org/10.1109/ICDE.2011.5767911
8. Koutra, D., Vogelstein, J.T., Faloutsos, C.: DELTACON: a principled massive-graph similarity function. In: Proceedings of the 2013 SIAM International Conference on Data Mining, pp. 162–170. SIAM (2013)
9. Lahav, N., Ksherim, B., Ben-Simon, E., Maron-Katz, A., Cohen, R., Havlin, S.: K-shell decomposition reveals hierarchical cortical organization of the human brain. New J. Phys. 18(8), 083013 (2016)
10. Li, R.H., et al.: Skyline community search in multi-valued networks. In: Proceedings of the 2018 International Conference on Management of Data, pp. 457–472. ACM (2018)
11. Li, R.H., Qin, L., Yu, J., Mao, R.: Influential community search in large networks. Proc. VLDB Endow. 8, 509–520 (2015). https://doi.org/10.14778/2735479.2735484
12. Li, R.H., Su, J., Qin, L., Yu, J.X., Dai, Q.: Persistent community search in temporal networks. In: 2018 IEEE 34th International Conference on Data Engineering (ICDE), pp. 797–808. IEEE (2018)
13. Lü, L., Zhou, T., Zhang, Q.M., Stanley, H.E.: The H-index of a network node and its relation to degree and coreness. Nat. Commun. 7, 10168 (2016)

14. Montresor, A., De Pellegrini, F., Miorandi, D.: Distributed k-core decomposition. IEEE Trans. Parallel Distrib. Syst. **24**(2), 288–300 (2013)

15. Morone, F., Del Ferraro, G., Makse, H.A.: The k-core as a predictor of structural collapse in mutualistic ecosystems. Nat. Phys. **15**(1), 95 (2019)

16. Nikolentzos, G., Meladianos, P., Limnios, S., Vazirgiannis, M.: A degeneracy framework for graph similarity. In: IJCAI, pp. 2595–2601 (2018)

17. Papadimitriou, P., Dasdan, A., Garcia-Molina, H.: Web graph similarity for anomaly detection. J. Internet Serv. Appl. **1**(1), 19–30 (2010)

18. Seidman, S.B.: Network structure and minimum degree. Soc. Netw. **5**(3), 269–287 (1983)

19. Shin, K., Eliassi-Rad, T., Faloutsos, C.: CoreScope: graph mining using k-core analysis–patterns, anomalies and algorithms. In: 2016 IEEE 16th International Conference on Data Mining (ICDM), pp. 469–478. IEEE (2016)

20. Wen, D., Qin, L., Zhang, Y., Lin, X., Yu, J.X.: I/O efficient core graph decomposition at web scale. In: 2016 IEEE 32nd International Conference on Data Engineering (ICDE), pp. 133–144. IEEE (2016)

21. Zhang, F., Zhang, W., Zhang, Y., Qin, L., Lin, X.: OLAK: an efficient algorithm to prevent unraveling in social networks. Proc. VLDB Endow. **10**(6), 649–660 (2017)

22. Zhang, F., Zhang, Y., Qin, L., Zhang, W., Lin, X.: When engagement meets similarity: efficient (k, r)-core computation on social networks. Proc. VLDB Endow. **10**(10), 998–1009 (2017)

23. Zhang, Y., Yu, J.X., Zhang, Y., Qin, L.: A fast order-based approach for core maintenance. In: 2017 IEEE 33rd International Conference on Data Engineering (ICDE), pp. 337–348. IEEE (2017)

Weighted Multi-label Learning
with Rank Preservation

Chong Sun[1,2], Weiyu Zhou[1,2(✉)], Zhongshan Song[1,2], Fan Yin[1,2],
Lei Zhang[1,2], and Jianquan Bi[1,2]

[1] School of Computer Science, South-Central University for Nationalities,
Wuhan 430074, China
2017110236@mail.scuec.edu.cn
[2] Hubei Manufacturing Enterprise Intelligent Management Engineering
Technology Research Center, Wuhan 430074, China

Abstract. As one of the central topic in the field of machine learning, multi-label learning gets widely applied in real life. The classical algorithm does not consider the relation of rank and weight between labels simultaneously, while correlation between labels own a certain impact on the quality of classification models, which makes the algorithm unable to be applied in some scenarios and the accuracy of the model is affected. To solve this problem, a new algorithm named weighted multi-label learning with rank preservation (abbrev. WMR) is proposed. WMR extends and optimizes the SVM-based multi-label learning algorithm by introducing two kinds of label pairs, which is called "related-unrelated" and "related-related" label pairs, to measure the rank and weight between labels. The experiment is based on the real datasets and compared to the RankSVM algorithm, and the experimental results show that WMR mines the correlation between labels fully and improve the quality of the classification model effectively.

Keywords: Multi-label learning · Correlation between labels · Weight · Rank

1 Introduction

The concept of multi-label learning comes from the ambiguity problem in document classification [1]. Traditional algorithms assume that each object has unique semantics, i.e. each object matches only one label. However, the "ambiguity" of objects is ubiquitous, and an object can be associated with multiple labels. Traditional multi-label learning algorithms are Binary Relevance [2], Label Powerset [3], boosting-based method [4], KNN-based method [5], etc.

In practical applications, the relative weight between labels is also one of the factors affecting the accuracy of the classification. As shown below, a news article on the Yahoo News page called "David Beckham to face trial over speeding offense", which can be considered as belonging to categories "sports news", "social news", or even "entertainment news", but when users searching for this news, it tends to search under the category of "sports news", some news websites may attribute it to "sports news" exclusively.

© Springer Nature Singapore Pte Ltd. 2019
H. Jin et al. (Eds.): BigData 2019, CCIS 1120, pp. 312–324, 2019.
https://doi.org/10.1007/978-981-15-1899-7_22

As the above example shows, each label in the label set owns a different degree for the description of the object, i.e. each label have a predetermined weight. Traditional algorithms have not considered the relation of rank and weight between labels at the same time, so that the accuracy of the model calculation is affected.

To solve the above problems, this paper proposes a weighted multi-label learning with Rank Preservation algorithm (WMR). WMR aims at the correlation between labels, using "related-unrelated", "related-related" label pairs to consider the relation of rank and weight between labels separately. For the "related-unrelated" label pairs, defining the rank of labels based on the distance from the sample points to the hyperplane. For the "related-related" label pairs, defining and calculating the weight between labels to decide which label is more important. WMR follows the SVM-like reasoning process, and calculates the weight probability between labels, such that the classifier gives the label sequence accurately.

This paper is organized as follows: the related work is introduced in Sect. 2, Sect. 3 presents the problem description and formal definition, the loss function and how to construct a classifier model is described in Sect. 4 and 5, experimental results are discussed in Sects. 6 and 7 is conclusion and future work (Fig. 1).

Fig. 1. News Page

2 Related Work

The classical algorithms for multi-label learning are mainly divided into two types: data transformation algorithm and method adaptation algorithm [6]. The data transformation algorithm converts the sample from multi-label to multiple single labels and learns with existing single label algorithms. Binary Relevance algorithm is to decompose the multi-label problem into multiple independent two-class problems. Each label is considered as a binary classifier, and each classifier is trained with the entire training data set. The optimized algorithm based on Binary Relevance is to add the copy and selection method, considering the probability factor of the label [7]. Classifier Chains algorithm [8] is a further improvement of Binary Relevance, which consider the correlation between labels. The basic idea is to transform the multi-label problem into a binary chain. Each sub-binary classifier is constructed by the previous prediction result. Based on the improved algorithms, Label Powerset, Random k-labelsets [9], Pruned Problem Transformation [10], Calibrated Label Ranking [11] algorithms consider the rank between labels. Another type, the method adaptation algorithm, adapting single label algorithms to multi-label problem. The Adaboost.MH and Adaboost.MR algorithms [4] are two algorithms that extended on Adaboost to minimize Hamming loss and ranking loss. The multi-label algorithm based on support vector machine [12] regards each label in the training set as an SVM classifier. The probabilistic generative model [13] generates different words for each label in the training set, which construct the label set, with the distribution learning of each mixed phrase. The conditional random fields method [14] is a graph model using parameters to present the co-occurrence of labels. In recent years, deep learning has been widely applied to multi-label problems. BP-MLL algorithm [15] is the first neural network-based multi-label learning algorithm, i.e. multi-label back propagation, which replace original loss function with a new error function to capture the features. Then, various scholars proposed multi-label algorithms: Boltzmann machine-based fully connected neural network, the combination of CNN and RNN structure [16–19].

In the classical algorithms, some consider the rank between labels, which improves the accuracy of the classifier. However, the weight between labels is not considered simultaneously, algorithms are limited to be applied in some scenarios. Therefore, this paper proposes a weighted multi-label learning with rank preservation algorithm, WMR calculates two types of label pairs separately to mine the relation of rank and weight between labels, which improves the quality of generated model effectively.

3   Problem Description and Formal Definition

3.1   Symbol Convention

Let $\mathcal{X} = \mathbb{R}^d$ denote the dimension space of instance, $y = \{1, \ldots, Q\}$ is the set of labels. Given a data set $D = \{(x_1, Y_1), (x_2, Y_2), \ldots, (x_m, Y_m)\}$, where, $|D| = m$, $x_i \in \mathcal{X}$, $Y_i \subset 2^y$. Multi-label classifier $h : \mathcal{X} \to 2^y$.

3.2   Formalization of the Problem

Assume that for any sample (x_i, Y_i) in D, the same probability distribution is obeyed, there exists a real-valued function $f_k = <w_k, x> + b_k$ $(k \in y)$. The goal of WMR is to output the classifier h, i.e. to learn Q decision functions $f_n(n \in y)$, $h = (f_1, f_2, \ldots, f_Q)$

from D. WMR finds the optimal solution for the loss function of a set of decision functions in the classifier.

4   Loss Function

Definition 1. (related-related) label pair. For sample (x_i, Y_i), label $R \in Y_i$, $R' \in Y_i$, then label pair formed by R and R' is (related-related) label pair, abbreviated as $(R - R')$.

Definition 2. (related-unrelated) label pair. For sample (x_i, Y_i), $\bar{Y}_i = y - Y_i$, label $R \in Y_i$, $U \in \bar{Y}_i$, then label pair formed by R and U is (related-unrelated) label pair, abbreviated as $(R - U)$.

4.1   Rank Between Labels

For sample (x_i, Y_i), WMR trains a set of decision functions f to form a classifier h, i.e. there exists function $f_k = \,<w_k, x_i> + b_k$, $k \in Y_i$, $w_k \in \mathbb{R}^d$ is the weight of k_{th} label, $b_k \in \mathbb{R}^d$ is the bias of k_{th} label. Decision model takes the Ranking Loss as the loss function, which is used to investigate the occurrence of sorting errors in the label sequence. The loss function $R(f)$ formula is written as follows.

$$\frac{1}{p}\sum_{i=1}^{p}\frac{|R(x_i)|}{|Y_i||\bar{Y}_i|} \tag{1}$$

where $R(x_i)$ is: $R(x_i) = \{(y_1, y_2)|f(x_i, y_1) \le f(x_i, y_2), (y_1, y_2) \in Y_i \times \bar{Y}_i\}$

4.2   Weight Between Labels

For sample (x_i, Y_i), there exists label pair $(R - U)$, with $f_R(x) = \,<w_R, x_i> + b_R$, $f_{R'}(x) = \,<w_{R'}, x_i> + b_{R'}$.

Suppose that in the label set Y, the label R is placed before R', then the weight of R should be larger than the weight of R' : $Weight_R > Weight_{R'}$. The probability that the label R is more relevant than the R' is represented by the weight probability $P_{R,R'}(f)$. Let the true weight probability between the label pairs be $\bar{P}_{R,R'}$, then the loss function $C_{R,R'}(f)$ that formed by cross entropy is written as follows.

$$C_{R,R'}(f) = -\bar{P}_{R,R'}\ln P_{R,R'}(f) - (1 - \bar{P}_{R,R'})\ln(1 - P_{R,R'}(f)) \tag{2}$$

5   Construct a Classification Model

WMR trains $(R - U)$ and $(R - R')$ label pair on each sample respectively. On the $(R - U)$ label pair, WMR defines the hyperplane, and construct a classifier based on the distance between the sample points and the plane, with minimizing the Ranking

Loss. On the $(R - R')$ label pair, WMR defines the weight probability between labels, and calculates the weight probability to determine the weight sequence of labels.

5.1 Linear Classifier of $(R - U)$ Label Pairs

Definition 3. decision boundary. Decision boundary of $(R - U)$ label pair is to divide the vector space into two sets, and the decision boundary $g(x_i)$ is expressed as:

$$< w_R - w_U, x_i > + b_R - b_U = 0 \tag{3}$$

where $(R - U)$ is: $R \in Y_i$, $U \in \bar{Y}_i$, $(R, U) \in Y_i \times \bar{Y}_i$

Definition 4. margin. Margin of the decision boundary based on $(R - U)$ label pair, and the margin $m(x_i)$ is expressed as:

$$\min_{R \in Y, U \in \bar{Y}} \frac{< w_R - w_U, x_i > + b_R - b_U}{\| w_R - w_U \|} \tag{4}$$

Specific to the training set is written as follows:

$$\min_{(x,Y) \in D} \min_{R \in Y, U \in \bar{Y}} \frac{< w_R - w_U, x_i > + b_R - b_U}{\| w_R - w_U \|} \tag{5}$$

Assuming that the elements in the label set are ordered, then (5) should be positive, i.e. $< w_R - w_U, x_i > + b_R - b_U > 0$. Normalizing the parameter w yields $< w_R - w_U, x_i > + b_R - b_U \geq 1$, then the maximization margin can be defined as the following optimization problem.

$$\max_{w_j, j=1,\dots Q} \min_{(x,Y) \in D} \min_{R \in Y, U \in \bar{Y}} \frac{1}{\| w_R - w_U \|^2} \tag{6}$$

s.t. $< w_R - w_U, x_i > + b_R - b_U \geq 1$ where $x_i \in \mathcal{X}$, $(R, U) \in Y_i \times \bar{Y}_i$

Ideally, (6) can be further optimized as :

$$\min_{w_j, j=1,\dots Q} \max_{1 \leq R < U \leq Q} \| w_R - w_U \|^2 \tag{7}$$

s.t. $< w_R - w_U, x_i > + b_R - b_U \geq 1$ where $x_i \in \mathcal{X}$, $(R, U) \in Y_i \times \bar{Y}_i$

To simplify (7), the maximization operator is replaced by a summation operator. Suppose that $\sum_{R=1}^{Q} \| w_R \|^2 = 0$, then (7) can be rewritten as:

$$\min_{w_j, j=1,\dots Q} \sum_{R,U=1}^{Q} \| w_R \|^2 \tag{8}$$

s.t. $< w_R - w_U, x_i > + b_R - b_U \geq 1$ where $x_i \in \mathcal{X}$, $(R, U) \in Y_i \times \bar{Y}_i$

For this constrained optimization goal, a slack variable is introduced, whose set is $\{\xi_{iRU} | 1 \le i \le m, (R, U) \in Y_i \times \bar{Y}_i\}$, parameter I is the adjustment parameter, with the loss function (1), the quadratic problem under linear constraint is written as follows.

$$\min_{w_j, j=1,\ldots,Q} \sum_{R=1}^{Q} \|w_R\|^2 + I \sum_{i=1}^{m} \frac{1}{|Y_i||\bar{Y}_i|} \sum_{(R,U) \in Y_i \times \bar{Y}_i} \xi_{iRU} \tag{9}$$

$$\text{s.t.} \quad \begin{array}{l} <w_R - w_U, x_i> + b_R - b_U \ge 1 - \xi_{iRU}, (R, U) \in Y_i \times \bar{Y}_i \\ \xi_{iRU} \ge 0 \end{array}$$

The specific algorithm of the linear classification model of (x_i, Y_i) is as described in Algorithm 1.

Algorithm 1. Training classification model
Input: training set D and parameters I
Output: Q parameters of the classification model: w, b

Step1. Initialize $w, b \leftarrow \varnothing$

Step2. For each $(R - U)$, $(R, U) \in Y_i \times \bar{Y}_i$

Step2.1. Define the decision boundaries $g(x_i)$

Step2.2.   Define the margin $m(x_i)$, calculate $op\_m(x_i) = \max_{w_R, b_R} m(x_i)$

Step2.3.   Define the loss function $R(f)$, optimize $op\_R(f) = \min_{w_R, b_R} R(f)$

Step2.4. $op = op\_m(x_i) + op\_R(f)$, quadratic optimization

Step2.5. $(w, b) = (w, b) \cup (w_R, b_R)$

End for

Step3. Return w, b

5.2   Weight Probability of $(R - R')$ Label Pair

Definition 5. weight probability. Indicates the probability that the label R is more relevant than R', parameter δ is the operator presents the weight ratio of labels. The formula is as follows.

$$P_{R,R'}(f) = \frac{\exp[\delta(f_R(x_i) - f_{R'}(x_i))]}{1 + \exp(f_R(x_i) - f_{R'}(x_i))} \tag{10}$$

where $P_{R,R'}(f)$ is: $P_{R,R'}(f) \subset [0, 1]$

If $Weight_R > Weight_{R'}$, $P_{R,R'}(f)$ should be close to 1 infinitely, if $Weight_R < Weight_{R'}$, $P_{R,R'}(f)$ approaches 0 infinitely, and if $Weight_R = Weight_{R'}$, $P_{R,R'}(f)$ equals 0.5 approximately.

The actual weight probability of (R, R') label pair is $\bar{P}_{R,R'}$, which calculates as:

$$\bar{P}_{R,R'} = \frac{1 + W_{R,R'}}{2} \tag{11}$$

where $W_{R,R'}$ is $W_{R,R'} = \begin{cases} 1, Weight_R > Weight_{R'} \\ 0, Weight_R = Weight_{R'} \\ -1, Weight_R < Weight_{R'} \end{cases}$

Then, the sum of loss function (2) is minimized by gradient descent method to obtain further optimized classification model parameters:

$$\min_{w_j, j=1,\ldots,Q} \sum_{R,R'=1}^{Q} C_{R,R'}(f) \tag{12}$$

Optimize the linear classification model parameters w of (x_i, Y_i) is described in Algorithm 2.

Algorithm 2. Optimize classification model parameters
Input: training set D and parameters w, δ
Output: Q parameters of the classification model: w
Step1. Foe each $(R - R')$, $R, R' \in Y_i$
  Step1.1. Define the weight probability $P_{R,R'}(f)$
  Step1.2. Calculate the true weight probability $\bar{P}_{R,R'}$
  Step1.3. Define the loss function $C_{R,R'}(f)$, calculate $w_k = \min_{w_R} C_{R,R'}(f)$

End for
Step2. Return w

5.3   Prediction of Label Sequence

The above introduces how to build a classification model. The following describes how to determine the size of the label sequence to output the final results.

Definition 6. size of label sequence. The size of label sequence $S(x)$ can be determined by the following rules:

$$S(x) = |\{f_k(x) > t(x)\}| \tag{13}$$

where $t(x)$ is the corresponding threshold function.

Let $t(x) = <w^*, f^*(x)> + b^*$, which is linear function, $f^*(x) = (f(x, y_1),$ $f(x, y_2), \ldots, f(x, y_m))^T \in \mathbb{R}^d$, $w^* \in \mathbb{R}^d$ is the weight vector, $b^* \in \mathbb{R}^d$ is the bias. Then the solution of $S(x)$ is the process of minimizing the loss function:

$$\min_{(w^*, b^*)} \sum_{i=1}^{m} (<w^*, f^*(x_i)> + b^* - s(x_i))^2 \tag{14}$$

where $s(x_i)$ is: $s(x_i) = \arg\min_{u \in R}(|\{y_j | y_j \in Y_i, f(x_i, y_j) \le u)\}| + |\{y_k | y_k \in \bar{Y}_i, f(x_i, y_k \ge u)\}|)$

Based on the solution of linear model and threshold function, the multi-label classifier is written as follows.

$$h(x) = \{y_j | <w_j, x> + b_j > <w^*, f^*(x)> + b^*, 1 \le j \le m\} \tag{15}$$

6 Experiment and Analysis

6.1 Data Set

The data set in this paper uses the yeast data set (Yeast Saccharomyces cerevisiae) [12], and each gene is divided into several categories according to their functions. The yeast data set includes 2417 samples, 103 attributes, and 14 categories. And 62% of the sample data is used as the training set and 38% is used as the test set. The details are as follows (Table 1).

Table 1. Description of datasets

Name	Instances	Attributes	Labels	Cardinality	Density	Train	Test
yeast	2417	103	14	4.237	0.303	1500	917

In addition, the yeast data set used in this paper is a manually preprocessed data set, and the label set is processed into a label sequence obeying a certain probability distribution. Three probability distributions are applied, assume that the weights in the label set obey the continuous random distribution in [0,1], uniform distribution in [0,1], and normal distribution. The parameter δ take three values: 0.37, 0.43, and 0.50, corresponding to the three intervals, which covers the whole weight ratio.

6.2 Experiment Environment

The experimental environment configuration is shown in Table 2. The experimental platform is MATLAB, and process of data distribution is performed with Python language.

Table 2. Experimental configuration environment

Operating system	Memory	CPU
Windows7	8G	Intel Core i7-6700, 3.40 GHz

6.3   Experimental Results and Analysis

WMR calculates the correlation between labels fully, which considers the relation of rank and weight between labels at the same time, to give the label sequence accurately. The following experimental results are displayed according to the three kinds of probability distributions.

In order to verify the validity of the WMR, this paper compares four evaluation indicators: 1-AveragePrecision, 2-OneError, 3-RankingLoss, and 4-MRR (Mean Reciprocal Rank) with the classic RankSVM algorithm.

The AveragePrecision indicator is to examine the case that the label, ranking before the label belonging to the sample, also belongs to the sample. The OneError indicator is to examine the case that in the label sequence, the first label dose not belong to the sample. The RankingLoss indicator is to investigate the wrong ranking of labels. The MRR indicator is to measure the effectiveness of algorithms, when the label is more relevant to the sample, the label ranks the top.

Random Distribution. The experimental results under above conditions are shown in Fig. 2.

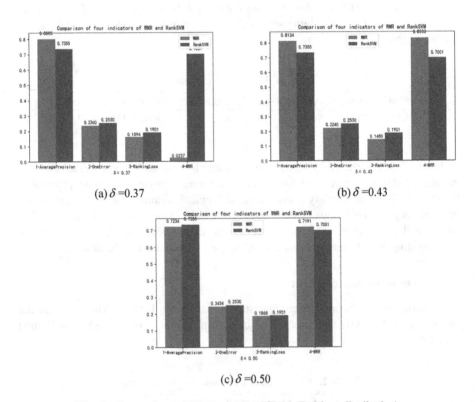

(a) δ =0.37

(b) δ =0.43

(c) δ =0.50

Fig. 2. Comparison of WMR and RankSVM (Random distribution).

As can be seen from the above figure, when the label weights obey the random probability distribution, WMR is 0.1% lower than RankSVM on the AveragePrecision, and the other three indicators are better than RankSVM. When $\delta = 0.37$, $\delta = 0.43$, WMR is apparently better than RankSVM with the four indicators, among them, WMR perform best when $\delta = 0.43$. the results show that when the weight ratio is large, i.e. the relation of weight is mined fully, WMR works best.

Uniform Distribution. The experimental results under above conditions are shown in Fig. 3.

Under the condition that the label weights obey the uniform probability distribution, WMR is obviously better than RankSVM with the four evaluation indicators, and the promotion of MRR is highest. When $\delta = 0.43$, WMR works best.

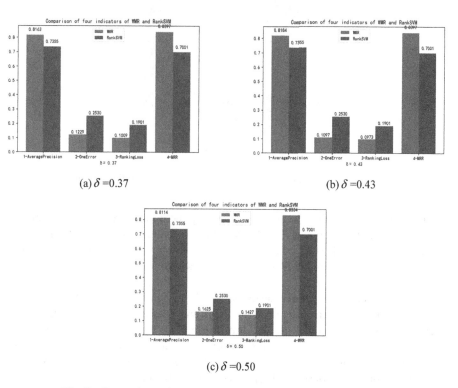

(a) δ =0.37

(b) δ =0.43

(c) δ =0.50

Fig. 3. Comparison of WMR and RankSVM (Uniform distribution).

Normal Distribution. The experimental results under above conditions are shown in Fig. 4.

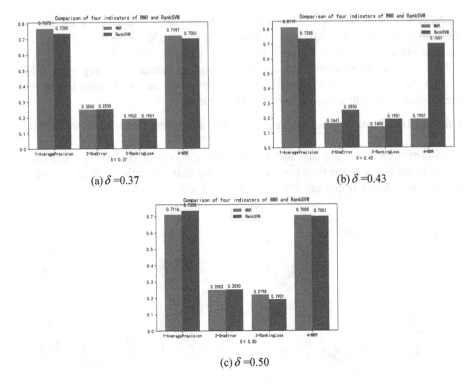

(a) δ =0.37
(b) δ =0.43

(c) δ =0.50

Fig. 4. Comparison of WMR and RankSVM (Normal distribution).

Under the condition that the label weights obey the normal probability distribution, when $\delta = 0.50$, WMR compares to RankSVM, the AveragePrecision is 0.02% lower and the RankingLoss is 0.02% higher, the other two indicators are more outstanding. When $\delta = 0.37$, $\delta = 0.43$, WMR performs better with four indicators, especially when $\delta = 0.43$.

As can be seen from above analysis, WMR performs significantly better than RankSVM when considering the relation of rank and weight between labels. Especially when the label weights obey the uniform distribution, the performance of WMR is optimal. The possible reason is that the uniform distributed weights are relatively easier to mine by WMR, which means that the label weight distribution has an impact on classifier performance. Besides, the parameter δ takes three values, there exists two results with little difference when δ takes the third value, in other cases, WMR performs better than RankSVM, which indicates that when the weight ratio is an intermediate value, the model works optimal. It illustrates that WMR mines the correlation between labels more comprehensively.

7 Conclusion

WMR calculates the weight probability between labels to consider the influence of the original label weights on classification results, which improves the accuracy of multi-label classifier effectively. Besides, WMR mines the correlation (rank and weight) between labels fully, so that the model owns superior practicability, which has been verified by experiments. However, the algorithm complexity has a certain relationship with the scale of data set, when the scale is large, how to reduce the complexity of the algorithm becomes the main task of next step.

Acknowledgement. This work is supported by the National Fund Major Project (17ZDA166), the Central University Basic Research Business Expenses Special Fund Project (CZY18015).

References

1. Maron, O.: Learning from ambiguity. Ph.D. dissertation, Department of Electrical and Computer Science, MIT, Cambridge, MA, June 1998
2. Boutell, M.R., Luo, J., Shen, X., Christopher, M.B.: Learning multi-label scene classification. Pattern Recogn. **37**(9), 1757–1771 (2004)
3. Read, J., Pfahringer, B., Holmes, G.: Multi-label classification using ensembles of pruned sets. In: ICDM 2008, 8th IEEE International Conference on Data Mining, Pisa, Italy, pp. 995–1000 (2008)
4. Schapire, R.E., Singer, Y.: BoosTexter: a boosting-based system for text categorization. Mach. Learn. **39**(2/3), 135–168 (2000)
5. Zhang, M.L., Zhou, Z.H.: ML-KNN: a lazy learning approach to multi-label learning. Pattern Recogn. **40**(7), 2038–2048 (2007)
6. Arnaiz-González, Á., Díez-Pastor, J.F., Rodríguez, J.J., García-Osorio, C.: Study of data transformation techniques for adapting single-label prototype selection algorithms to multi-label learning. Expert Syst. Appl. **109**, 114–130 (2018)
7. Chen, W., Yan, J., Zhang, B., Chen, Z., Yang, Q.: Document transformation for multi-label feature selection in text categorization. In: 7th IEEE International Conference on Data Mining, pp. 451–456 (2007)
8. Read, J., Pfahringer, B., Holmes, G., Frank, E.: Classifier chains for multi-label classification. Mach. Learn. **85**, 333 (2011)
9. Tsoumakas, G., Katakis, I., Vlahavas, I.: Random k-labelsets for multilabel classification. IEEE Trans. Knowl. Data Eng. **23**, 1079–1089 (2011)
10. Read, J.: A pruned problem transformation method for multi-label classification. In: Proceedings of 2008 New Zealand Computer Science Research Student Conference, pp. 143–150 (2008)
11. Furnkranz, J., Hullermeier, E., Mencia, E.L.: Multi-label classification via calibrated label ranking. Mach. Learn. **73**(2), 133–153 (2008)
12. Elisseeff, A., Weston, J.: Kernel methods for multi-labelled classification and categorical regression problems, BIOwulf Technologies, Technical report (2001)
13. McCallum, A.: Multi-label text classification with a mixture model trained by EM. In: AAAI 1999 Workshop on Text Learning, pp. 1–7 (1999)

14. Ghamrawi, N., McCallum, A.: Collective multi-label classification. In: Proceedings of the 14th ACM International Conference on Information and Knowledge Management, pp. 195–200 (2005)
15. Zhang, M.-L., Zhou, Z.-H.: Multilabel neural networks with applications to functional genomics and text categorization. IEEE Trans. Knowl. Data Eng. **18**(10), 1338–1351 (2006)
16. Guo, Y., Gu, S.: Multi-label classification using conditional dependency networks. In: IJCAI 2011: 24th International Conference on Artificial Intelligence, pp. 1300–1305. IJCAI/AAAI (2011)
17. Berger, M.J.: Large scale multi-label text classification with semantic word vectors, pp. 1–8. Technical report, Stanford, CA 94305 (2014)
18. Kurata, G., Xiang, B., Zhou, B.: Improved neural network-based multi-label classification with better initialization leveraging label co-occurrence, San Diego, California, 12–17 June 2016, pp. 521–526. Association for Computational Linguistics (2016)
19. Chen, G., Ye, D., Xing, Z., Chen, J., Cambria, E.: Ensemble application of convolutional and recurrent neural networks for multi-label text categorization. https://doi.org/10.1109/IJCNN.2017.7966144

Orthogonal Graph Regularized Nonnegative Matrix Factorization for Image Clustering

Jinrong He[1,2,3], Dongjian He[1,3(✉)], Bin Liu[1,3], and Wenfa Wang[2]

[1] Key Laboratory of Agricultural Internet of Things,
Ministry of Agriculture and Rural Affairs, Northwest A&F University,
Yangling 712100, Shaanxi, China
hdjl68@nwsuaf.edu.cn
[2] College of Mathematics and Computer Science, Yan'an University,
Yan'an 716000, China
[3] Shaanxi Key Laboratory of Agricultural Information Perception and Intelligent
Service, Northwest A&F University, Yangling 712100, Shaanxi, China

Abstract. Since high-dimensional data can be represented as vectors or matrices, matrix factorization is a common useful data modeling technique for high-dimensional feature representation, which has been widely applied in feature extraction, image processing and text clustering. Graph regularized nonnegative matrix factorization (GNMF) incorporates the non-negativity constraint and manifold regularization simultaneously to achieve a parts-based meaningful high-dimensional data representation, which can discover the underlying local geometrical structure of the original data space. In order to reduce the redundancy between bases and representations, and enhance the clustering power of NMF, three orthogonal variants of GNMF are proposed, which incorporates the orthogonal constraints into GNMF model. The optimization algorithms are developed to solve the objective functions of Orthogonal GNMF (OGNMF). The extensive experimental results on four real-world face image data sets have confirmed the effectiveness of the proposed OGNMF methods.

Keywords: Image representation · Nonnegative Matrix Factorization · Manifold regularization · Orthogonal projection

1 Introduction

Non-negative Matrix Factorization (NMF) is a popular useful method for learning low-dimensional compact meaningful representations of high-dimensional dataset, which is very helpful for clustering interpretation. By decomposing original high-dimensional non-negative data matrix X whose columns are samples into two low-dimensional non-negative factors U and V, namely basis matrix and coefficient matrix, such that $X \approx UV^T$. In this way, the dimensionality of original data can be reduced. The non-negative constraints on the elements of data matrix can lead to part-based data representation in the process of matrix factorization, which is in accordance with the cognition way of the human beings [1]. Due to its meaningful physiological interpretation abilities, NMF has been used in a variety of areas during the past years, including face recognition [5, 6], document clustering [4], microarray data analysis [2, 3], and more.

© Springer Nature Singapore Pte Ltd. 2019
H. Jin et al. (Eds.): BigData 2019, CCIS 1120, pp. 325–337, 2019.
https://doi.org/10.1007/978-981-15-1899-7_23

Since classical NMF algorithms aim to minimize the L_2 norm based distance between the original high-dimensional data matrix and the products of its low-rank approximations, which can be proved theoretically that it is optimal when the original data samples contaminated by additive Gaussian noise, and they may fail when the corruptions or outliers are not satisfied with the noise assumption. To overcome the aforementioned limitations of classical NMF models, many NMF variants have been proposed. Hypersurface cost based NMF (HCNMF [2]) used hypersurface objective function between the original high-dimensional data matrix and its low-rank approximations to improve the robustness of classical NMF. While L_1-NMF [3] proposed by E. Lam used L_1-norm based loss function to model the noise with Laplace distribution. Moreover, L_1-norm regularized Robust NMF (RNMF- L_1 [4]) assumed that the original high-dimensional data matrix can be factorized into a low-rank NMF approximation and a sparse error matrix, and $L_{2,1}$-NMF [5] used $L_{2,1}$-norm based loss function to eliminate the effect of the noise with large magnitude from dominating the NMF objective function, since the L_1 norm of large value will not be amplified. Robust capped norm NMF (RCNMF [6]) used pre-defined capped norm instead of L_1 norm to design the loss function, which can eliminate the effect of outliers by confining their proportions in the NMF objective function. When the subspace of a high-dimensional data samples is contaminated by outliers, Guan et al. proposed a Truncated Cauchy NMF, in which outliers are modeled by truncated Cauchy loss, then a half-quadratic optimization algorithm is designed to optimize the model [7].

Since NMF methods fails to discover the geometric structure of high-dimensional data samples, Cai et al. incorporated the manifold learning into NMF framework, and propose graph regularized NMF (GNMF [8]) model, which combines the graph embedding technique into the original NMF model for generating a compact low-dimensional data representation, then the latent intrinsic local geometric structure and semantic components can be extracted effectively. GNMF achieved better performance for high-dimensional data clustering than the original NMF model. In this framework, graph construction is the core problem, which affects the clustering performance seriously. The traditional data graph can only connect two vertices with one edge and neglect the high-order information contained in the high-dimensional data set. To deal with such problems, Zeng et al. Introduced the definition of hypergraph embedding to NMF model and proposed a hypergraph regularized NMF (HNMF [9]), which used hypergraph of data set to represent the local geometric structure of data manifold, then the underlying high-order information can be used to generate better data representations. Since the sparse hypergraph can make use of the merits of both the hypergraph models and sparse learning, Huang et al. proposed Sparse Hypergraph regularized NMF (SHNMF [10]) to exploit the high-order information, geometric and discriminant structure for data representation. Motivated by robust NMF model, Wang et al. combined hypergraph regularization technique with $L_{2,1}$-NMF objective function for better extracting the spectral-spatial structure of hyperspectral image data [11]. In order to depict complex relations of samples, Wang et al. encoded different sample relations into multiple graphs and proposed Multiple Graph regularized NMF (MGNMF [12]). By learning the optimal combination of graphs, MGNMF can better discover the underlying intrinsic manifold structure for data representation. However, the process of multiple graphs construction is very time consuming. Motivated by recent studies on sample diversity learning [13], Wang et al. proposed GNMF with Sample Diversity (GNMFSD [14]), which incorporated supervised prior knowledge, such as label information, into the data graph construction process to represent the intrinsic global geometrical structures and discriminant structure of the original data space, then the discriminant abilities of the basis vectors is enhanced. Furthermore, in order to make full use of higher-order information contained in the data graph to improve the data clustering performance, Wu et al. proposed Mixed Hypergraph regularized NMF (MHGNMF [15]) method, which projects the nodes within the identical hyper-edge onto the same latent space.

By imposing **orthogonality constraints** on NMF, Ding et al. proposed orthogonal NMF (ONMF) method [16], which have been shown to work remarkably well for clustering tasks. Pompili et al. show that ONMF is mathematically equivalent to the model of weighted spherical k-means [17]. Motivated by recent studies in orthogonal projection and manifold learning, we proposed a novel algorithm, called Orthogonal GNMF (OGNMF), which combining orthogonal constraints and manifold regularization in a parts-based low-dimensional space. The main contributions in the paper are summarized as follows:

(1) Three types of orthogonal constraints are added into GNMF model, including U orthogonal, V orthogonal and bi-orthogonal. In this way, the potential geometrical structural information can be preserved during the process of data representation, which can effectively improve the discriminant ability of data clustering.
(2) The multiplicative iterative updating rules derived from original NMF are used to optimize the proposed OGNMF models.
(3) Comprehensive experiments on four real world face image datasets are conducted to confirm the effectiveness of the proposed OGNMF methods, and demonstrate its advantage over other state-of-the-art methods.

The rest of the paper is organized as follows. Section 2 briefly introduces related works, such as NMF, ONMF and GNMF. In Sect. 3, we propose three types of OGNMF models and their optimization algorithms. Extensive image clustering experiments are conducted in Sect. 4. At last, conclusions are drawn in Sect. 5.

2 Related Works

2.1 NMF

NMF aims to express each data sample x_i as a linear combination of the basis vectors u_i, which are columns of U, namely $x_i = U v_j^T$, where v_j^T is the jth column of V^T. Such a matrix factorization model can be achieved by minimizing the following objective function:

$$\min_{U,V} \quad \|X - UV^T\|_F^2$$
$$s.t. \quad u_{ij} \geq 0, v_{ij} \geq 0 \tag{1}$$

where $X = [x_1, x_2, \cdots, x_n] \in R^{d \times n}$ denotes the original data matrix, which contains n data samples with the dimensionality d, and $U \in R^{d \times r}$ denotes the basis matrix,

$V \in R^{n \times r}$ denotes the coefficient matrix, where r is the reduced target dimensionality of projected low dimensional data. The optimal solution of NMF model (1) can be obtained by using the following iterative procedure [1]:

$$u_{ij} = u_{ij} \frac{(XV)_{ij}}{(UV^T V)_{ij}} \tag{2}$$

$$v_{ij} = v_{ij} \frac{(X^T U)_{ij}}{(VU^T U)_{ij}} \tag{3}$$

Since the objective function in NMF model (1) is not convex on both U and V, the NMF iterative updating rules (2) and (3) can only find the local minimal solution of the objective function. The convergence proof of the iterative optimization algorithm can be find in [1].

2.2   ONMF

ONMF is a variant of NMF, which approximate the original data matrix with the product of two low-rank non-negative matrices, one of which has orthonormal columns. For example, when the coefficient matrix V has orthonormal columns, the corresponding optimization model is

$$\min_{U,V} \quad \|X - UV^T\|_F^2$$
$$s.t. \quad \begin{array}{c} u_{ij} \geq 0, v_{ij} \geq 0 \\ V^T V = I \end{array} \tag{4}$$

Since the column vectors of coefficient matrix V is orthonormal, ONMF can generate sparser part-based decompositions of data with smaller basis vectors, that are easier to interpret.

The updating rules to optimize the above objective function (4) are given as follows

$$u_{ij} = u_{ij} \frac{(XV)_{ij}}{(UV^T V)_{ij}} \tag{5}$$

$$v_{ij} = v_{ij} \frac{(X^T U)_{ij}}{(VU^T XV)_{ij}} \tag{6}$$

2.3   GNMF

GNMF incorporates the manifold learning into the classical NMF model to discover the local geometric structure, in which the low-dimensional compact representations of data manifold have better discriminant ability. Let $G(X, W)$ denote a data graph whose vertex set is $X = \{x_1, x_2, \cdots, x_n\}$ and weight matrix is W, in which the entry W_{ij}

denotes the similarity between sample x_i and x_j. There several ways to compute the similarity weight matrix W, and we use the Gauss kernel function as follows:

$$W_{ij} = \begin{cases} e^{-\frac{\|x_i-x_j\|_2^2}{t}} & x_i \in N_k(x_j) \vee x_j \in N_k(x_i) \\ 0 & otherwise \end{cases} \tag{7}$$

where $N_k(x_i)$ is the sample points set that contains the k nearest neighbors of x_i, and t is heat kernel parameter. Since G is an undirected data graph, its weight matrix W is symmetric and the diagonal entries of W are zero.

GNMF aims to learning non-negative basis vectors of high-dimensional data space, which can not only minimize the reconstruction error between the original data matrix and its low-rank approximations, but also preserve the similarities between pairwise samples that encoded in neighborhood graph. Therefore, the optimal non-negative basis matrix U can be obtained by minimizing the following objective function:

$$\min_{U,V} \quad \|X - UV^T\|_F^2 + \lambda tr(V^T LV)$$
$$s.t. \quad u_{ij} \geq 0, v_{ij} \geq 0 \tag{8}$$

where $L = D - W$ is the graph Laplacian matrix which encoded the information of pairwise similarities between data samples, and D is a diagonal matrix whose entry is $D_{ii} = \sum_{j=1}^{n} W_{ij}$. The λ is a regularization parameter which is used to tradeoff the data matrix reconstruction error and graph embedding term. Since the graph embedding techniques imposed larger punishment on differences between two samples with larger similarity in the latent low-dimensional embedding space, the performance of data clustering based on GNMF model can be effectively enhanced.

Similar to Eqs. (2) and (3), the iterative updating rules to solve (8) are presented as follows:

$$u_{ij} = u_{ij} \frac{(XV)_{ij}}{(UV^T V)_{ij}} \tag{9}$$

$$v_{ij} = v_{ij} \frac{(X^T U + \lambda WV)_{ij}}{(VU^T U + \lambda DV)_{ij}} \tag{10}$$

3 Proposed Methods

OGNMF are extensions of classical NMF, which can be viewed as manifold regularized ONMF or orthogonal constrained GNMF. Since the orthogonal constraints can be imposed on basis matrix U or coefficient matrix V, or both of them, there are three variants of OGNMF.

3.1   OGNMF-U

The orthogonal constraint on the basis matrix of GNMF model can improve the data exclusivity across the different classes, thus it can improve the data clustering performance [16]. Incorporating orthogonal constraint to GNMF model (8), we have the following OGNMF-U model:

$$
\begin{aligned}
&\min_{U,V} \quad \|X - UV^T\|_F^2 + \lambda tr(V^T LV) \\
&s.t. \quad \begin{aligned} &u_{ij} \geq 0, v_{ij} \geq 0 \\ &U^T U = I \end{aligned}
\end{aligned}
\tag{11}
$$

The objective function in (11) can be reformulated in trace from:

$$
\begin{aligned}
\varepsilon &= tr((X - UV^T)(X - UV^T)^T) + \lambda tr(V^T LV) \\
&= tr(XX^T) - 2tr(XVU^T) + tr(UV^T VU^T) + \lambda tr(V^T LV)
\end{aligned}
\tag{12}
$$

Let ϕ_{ik} and φ_{jk} be the Lagrangian multipliers, then the Lagrangian function of optimization model (11) is

$$
\begin{aligned}
\ell =& tr(XX^T) - 2tr(XVU^T) + tr(UV^T VU^T) + \lambda tr(V^T LV) \\
&+ tr(\phi U^T) + tr(\varphi V^T)
\end{aligned}
\tag{13}
$$

Its partial derivatives with basis matrix U and coefficient matrix V is

$$
\frac{\partial \ell}{\partial U} = -2XV + 2UV^T V + \phi
\tag{14}
$$

$$
\frac{\partial \ell}{\partial V} = -2X^T U + 2VU^T U + 2\lambda LV + \varphi
\tag{15}
$$

where $L = D - W$ is graph Laplacian matrix.

Suppose that the gradient of the objective function ε in (11) can be decomposed into a positive part and a negative part, which has the following form [18]

$$
\nabla \varepsilon = [\nabla \varepsilon]^+ - [\nabla \varepsilon]^-
\tag{16}
$$

Then the derivatives of the objective function (12) with respect to basis matrix U with coefficient matrix V fixed can be given by

$$
\nabla_U \varepsilon = [\nabla_U \varepsilon]^+ - [\nabla_U \varepsilon]^- = UV^T X^T U - XV
\tag{17}
$$

With these gradient calculations, the multiplicative update rules for basis matrix U has the form

$$u_{ij} = u_{ij} \frac{(XV)_{ij}}{(UV^TX^TU)_{ij}} \tag{18}$$

$$v_{ij} = v_{ij} \frac{(X^TU + \lambda WV)_{ij}}{(VU^TU + \lambda DV)_{ij}} \tag{19}$$

The OGNMF-U algorithm is summarized in Algorithm 1.

Algorithm 1. OGNMF-U

Input: Non-negative matrix X, regularization parameters γ, reduced dimensionality r
Output: Coefficient matrix U, basis matrix V
Procedure:
  Step 1. Initialize matrices U and V.
  Step 2. Update U according to equation (18).
  Step 3. Update V according to equation (19).
  Step 4. If not convergence or number of iterations < Threshold:
      Go to Step 2.
      Else return U, V.

3.2 OGNMF-V

Combining OGNMF model (4) and GNMF model (8), we have the following OGNMF-V model:

$$\min_{U,V} \quad \|X - UV^T\|_F^2 + \lambda tr(V^TLV)$$
$$s.t. \quad \begin{array}{l} u_{ij} \geq 0, v_{ij} \geq 0 \\ V^TV = I \end{array} \tag{20}$$

Similar to optimization techniques in Sect. 3.1, the derivatives of the objective function (12) with respect to basis matrix U with V fixed can be given by

$$\nabla_V \varepsilon = [\nabla_V \varepsilon]^+ - [\nabla_V \varepsilon]^- = VU^TXV - X^TU \tag{21}$$

For the V-orthogonal optimization problem (20), the update rules are

$$u_{ij} = u_{ij} \frac{(XV)_{ij}}{(UV^TV)_{ij}} \tag{22}$$

$$v_{ij} = v_{ij} \frac{(X^TU + \lambda WV)_{ij}}{(VU^TXV + \lambda DV)_{ij}} \tag{23}$$

3.3   OGNMF

OGNMF imposes the non-negativity constraints in learning both the basis matrix and coefficient matrix, which involves the following optimization model:

$$\min_{U,V} \quad \|X - UV^T\|_F^2 + \lambda tr(V^T L V)$$

$$u_{ij} \geq 0, v_{ij} \geq 0$$
$$s.t. \quad U^T U = I \tag{24}$$
$$V^T V = I$$

The orthogonal constraints in (24) on the basis matrix U and coefficient matrix V in the OGNMF model simultaneously can enhance the capability of clustering rows and columns, which are called bi-orthogonal constraints in NMF. Combining update rules (25) in OGNMF-U and (26) in OGNMF-V, the update rules of OGNMF are given as follows

$$u_{ij} = u_{ij} \frac{(XV)_{ij}}{(UV^T X^T U)_{ij}} \tag{25}$$

$$v_{ij} = v_{ij} \frac{(X^T U + \lambda WV)_{ij}}{(VU^T XV + \lambda DV)_{ij}} \tag{26}$$

The procedure of OGNMF-V and OGNMF are same as Algorithm 1, other than the updating rules of U and V.

4   Experimental Results

Previous works on data clustering tasks have confirmed the superiority of the orthogonal constraints and local similarity information imposed on NMF. In the experiments, the clustering performances of the proposed OGNMF-U, OGNMF-V and OGNMF are compared with the related methods, such as NMF, ONMF and GNMF.

4.1   Experimental Setting

To investigate the image clustering performances, two popular evaluation metric are used in the experiments, i.e., the data clustering accuracy (AC) and the normalized mutual information (NMI) [8], that are defined as follows

$$AC = \frac{\sum_{i=1}^{n} \delta(l_i, \tau_i)}{n}$$

where δ is a cluster label indicator function, which equals one if the two entries have the same value and equals zero otherwise. l_i and τ_i denote the true label and predicted label.

$$NMI = \frac{MI(C,C')}{\max(H(C),H(C'))}$$

where C is the set of ground truth clusters, and C' is predicted clusters by the clustering method. $H(\cdot)$ is the entropy of a set, and the mutual information $MI(C,C')$ between two sets of clusters C and C' is defined as follows

$$MI(C,C') = \sum_{c_i \in C, c'_j \in C'} p\left(c_i, c'_j\right) \log \frac{p\left(c_i, c'_j\right)}{p(c_i)p\left(c'_j\right)}$$

where $p(c_i)$ and $p\left(c'_j\right)$ are the probabilities of a data sample belonging to the clusters c_i and c'_j, respectively. The joint probability $p\left(c_i, c'_j\right)$ represents the chance that the selected data sample belongs to the clusters c_i as well as c'_j simultaneously. Obviously, the larger AC and higher NMI indicate a better clustering performance [8].

For GNMF and OGNMF methods, the neighborhood parameter k is set as 5, which is suggested in [8]. The regularization parameter in model (8), (11), (20) and (24) are empirically set as 100. After the low-dimensional data representations were produced by matrix factorization methods, the k-means clustering method can be performed in the low-dimensional feature space. Since k-means method is sensitive to the initial centers, the average results on 20 rounds k-means are reported. All the experiments in this paper are run in MATLAB R2015b on Win7, RAM 8G, CPU 3.4 GHz.

4.2 Datasets Description

All the experiments were carried out on the four face image datasets, including UMIST[1], FERET[2], YaleB[3] and ORL[4]. Some statistics of the four face image datasets are summarized in Table 1.

UMIST. The UMIST face dataset consists of 575 images with 20 different persons, the number of which changes from 19 to 36. All images were taken against different angles from left profile to right profile.

FERET. This dataset comes from U.S. military. The data used in the experiments is a subset of the original FERET dataset, including 1400 images of 200 different people, and each people has seven sample images with different poses, expressions and illumination conditions.

---

[1] http://www.sheffield.ac.uk/eee/research/iel/research/face.

[2] http://www.nist.gov/itl/iad/ig/colorferet.cfm/.

[3] http://vision.ucsd.edu/~leekc/ExtYaleDatabase/ExtYaleB.html.

[4] http://www.uk.research.att.com/facedatabase.html.

YaleB. The YaleB face image data set is an extension of traditional Yale face image data set, including 16128 images with 9 poses and 64 illumination conditions. In our experiments, we just use 38 individuals, each of which has 64 illumination conditions.

ORL. The ORL face image data set has 400 images of 10 persons, each of which contains 40 face images that obtained at different conditions, including shooting time, illuminations, expressions and some other facial details, such as with/without glasses.

Table 1. The description of face datasets used in the experiments.

Datasets	#Samples	#Dimensionality	#Classes
UMIST	575	2576	20
FERET	1400	1600	200
YaleB	2414	1024	38
ORL	400	1024	40

4.3   Results Discussion

The image clustering performance comparing all algorithms on the four face image data sets are shown in Tables 2 and 3, where the clustering performance evaluation metrics AC and NMI reported. The numbers in the parenthesis were the optimal reduced dimensionality. As can be shown, the proposed orthogonal GNMFs have better clustering performances than the others.

Moreover, the clustering performances under different reduced dimensionalities are shown in Figs. 1, 2, 3 and 4. The results indicated that the clustering performance of each method was sensitive to the reduced dimensionality. Manifold regularized methods outperform the ONMF method, since the low-dimensional representations produced by them captured the local similarity of the original high-dimensional image vectors, which is very helpful on clustering tasks [8]. The orthogonal constraint on U performed better on UMIST and YaleB datasets, while the orthogonal constraint on V performed better on FERET and ORL datasets.

Table 2. The optimal clustering results (AC%) on face datasets.

Datasets	NMF	ONMF	GNMF	OGNMF-U	OGNMF-V	OGNMF
UMIST	24.00(94)	19.14(11)	21.07(92)	21.86(86)	**24.57(36)**	22.14(63)
FERET	48.35(77)	33.91(8)	62.96(19)	**65.39(49)**	56.17(67)	58.43(12)
YaleB	16.20(91)	6.96(1)	32.97(18)	**34.80(62)**	21.33(86)	15.04(84)
ORL	56.50(76)	36.75(27)	54.75(20)	56.75(67)	57.50(99)	**58.50(41)**

Table 3. The optimal clustering results (NMI%) on face datasets.

Datasets	NMF	ONMF	GNMF	OGNMF-U	OGNMF-V	OGNMF
UMIST	64.62(5)	63.19(11)	63.99(2)	64.59 (86)	**65.50(74)**	64.28(18)
FERET	66.86(79)	48.13(8)	79.71(36)	**81.86 (35)**	73.50(97)	75.64(23)
YaleB	27.63(86)	8.64(1)	45.33(18)	**47.73 (94)**	33.14(90)	25.58(92)
ORL	74.18(59)	59.31(27)	72.90(95)	73.19 (79)	74.09(99)	**74.35(81)**

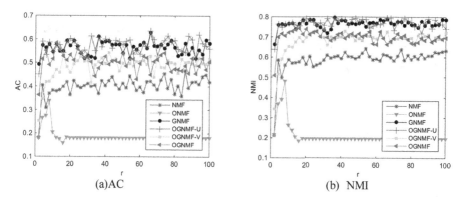

(a)AC   (b) NMI

Fig. 1. The performance comparison on UMIST face image clustering.

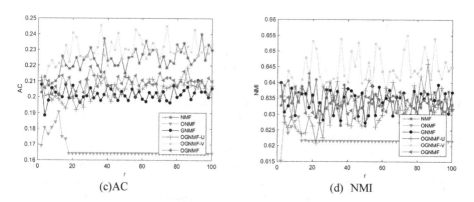

(c)AC   (d) NMI

Fig. 2. The performance comparison on FERET face image clustering.

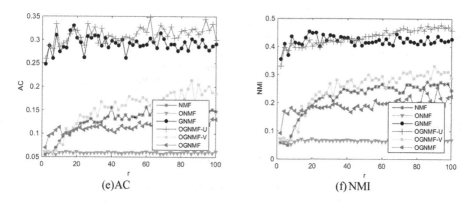

(e)AC   (f) NMI

Fig. 3. The performance comparison on YaleB face image clustering.

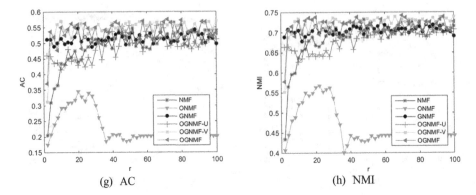

(g) AC   (h) NMI

Fig. 4. The performance comparison on ORL face image clustering.

5 Conclusion

This paper introduced a data representation method by combining GNMF model and ONMF model, which is called orthogonal GNMF (OGNMF). In the proposed OGNMF model, the orthogonal constraints imposed on the basis matrix U or coefficient matrix V of NMF model can not only make full use of geometrical structural information underlying the data manifold, but also enhance the generalization performance and robustness of the proposed methods.

In future, to cope with the large-scale dataset, we will attempt to build an incremental learning framework for OGNMF. Besides, motivated by recent works on dual graph regularization techniques in dimensionality reduction and manifold learning, we will also consider advantages of dual graph regularization technique to improve the data representation performances of NMF.

Acknowledgement. This work was partially supported by National Natural Science Foundation of China (61902339, 61602388), China Postdoctoral Science Foundation (2018M633585, 2017M613216), Natural Science Basic Research Plan in Shaanxi Province of China (2018JQ6060, 2017JM6059), Fundamental Research Funds for the Central Universities (2452019064), Key Research and Development Program of Shaanxi (2019ZDLNY07-06-01), and the Doctoral Starting up Foundation of Yan'an University (YDBK2019-06).

References

1. Lee, D.D., Seung, H.S.: Learning the parts of objects by nonnegative matrix factorization. Nature **401**(6755), 788–791 (1999)
2. Hamza, A.B., Brady, D.J.: Reconstruction of reflectance spectra using robust non-negative matrix factorization. IEEE Trans. Sig. Process. **54**(9), 3637–3642 (2006)
3. Lam, E.Y.: Non-negative matrix factorization for images with Laplacian noise. In: IEEE Asia Pacific Conference on Circuits and Systems, pp. 798–801 (2008)
4. Zhang, L., Chen, Z., Zheng, M., He, X.: Robust non-negative matrix factorization. Front. Electr. Electron. Eng. China **6**(2), 192–200 (2011)

5. Kong, D., Ding, C., Huang, H.: Robust non-negative matrix factorization using L21-norm. In: Proceedings of the 20th ACM International Conference on Information and Knowledge Management, pp. 673–682 (2011)

6. Gao, H., Nie, F., Cai, W., Huang, H.: Robust capped norm nonnegative matrix factorization. In: Proceedings of the 24th ACM International Conference on Information and Knowledge Management, Melbourne, Australia, 19–23 October 2015

7. Guan, N., Liu, T., Zhang, Y., et al.: Truncated cauchy non-negative matrix factorization for robust subspace learning. IEEE Trans. Pattern Anal. Mach. Intell. **41**(1), 246–259 (2019)

8. Cai, D., He, X., Han, J., et al.: Graph regularized nonnegative matrix factorization for data representation. IEEE Trans. Pattern Anal. Mach. Intell. **33**(8), 1548–1560 (2011)

9. Zeng, K., Yu, J., Li, C., You, J., Jin, T.: Image clustering by hyper-graph regularized nonnegative matrix factorization. Neurocomputing **138**(11), 209–217 (2014)

10. Huang, S., Wang, H., Ge, Y., et al.: Improved hypergraph regularized nonnegative matrix factorization with sparse representation. Pattern Recogn. Lett. **102**(15), 8–14 (2018)

11. Wang, W., Qian, Y., Tang, Y.Y.: Hypergraph-regularized sparse NMF for hyper-spectral unmixing. IEEE J. Sel. Top. Appl. Earth Obs. Remote Sens. **9**(2), 681–694 (2016)

12. Wang, J.Y., Bensmail, H., Gao, X.: Multiple graph regularized nonnegative matrix factorization. Pattern Recogn. **46**(10), 2840–2847 (2013)

13. Xu, Y., Li, Z., Zhang, B., Yang, J., You, J.: Sample diversity, representation effectiveness and robust dictionary learning for face recognition. Inform. Sci. **375**(1), 171–182 (2017)

14. Wang, C., Song, X., Zhang, J.: Graph regularized nonnegative matrix factorization with sample diversity for image representation. Eng. Appl. Artif. Intell. **68**(2), 32–39 (2018)

15. Wenhui, W., Sam Kwong, Yu., Zhou, Y.J., Gao, W.: Nonnegative matrix factorization with mixed hypergraph regularization for community detection. Inf. Sci. **435**(4), 263–281 (2018)

16. Ding, C., Li, T., Peng, W., Park, H.: Orthogonal nonnegative matrix t-factorizations for clustering. In: Proceedings of the 12th ACM SIGKDD International Conference on Knowledge Discovery and Data Mining, pp. 126–135. ACM (2006)

17. Li, B., Zhou, G., Cichocki, A.: Two efficient algorithms for approximately orthogonal nonnegative matrix factorization. IEEE Sig. Process. Lett. **22**(7), 843–846 (2015)

18. Yoo, J., Choi, S.: Orthogonal nonnegative matrix factorization: multiplicative updates on Stiefel manifolds. In: Fyfe, C., Kim, D., Lee, S.-Y., Yin, H. (eds.) IDEAL 2008. LNCS, vol. 5326, pp. 140–147. Springer, Heidelberg (2008). https://doi.org/10.1007/978-3-540-88906-9_18

Big Data Application

Characteristics of Patient Arrivals and Service Utilization in Outpatient Departments

Yonghou He[1], Bo Chen[2], Yuanxi Li[3], Chunqing Wang[1], Zili Zhang[1], and Li Tao[1(✉)]

[1] School of Computer and Information Science,
Southwest University, Chongqing, China
{holder523,wangchunqing}@email.swu.edu.cn, {zhangzl,tli}@swu.edu.cn
[2] The First Affiliated Hospital of Chongqing Medical University, Chongqing, China
195123624@qq.com
[3] Department of Computer Science, Hong Kong Baptist University,
Kowloon Tong, Hong Kong, China
csyxli@comp.hkbu.edu.hk

Abstract. The characteristics of patient arrivals and service utilization are the theoretical foundation for modeling and simulating healthcare service systems. However, some commonly acknowledged characteristics of outpatient departments (e.g., the Gaussian distribution of the patient numbers, or the exponential distribution of diagnosis time) may be doubted because many outpatients make prior appointment before they come to a hospital in recent years. In this study, we aim to discover the characteristics of patient arrivals and service utilization in five outpatient departments in a big and heavy load hospital in Chongqing, China. Based on the outpatient registration data from 2016 to 2017, we have the following interesting findings: (1) the variation of outpatient arrival numbers in each day is non-linear and can be characterized as pink noise; (2) the distribution of daily arrivals follows a bimodal distribution; (3) the outpatient arrivals in distinct departments exhibit different seasonal patterns; (4) the registration intervals of outpatient arrivals and the doctors' diagnosis time in all the departments except the Geriatrics department exhibit a power law with cutoff distribution. These empirical findings provide some new insights into the dynamics of patient arrivals and service utilization in outpatient departments and thus enable us to make more reasonable assumptions when modeling the behavior of outpatient departments.

Keywords: Characteristics of outpatient arrivals · Service utilization · Statistical analysis · Power spectrum analysis

1 Introduction

The characterization of patient arrivals and service utilization is vital in modeling and simulating healthcare service systems to predict patient arrivals for

© Springer Nature Singapore Pte Ltd. 2019
H. Jin et al. (Eds.): BigData 2019, CCIS 1120, pp. 341–350, 2019.
https://doi.org/10.1007/978-981-15-1899-7_24

better resource allocations in hospitals. Based on empirical observations, existing studies usually assume that the number of patient arrivals in outpatient departments follows a Gaussian distribution [1], the time interval between two consecutive patients follows a Poisson distribution [2,3], and the diagnosis time (often referred as service time) obeys an exponential distribution [4]. However, these well-acknowledged and widely-utilized characteristics of patient arrivals and service utilization are observed from outpatients without a prior appointment [5]. It may not be the case in most large and heavy load hospitals in China, because a certain portion of outpatients make a prior appointment using online or offline appointment systems [6,7]. Thus, the patient arrivals may exhibit different features such as non-linear characteristics [8].

Besides, the patterns of arrivals in different outpatient departments may not be the same because the concerned disease in each department depends on various causes. For instance, the number of patient arrivals of a Respiratory medicine department may be influenced by weather and air quality [9–11], whereas the arrivals of other departments, e.g., a Neurology department, may not be affected by the same factors at the same level [12,13]. Therefore, it would be valuable to reexamine the patterns of patient arrivals and service utilization in different outpatient departments with both appointment and walk-in patients.

In this study, we analyzed the outpatient registration and payment records in a large general hospital in Chongqing, China, to discover temporal patterns of patient arrivals and diagnosis in five major outpatient departments, including Respiratory medicine department (RESP), Neurology department (NEUR), Geriatrics department (GER), Dermatology department (DERM), and Cardiovascular medicine department (CVS). A power spectrum analysis on the outpatient arrivals has shown that the number of daily outpatient arrivals is nonlinear and it could be characterized as pink noise, instead of obeying random walks. Thus, it should not be assumed as a Gaussian process. Furthermore, by statistical analysis on the number of outpatient arrivals in different time scales, we found that the daily outpatient arrivals exhibit a bimodal pattern but with different peaks in the five outpatient departments. Furthermore, the monthly arrivals of different outpatient departments exhibit distinct seasonal patterns. Besides, both of the time intervals between two consecutive registered outpatients and the diagnosis time for each outpatient have been found to exhibit power-law with cutoff distributions in all the five departments except the Geriatrics department. The results suggested that the arrivals of mixed appointment and walk-in outpatients, as well as the service process in outpatient departments, should not be described as a Poisson process.

The remainder of the paper is organized as follows. Section 2 describes the data set and the analytical methods used for mining the patterns of outpatient arrivals, as well as the diagnosis behavior of outpatient doctors. Section 3 presents the results and discussions of empirical data analysis. We conclude this paper in Sect. 4.

2 Method

The empirical data studied in this paper comes from a large 3A (Class Three/Grade A) general hospital in Chongqing, China. The data contains the registration and payment records from 2016 to 2017 in five outpatient departments, which include Respiratory medicine department, Neurology department, Geriatrics department, Dermatology department, and Cardiovascular medicine department. After removing invalid and erroneous records, the total number of valid records is more than 2 millions. Specifically, each patient record p_i contains four attributes, including the registration time r_{p_i}, the bill payment time b_{p_i}, the diagnosis department d_{p_i}, and the serving doctor s_{p_i}. In order to analyze the characteristics of outpatient arrivals in various departments at different time scales, the raw data of each department has been processed into four time series data to present the number of corresponding outpatient arrivals in a specific hour N_h, in a specific day N_d, in a specific week N_w, and in a specific month N_m, respectively. The registration interval between two consecutive registered outpatients, $\hat{t}_{p_i,p_{i+1}}$, is calculated based the registration time of the two outpatients, i.e., $\hat{t}_{p_i,p_{i+1}} = r_{p_{i+1}} - r_{p_i}$. We ignored the registration intervals that are larger than ten hours because such big intervals often indicate that the two outpatients register in two different days. To estimate the time of diagnosis d_{p_i} for an outpatient p_i, we assume that d_{p_i} equals to the time interval between payment time of p_i and the payment time of the previous patient p_j who receives services at the same doctor at busy hours, or the length of the time between the registration time r_{p_i} and the payment time b_{p_i} of patient p_i at off-peak hours, i.e., $d_{p_i} = \min(b_{p_i} - b_{p_j}, b_{p_i} - r_{p_i})$.

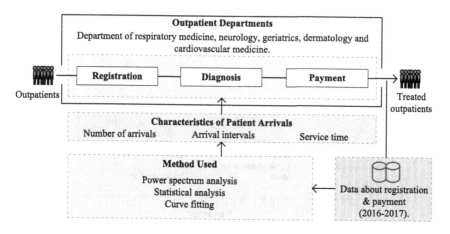

Fig. 1. A research framework of this study

Based on the processed data, we employed several statistical analyses, curve fitting, and graph plotting methods to analyze the temporal patterns of outpatient arrivals and the service utilization. To explore the nonlinear characteristics of the patient arrivals, we regarded the time series of outpatient arrivals at different time scales (e.g., N_d) as sequences of signals. Then, we used the power spectrum analysis, an effective method to discriminate the nonlinear characteristics of signals [8, 14] for the data analysis. In general, if the spectral density of signals obeys the power law, $s(f) = f^{-\beta}$, where f is the frequency and β is the scaling index, it means that the signal has a scaling property. Specifically, if $0 < \beta < 2$, the time series can be characterized as pink noise or a $\frac{1}{\beta}$ noise [15], which is the main characteristics of a nonlinear sequence. In this paper, the fast Fourier transform [16] method is used for analyzing the power spectrum of the time series of outpatient arrivals. The framework of empirical analysis and the corresponding methods is summarized in Fig. 1.

3   Results and Discussions

3.1   Temporal Characteristics of Outpatient Arrivals

Patterns of Daily Outpatient Arrivals. The results of the power spectrum analysis on the daily series of outpatient arrivals in the five departments are represented in Fig. 2. As shown in Fig. 2, the spectral density $s(f)$ of each outpatient arrival series has a power-law relationship with the corresponding frequency f. The scaling indexes are $\beta_R = 0.76$, $\beta_G = 0.70$, $\beta_D = 0.67$, $\beta_N = 0.85$, and $\beta_C = 0.95$ for the Respiratory medicine department, Geriatric department, Dermatology department, Neurology department, and Cardiovascular medicine department, respectively.

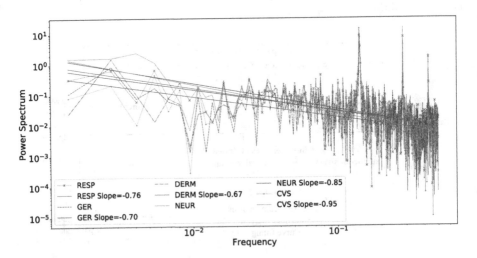

Fig. 2. The power spectrum analysis results on daily arrivals of five outpatient departments

As all the scaling indexes of daily arrival series are within the interval $(0, 1)$, this indicates that all the daily series of outpatient arrivals can be characterized as pink noise. This reveals that the series of daily arrivals do not obey random walks. Thus the number of daily outpatient arrivals should be better to be not assumed as a Gaussian distribution. Specifically, as shown in Fig. 3, the daily outpatient arrivals of all the five departments follow bimodal distributions. In the context of a mixture of walk-in and appointment outpatient, the finding differs from the existing studies in which the daily patient arrivals (mainly walk-in) are empirically found or are assumed to follow Gaussian distributions.

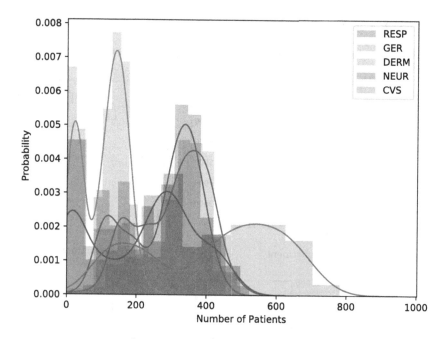

Fig. 3. The density distribution of daily outpatient arrivals

Seasonal Patterns of Outpatient Arrivals. Figure 4 presents the temporal pattern of monthly outpatient arrivals. From this figure, we can see that the monthly outpatients of the five departments have distinct seasonal patterns. The Respiratory medicine department has a relatively smaller number of patient arrivals in the months of January, February, and July to October, while it has a larger number of arrivals in the rest months. In the Neurology department, there are more patient arrivals from May to December. Different from the previously introduced two departments, the Geriatrics and Cardiovascular medicine departments have relatively more patients from March to June, whereas more patients come to the Dermatology department from March to September.

This finding implies that the number of outpatient arrivals of each month is not truly random, but has some particular patterns. These patterns may be caused by certain impact factors relating to specific time or seasons in a year,

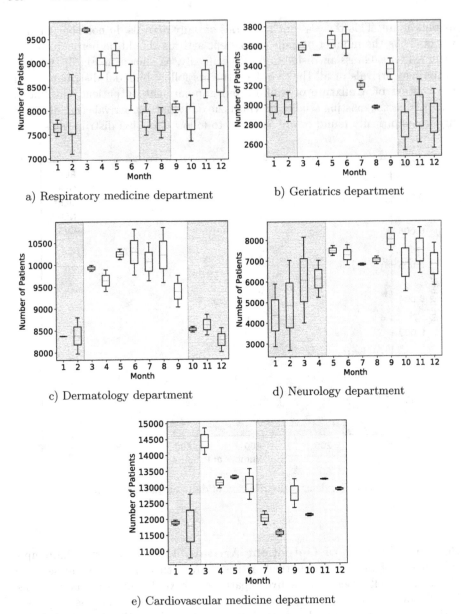

a) Respiratory medicine department

b) Geriatrics department

c) Dermatology department

d) Neurology department

e) Cardiovascular medicine department

Fig. 4. The patterns of monthly arrivals of five outpatient departments

such as temperature, air quality, and pollen. This suggests that if we attempt to model outpatient arrivals, the number of patients at different time scales (e.g., a day or a month) should not be simply generated from a specific distribution, but should take into account the relationship between the number of patients and the time of the year.

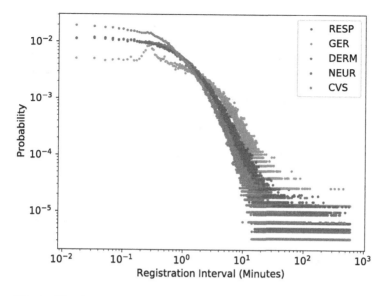

Fig. 5. The density distribution of outpatients' registration intervals

3.2   Patterns of Registration Intervals

The statistical characteristics of registration intervals between every two outpatients of the five departments are shown in Fig. 5. According to Fig. 5, we can see that most of the outpatients register into the hospital at short intervals, normally within one or a few minutes, and a small portion of outpatients register into the hospital at intervals of more than ten minutes. Based on the curve-fitting test method provided by Alstott [17], the densities of registration intervals ($p < 0.001$) in all the five departments except the Geriatrics department, are shown to be better characterized as power-law with cutoff distributions than exponential distributions, stretched exponential distributions, and log-normal distributions. For the registration intervals in the Geriatrics department, the result of Alstott's test [17] shows that both of the power-law with cutoff distribution and the log-normal distribution fit the data well. These findings suggest that it should be better to not simply assume the registration intervals of outpatients s follow an exponential distribution when we model the behavior of outpatient departments.

3.3   Patterns of Service Utilization

The patterns of service utilization, as represented by the distribution of doctors' diagnosis time for patients, are shown in Fig. 6. The curve fitting results based on Alstott's test method [17] reveal that the density distributions of doctors' diagnosis time in all of the five outpatient departments are well characterized by power-law with cutoff distributions. This finding reveals that the pattern of service time of outpatient departments that have both appointment and walk-in patients does not consistent with the common assumption, i.e., the service time

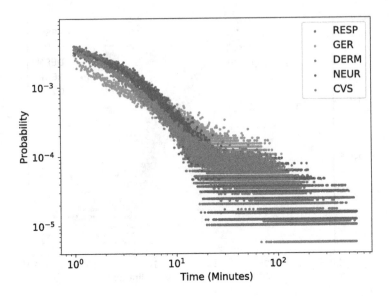

Fig. 6. The probability density distributions of doctors' diagnosis time

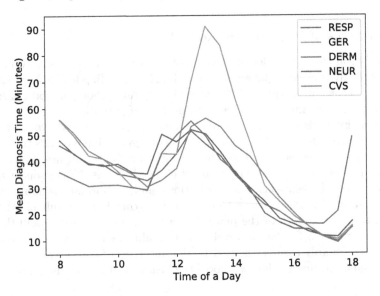

Fig. 7. The mean diagnosis time of doctors in working hours

of patients follows an exponential distribution. The trends of mean diagnosis time of working hours in a day (as shown in Fig. 7) further reveals that the diagnosis time continuously declines from 8 am to around 11:30 am, and from 1 pm to around 5:30 pm. This result reveals that doctors continuously lifting the diagnosis efficiency in work hours.

4 Conclusion

In this paper, we have empirically examined the characteristics of outpatient arrivals and service utilization in five outpatient departments in which there are both appointment and walk-in patients. By applying the statistical analysis and power spectrum analysis, we have found that (1) the daily outpatient arrivals in all the departments are nonlinear and can be characterized as a pink noise; (2) the distribution of daily patient arrivals follows a bimodal distribution; (3) the monthly outpatient arrivals present different seasonal patterns in distinct outpatient departments; (4) the registration intervals and diagnosis time in almost all the departments are better fitted by a power law with cutoff distribution, rather than the commonly acknowledged exponential distribution. The empirical findings in this study suggest that when modeling outpatient departments with a mixture of appointment and walk-in arrivals, it is better to assume that the number of patients in each day follows a bimodal distribution and should be seasonal adjusted. In addition, we also suggest that the registration intervals between two outpatients and the diagnosis time for each patient should follow a power law with cutoff distribution when modeling the diagnosis service of outpatient departments.

Acknowledgment. This work is supported by Fundamental Research Funds for the Central Universities (XDJK2018C045 and XDJK2019D018).

References

1. Kim, S.H., Whitt, W., Cha, W.C.: A data-driven model of an appointment-generated arrival process at an outpatient clinic. INFORMS J. Comput. **30**(1), 181–199 (2018)
2. Biswas, S., Arora, H., et al.: On an application of geiger-muller counter model (type-ii) for optimization relating to hospital administration. Acta Med. Int. **4**(2), 16 (2017)
3. Peter, P.O., Sivasamy, R.: Queueing theory techniques and its real applications to health care systems-outpatient visits. Int. J. Healthcare Manag. 1–9 (2019)
4. Vass, H., Szabo, Z.K.: Application of queuing model to patient flow in emergency department. case study. Proc. Econ. Financ. **32**, 479–487 (2015)
5. Ghamsari, B.N.: Modeling and Improving Patient flow at an Emergency Department in a Local Hospital Using Discrete Event Simulation. Ph.D. thesis, University of Minnesota (2017)
6. Yang, P.C., et al.: Features of online hospital appointment systems in Taiwan: a nationwide survey. Int. J. Environ. Res. Public Health **16**(2), 171 (2019)
7. Zhang, M., Zhang, C., Sun, Q., Cai, Q., Yang, H., Zhang, Y.: Questionnaire survey about use of an online appointment booking system in one large tertiary public hospital outpatient service center in china. BMC Med. Inform. Decis. Making **14**(1), 49 (2014)
8. Zhang, L., Liu, Z.: Empirical analysis of nonlinear characteristics on the patient flows. J. Syst. Manag. **25**(3), 527–531 (2016)
9. D'amato, G., et al.: Climate change and air pollution: effects on respiratory allergy. Allergy Asthma Immunol. Res. **8**(5), 391–395 (2016)

10. Kelly, F.J., Fussell, J.C.: Air pollution and public health: emerging hazards and improved understanding of risk. Environ. Geochem. Health **37**(4), 631–649 (2015)
11. Strosnider, H.M., Chang, H.H., Darrow, L.A., Liu, Y., Vaidyanathan, A., Strickland, M.J.: Age-specific associations of ozone and fine particulate matter with respiratory emergency department visits in the united states. Am. J. Respir. Critical Care Med. **199**(7), 882–890 (2019)
12. Bao, X., Tian, X., Yang, C., Li, Y., Hu, Y.: Association between ambient air pollution and hospital admission for epilepsy in Eastern China. Epilepsy Res. **152**, 52 (2019)
13. Mukamal, K.J., Wellenius, G.A., Suh, H.H., Mittleman, M.A.: Weather and air pollution as triggers of severe headaches. Neurology **72**(10), 922–927 (2009)
14. Ren, J., Wang, W.X., Yan, G., Wang, B.H.: Emergence of cooperation induced by preferential learning. arXiv preprint physics/0603007 (2006)
15. Lalwani, A.: Long-range correlations in air quality time series: effect of differencing and shuffling. Aerosol Air Qual. Res. **16**(9), 2302–2313 (2016)
16. Welch, P.: The use of fast fourier transform for the estimation of power spectra: a method based on time averaging over short, modified periodograms. IEEE Trans. Audio Electroacoust. **15**(2), 70–73 (1967)
17. Alstott, J., Bullmore, E., Plenz, D.: Powerlaw: a python package for analysis of heavy-tailed distributions. PloS One **9**(1), e85777 (2014)

College Academic Achievement Early Warning Prediction Based on Decision Tree Model

Jianlin Zhu[1(✉)], Yilin Kang[1], Rongbo Zhu[1], Dajiang Wei[1],
Yao Chen[1], Zhi Li[1], Zhentong Wang[1], and Jin Huang[2]

[1] College of Computer Science, South-Central University for Nationalities,
Wuhan City 430074, China
{Jianlin.Zhu,ylkang,rbzhu}@mail.scuec.edu.cn,
22252724@qq.com, 136065796@qq.com,
chenyao198516@hotmail.com
[2] School of Mathematics and Computer Science, Wuhan Textile University,
Wuhan, China
derick0320@foxmail.com

Abstract. With the transformation of higher education from elitism to popularity, the problem of academically at-risk students has aroused widespread concern in universities and society. Many colleges and universities at home and abroad have implemented the academic achievement early warning system for academically at-risk students, that is, to intervene in the crisis of learning problems and academic difficulties that have occurred or are about to occur during the school period, and take targeted preventive and remedial measures to help solve the academic difficulties. In this paper, based on the education big data, we model the academic achievement early warning prediction as multi-class classification problem and try to give academic achievement warning to students as early as possible. The data pre-precessing and feature engineer is introduced, the decision tree model is adopted to do the academic achievement warning prediction. The prediction model provides decision-making assistance to educational and teaching managers, and provides early warning to students for giving timely protection and intervention to warn them.

Keywords: Academic crisis · Academic achievement early warning · Decision tree

1 Introduction

The policy of enrollment expansion of higher education started in 1999 marked the beginning of the transformation of higher education from elitism to popularization in China. The source structure of students in colleges and universities has changed greatly in the past 20 years. With the rapid development of society, the problems of students with learning difficulties or whose academic level can not meet the requirements of schools (Academically At-Risk Students [1, 2]) have gradually emerged and become increasingly prominent. With the development of information age, the academic achievement early warning mechanism based on educational data can better meet the

© Springer Nature Singapore Pte Ltd. 2019
H. Jin et al. (Eds.): BigData 2019, CCIS 1120, pp. 351–365, 2019.
https://doi.org/10.1007/978-981-15-1899-7_25

requirements of higher education management. Through the implementation of academic achievement early warning mechanism [1], students can get the continuous attention and guidance from schools and families, and also can enhance students' learning autonomy. Students can get feedback from their own stage learning results through this mechanism, and make necessary adjustments and responses according to the feedback results, so as to minimize the emergence of crisis events.

Academically at-risk students have always been the focus of educational researchers, parents and universities. Concerning the concept of students in academic crisis, different researchers have different definitions. Reis and McCoach's [2] definition of students in academic crisis integrates the views of many researchers. They call students with academic crisis "underachievement" or "underachievement with ability". underachievement refers to the students whose expected results (obtained through standard academic achievement test scores, cognitive and psychological intelligence tests) differ greatly from the actual results (obtained through class performance and teacher evaluation). Gifted underachievement refers to those who have the ability to achieve high scores in expected performance tests (through standard academic performance test scores, cognitive and mental intelligence tests). Generally speaking, the academic crises ("those who have the ability to fail to meet the criteria") refer to the students whose learning potential and actual performance are quite different. Reis and McCoach argue that the underachievement are not caused by learners' learning disabilities, they believe that this phenomenon will only last for a period of time, so they define the subject of academic crisis as the underachievement with ability. Kerry Wimshurst and Richard Wortley [3] call the academic crisis "Academic Failure". Generally speaking, there are two basic conditions for the definition of students in academic crisis abroad: first, they have basic or normal learning ability, and there is no problem of intellectual factors; second, they have poor academic performance and can not meet certain learning standards, especially can not successfully complete undergraduate academic tasks. In the domestic research field, students with academic crisis generally refer to the normal level of sense organs and intelligence. Due to various subjective and objective factors, their academic performance declines, they can not complete their studies according to the school syllabus, fail the course examination and fail to reach a certain number of credits, thus affecting the graduation, even affecting the mental health and development of College students. In this project book, the problem is briefly described as the problem of students with learning difficulties. Students with learning difficulties in Colleges and universities refer to students with normal intelligence but certain obstacles or difficulties in learning, whose academic achievements are obviously behind others and fail to meet the basic requirements of the syllabus.

Aiming at the early warning needs of process management for academically at-risk students in colleges and universities, according to the number of failed courses per semester, this paper divides the academic achievement early warning level into five levels: *no warning*(zero courses failed), *blue warning*(1 courses failed), *yellow warning*(2 courses failed), *orange warning*(3 courses failed) and *red warning*(more than 4 courses failed). Through the basic data of students' entrance information (including students' basic information, college entrance examination achievement, psychological data, etc.), we build relevant features (performance-related characteristics, psychological characteristics, local educational economic level characteristics, etc.)

after data pre-processing. We adopt the decision tree model to predict the students' academic achievement warning level. Only by providing assistance and guidance to academically at-risk students as early as possible, the implement of the academic achievement early warning mechanism is more effectively, and ultimately achieve the goal of helping students successfully complete their studies, and improving the quality of teaching.

The contents can be summarized as follows:

(1) The raw dataset is the enrollment information, academic achievement, e-card data of the Grade 2012 students from South-Central University for Nationalities. This paper mainly focus the enrollment information with the purpose to predict the academic achievement warning level as early as possible.
(2) Data pre-processing includes data acquisition, data cleaning, data encoding, data standardization. The feature engineering takes a lot of time to construct the related features.
(3) Decision tree model is used to predict the academic achievement warning level, different evaluation indicators are given. Discussions on the shortcomings of existing work and future work is discussed.

2   Related Work

With the development of big data analysis in the field of education, many excellent systems are designed to solve the problems of students with learning difficulties. For example, Intelligent Instruction System (ITS) [4] is based on the log data of interaction with students to diagnose personalized knowledge, analyze students' knowledge mastery, and find students' weaknesses, so as to help students acquire knowledge and skills better adaptively [5, 6]. Carnegie Learning's "cognitive guidance" system is a typical ITS [7], which formulates follow-up questions based on students' answers to previous questions. Many large-scale online classroom platforms, such as edX, Coursera, Udacity abroad and online school in China, focus on the serious problem of high dropout rate in online education. Based on the demographic data of students and behavior data of students' registration courses, watching videos, completing homework after class and participating in forum discussions, the aim is to find out the important factors affecting students' dropout, and then to formulate corresponding actions. Pre-strategies guide students to reduce the dropout rate of online education [8]. Xiao et al. [9] do the research on prediction of MOOC course achievement based on behavior sequence characteristics. Huang et al. [10] propose a Test-aware Attention-based Convolutional Neural Network (TACNN) framework to automatically solve this Question Difficulty Prediction (QDP) task for READING problems (a typical problem style in English tests) in standard tests. Su [11] propose a Exercise-Enhanced Recurrent Neural Network (EERNN) framework for student performance prediction by taking full advantage of both student exercising records and the text of each exercise.

Wu et al. [12] A person's career trajectory is composed of his/her work or learning experience (institutions) in different periods. Understanding the career trajectory of people, especially scholars, can help the government formulate more scientific strategies to allocate resources and attract talents, help companies formulate wise recruitment plans. Defu Lian et al. [13] studied the problem of predicting academic performance based on the history of student book borrowing, and proposed a supervised dimensionality reduction algorithm for collaborative learning performance prediction and Book recommendation. Huaxiu Yao et al. [14] attempted to build a campus social network based on behavioral records that appeared in multiple locations, and verified the relationship between campus social network and academic performance, indicating that students' academic performance is related to their circle of friends. Based on the influence of campus social network on academic performance, a new label propagation algorithm based on multiple networks is proposed to predict academic performance. Nie et al. [15] extracted four behavioral characteristics based on students' campus behavior, and proposed a data-driven career choice prediction framework. It was found that the extracted vocational skills, behavioral regularity and economic status were significantly related to career choice. Cao et al. [16] put forward the index of orderliness to measure every student's daily life on campus (such as diet and shower). Based on the above research foundation, educational administrators can better guide students' campus life and implement effective intervention measures in the early stage when necessary. Bairui Tao, etc. [17] design of early warning system for student's poor academic performance based on SVM improved by KFCM.

Many colleges and universities at home and abroad have implemented the academic early warning system for academically at-risk students, that is, to intervene in the crisis of learning problems and academic difficulties that have occurred or are about to occur during the school period, and take methods to help solve the academic difficulties. Early warning is to prevent, predict the adverse consequences that will occur in advance, and give warning to the main body in advance. Academic early warning mainly aims at supervising and warning students' academic completion. Dinovi [18] did the research to investigate the establishment of an early warning system and subsequent intervention with college freshmen, which addressed both the academic viability and retention of first-year students. The academic early warning is not a simple information communication mechanism, but a measure of management and education for students; academic early warning is not a punishment for students, but a kind warning and care for students.

3 Data Pre-processing

There are many ways to get raw dataset in the data acquisition step, such as database, Internet, traditional media, etc. In our paper, our raw dataset is got from database and website mentioned in chapter one. In this chapter, we focus on the data pre-processing flow.

3.1 Data Pre-processing Flow

Common characteristics of the raw dataset got from databases, Internet and other data sources in the real world is listed as follows:

- Incorrect
  - Incorrect attribute values
  - An outlier that contains errors or deviations from expectations
  - A lot of fuzzy information
- Incomplete
  - Some data attributes are missing or uncertain
  - Missing the necessary data
- Inconsistency
  - The original system is obtained from various practical application systems, and the data structure is quite different.

Without high-quality data, there will be no high-quality results. High-quality decision-making must rely on high-quality data. The meaning of data quality is correctness, consistency, completeness, reliability. Data pre-processing is one of the most important steps in the analysis and mining of large dataset. The general flow of data pre-processing is shown in Fig. 1, which can be processed according to the needs.

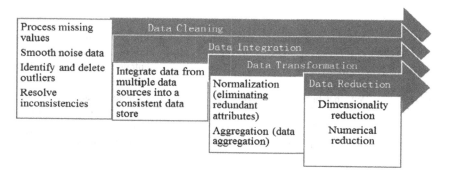

Fig. 1. Data pre-processing flow

3.2 Data Cleaning

Data cleaning task includes filling the missing value, recognizing outliers and smoothing noise data, correcting inconsistent data. There many methods could be adopted to deal with the "dirty" data, such as deletion and interpolation method for the missing value; binning, regression and clustering for the noise data.

3.3 Data Integration

The data come from the relevant departments of SCUEC, including the admission data of the Grade 2012 students. Initial data comes from different software management systems of the university, which need to be merged with the key of school number, discarding redundant and overlapping data, and finally stored in a unified data table. The enrollment features are extracted for academic achievement early warning, which are shown in Table 1.

Table 1. The features built from raw data

Feature	Meaning	Feature	Meaning
sno	Student's ID	ks_is_jf	Whether has adding points
male	Gender	jydf	Provincial Basic Education Score
female	Gender	csdm	City code
zybm	Major codes	kslb	Exam Type Code
xybm	College code	kldm	Class code
is_poor	Whether poor or not	csdj	City level
poor_level	Poor level	ybfsx_wk	College Entrance Examination Liberal Arts First Line
is_psy_warning	Whether Psychological warning or not	ebfsx_wk	College Entrance Examination Liberal Arts Second Line
is_qu	Whether birth place is district	ssmzlqf_wk	Minority Nationality College Entrance Examination Liberal Arts Minimum Entrance Score
is_shi	Whether birth place is city	hzlqf_wk	The lowest admission score of Liberal Arts in college entrance examination of Han nationality
is_xian	Whether birth place is county	ybfsx_lk	College Entrance Examination Science First Line
zf_yb_d	The different score between total score and first line score	ebfsx_lk	College Entrance Examination Science second Line
zy_fs_pm	Professional Ranking of College Entrance Examination Scores	ssmzlqf_lk	Minority Nationality College Entrance Examination Science Minimum Entrance Score
zfdj	Grades of College Entrance Examination Scores	hzlqf_lk	The lowest admission score of Science in college entrance examination of Han nationality
tdf	The total score of the college entrance examination with adding points	zf_lqf_d	The difference between the college entrance examination score and the first line score

(continued)

Table 1. (*continued*)

Feature	Meaning	Feature	Meaning
ncyj	Rural freshmen	zf_pjf_d	The difference between the total score of college entrance examination and the average score of the major
ncwj	Past rural students	zf	The total score of the college entrance examination
czyj	Urban freshmen		
czwj	Past urban freshmen	NumOfFailedCourse	Academic achievement warning level

At the same time, one-hot encoding is applied to the some characteristics, such as students' gender. The one-Hot Encoding is suitable for discrete features, and the features with n attribute values are transformed into N binary features. The effect of one-hot encoding for gender characteristics is shown in Table 2.

Table 2. Gender one-hot encoding results

Sno	Original data	One-hot encoding	
	Gender	Female	Male
2012210001	Female	1	0
2012210024	Male	0	1
…	…	…	…

Achievement Early Warning Mechanism is designed according to the distribution of the number of failed courses in the first semester. In order to improve the classification effect, it is divided into five categories, which are mainly distributed as shown in the statistics of failing subjects in Table 3.

Table 3. Statistics of failed courses in the first semester

The number of failed courses in the first semester	Numbers	Academic achievement early warning
0	4647	No Warning
1	761	Blue Warning
2	225	Yellow Warning
3	59	Orange Warning
4	23	Red Warning
Total	**5715**	

3.4   Data Standardization

Data standardization is the most basic part of data mining and data analysis. Because different evaluation indicators, i.e. different characteristics, often have different numerical sizes and modes of consideration, the meanings of numerical values may vary greatly, and the results of data analysis may be affected without data standardization. In order to eliminate the difference between indicators and the influence of the range of values, it is necessary to standardize the processing, scaling the data in proportion, so that it can be evenly and reasonably distributed to a prescribed numerical area, that is, between the two digital ranges. For example, scaling the attribute value of the score into [− 1, 1] or [0,1], the smaller the value, the simpler the processing, and the easier to run the algorithm.

Because the measurement units of each feature of the original data are different, it is necessary to map its value to a certain value interval by functional transformation. The standardization of data is to scale the data to a smaller specific interval. The standardizing the pre-processing function StandardScaler in Sklearn library are described in Formula (1), which is used for data standardization of each feature variable. $\bar{X}$ means the variable's mathematical expectation, and σ means standard deviation.

$$x^* = \frac{X - \bar{X}}{\sigma} \tag{1}$$

4   Decision Tree Model for Achievement Early Warning Prediction

4.1   Decision Tree Model

Decision tree method originated in the 1960s. It is a learning system CLS (concept learning system) established by Hunt et al. [19] when they studied human conceptual modeling. By the end of 1970s, J. Ross Quinlan proposed ID3 algorithm [20], introduced some ideas of information theory, proposed using information gain as a measure of feature discrimination ability, selected attributes as nodes of decision tree, and embedded the method of tree building in an iterative program. At that time, his main purpose was to reduce the depth of the tree, but he neglected the study of the number of leaves. In 1993, Quinlan himself developed C4.5 algorithm based on ID3 algorithm [21]. The new algorithm has made great improvements in missing value processing, pruning technology and derivation rules of predictive variables.

Decision tree is a tree-structured classifier, as Fig. 2 show, which determines the final category of classification points by inquiring the attributes of classification points in sequence. Generally, a decision tree is constructed according to the information gain or other indicators of features. When classifying, we only need to judge according to the nodes in the decision tree, and then we can get the category of the sample. Decision tree is essentially to find a partition of feature space, aiming at building a decision tree with good fitting of training data and low complexity.

Basic concepts related to ID3 classification algorithm: Information Entropy and Information gain. Entropy (also known as information entropy) is used to measure the amount of information of an attribute. Assuming S is a training set, the target attribute

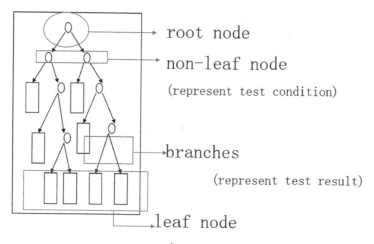

root node

non-leaf node

(represent test condition)

branches

(represent test result)

leaf node

(represent classificaition labels
obtained after classification)

Fig. 2. The structure of decision model

C of S has m possible class label values, C = {C_1, C_2,... C_m}, assuming that in training set S, the probability of Ci appearing in all samples is Pi (i = 1,2,3,... m), then the information entropy contained in the training set S is defined as in Formula (2):

$$Entropy(S) = Entropy(p_1, p_2, \cdots, p_m) = -\sum_{i=1}^{m} p_i \log_2 p_i \qquad (2)$$

The smaller the entropy is, the purer the distribution of the target attributes is. On the contrary, the larger the entropy is, the more confused the distribution of the target attributes is.

Information gain: is the difference between the degree of impurity (entropy) of the sample data set before partition and the degree of impurity (entropy) of the sample data set after partition. Assuming that the sample data set before partitioning is S, and the sample set S is partitioned by attribute A, the information gain Gain (S, A) of S is the entropy of the sample set S minus the entropy of the sample subset after partitioning S by attribute A, as show in Formula (3):

$$Gain(S, A) = Entropy(S) - Entropy_A(S) \qquad (3)$$

Entropy of sample subset divided by attribute A is defined as follows: Assuming that attribute A has k different values, S is divided into k sample subsets {S1, S2, ... Sk}, then the information entropy of the sample subset divided by attribute A is defined in Formula (4) :

$$Entropy_A(S) = \sum_{i=1}^{k} \frac{|S_i|}{|S|} Entropy(S_i) \qquad (4)$$

Where $|$ Si $|$ (i, = 1, 2,... K) is the number of samples contained in the sample subset Si, $|$ S $|$ is the number of samples contained in the sample subset S. The greater the information gain, the more pure the subset of samples partitioned by attribute A is, the better the classification is.

4.2   Model Evaluation

This paper classifies the academic achievement early warning grades into five categories and evaluates the results by using Micro F1 and ROC curves. F1 is the weighted harmonic mean of Precision and Recall, and both are considered equally important. Conventional classification evaluation indexes [22] are shown in Formulas (5), (6), (7) and (8) as follows:

$$Precision = \frac{TP}{TP + FP} \tag{5}$$

$$Recall = \frac{TP}{TP + FN} \tag{6}$$

$$Accuracy = \frac{TP + TN}{TP + FP + TN + FN} \tag{7}$$

$$F1 = 2 * \frac{Recall * Precision}{Recall + Precision} \tag{8}$$

Among them, regarding TP, TN, FP and FN, as shown in Table 4, confusion matrix below, under the binary classification problem, a few classes are Positive and most classes are Negative:

Table 4. Confusion matrix

Actual	Predicted	
	Yes	No
Yes	TP (True Positives)	FN (False Negatives)
No	FP (False Positives)	TN (True Negatives)

Since the academic achievement early warning prediction problem is multi-classification, it can be assessed using Micro F1 or Macro F1. The multi-classification evaluation is divided into several two-classification evaluations. The TP, FP and FN of several two-classification evaluations are added together to calculate the accuracy (Precision) and the recall rate (Recall). The F1 score calculated by the accuracy (Precision) and the recall rate (Recall) is Micro F1. The evaluation of multiple classifications is divided into several two classifications. The F1 score of each two classifications is calculated and the average value of the sum is calculated, which is Macro F1. Generally speaking, the higher the Macro F1 and Micro F1 are, the better the classification effect is. Macro F1 is greatly influenced by the categories with fewer samples. So this paper chooses Micro F1 to evaluate [23].

As for the ROC curve as shown in Fig. 3, the abscissa of the ROC curve is False Positive Rate (FPR). The larger the FPR, the more the actual negative classes in the positive classes are predicted; and the larger the TPR, the more the actual positive classes in the positive classes are predicted. FPR and TPR formulas are shown in Eqs. (9) and (10) as follows:

$$FPR = \frac{FP}{FP + NP} \tag{9}$$

$$TPR = \frac{TP}{TP + FN} \tag{10}$$

So ideally, the perfect case is TPR 1, FPR 0, which is the point in the sample ROC curve of Fig. 3 below. In fact, the closer to the upper left corner of the graph, the better the effect. At the same time, the larger the area under the curve, which is also defined as AUC (Area Under Curve), generally ranging from 0.5 to 1. AUC value is a probability value, i.e. random selection of a positive sample and a negative sample, the ranking probability of selected positive samples is higher than that of randomly selected negative samples. The larger the AUC value is, the more likely the classification algorithm will rank the positive samples ahead of the negative samples, that is, the better the classification effect [24].

Because in the actual data set, there is often a kind of imbalance phenomenon, that is, the proportion of samples is not balanced, and the distribution of samples in the test data may change with time. However, the ROC curve has a good characteristic: when the distribution of samples in the test set changes, the ROC curve can not be maintained. Change [25]. So using ROC curve evaluation can reduce the impact of sample distribution changes on it to a certain extent.

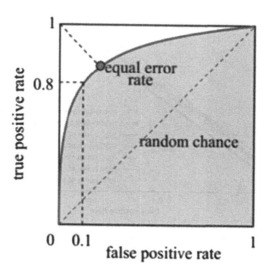

Fig. 3. Sample diagram of ROC curve

4.3   Model Evaluation and Comparison

According to the statistical table of the classification effect in Table 5, we can see that the effect of No Warning is the best. This kind of sample is the majority in the original data. However, in order to find out more students who may have red warning, the future work will focus on class imbalance challenge [26–28]. As shown in Table 5 below (support is the number of corresponding classifications in the test set):

Table 5. Statistics of classification results

Academic achievement early warning	The result of decision tree model			
	Precision	Recall	**f1 score**	Support
No warning	0.84	0.85	0.84	1384
Blue	0.21	0.20	0.21	240
Yellow	0.09	0.08	0.09	72
Orange	0.08	0.08	0.08	12
Red	0.00	0.00	0.00	7
Macro avg	0.24	0.24	0.24	1715
Weighted avg	0.71	0.72	0.71	1715

Figure 4 shows the ROC curve of the decision tree model.

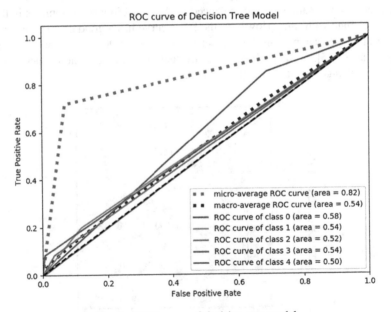

Fig. 4. ROC curve of decision tree model

Combining features, the top five features with the highest correlation are listed as shown in Table 6 below. From this, we can see that under the current data set, the results of college entrance examination are the most effective for the training of the model, so we can think that the academic performance determines the early warning grade of the first semester to a certain extent.

Table 6. Feature correlation ranking

Feature	Importance
tdf (The total score of the college entrance examination with adding points)	0.0805204308576356
zy_fs_pm (Professional Ranking of College Entrance Examination Scores)	0.0786424945260908
Zf (The total score of the college entrance examination)	0.0763173608678484
zf_pjf_d (The difference between the total score of college entrance examination and the average score of the major)	0.0742464268120314
zybm (Major codes)	0.0685766593741163

5 Conclusion

This paper focuses on the problem of students with learning difficulties or their academic level can not meet the school requirements. Based on the big data of education, under the guidance of machine learning theory, this paper studies the academic achievement early warning prediction. Although this paper has realized the academic early warning prediction model for college students, there are still many shortcomings summarized as follows, which need further study and exploration.

- The performance of the model needs to be further improved. In this paper, decision tree model is adopted. The future work will focus on the multi-class imbalance challenge.
- The feature engineering needs to be further improved. Data and features determine the upper limit. For existing data, there are problems of high dimension and data redundancy. The processing of features is far from enough. In the future, feature engineering can be emphasized to further improve the performance of the model.

Acknowledgments. This work is supported by "the Fundamental Research Funds for the Central Universities", South-Central University for Nationalities (CZQ18008, CZP17008), Undergraduate Teaching Quality Engineering (JYZD18035), Natural Science foundation of Hubei Province (BZY16011, BZY18008). We thank relevant departments of South-Central University for Nationalities for providing related data and supporting this work.

References

1. Jayaprakash, S.M., Moody, E.W., Lauría, E.J.M., et al.: Early alert of academically at-risk students: An open source analytics initiative. J. Learn. Anal. **1**(1), 6–47 (2014)
2. Reis, S.M., McCoach, D.B.: The underachievement of gifted students: what do we know and where do we go? Gifted Child Q. **44**(3), 152–170 (2000)
3. Wimshurst, K.J., Wortley, R.K.: Academic success and failure: student characteristics and broader implications for research in higher education. In: Effective Teaching and Learning (ETL) Conference (2004)
4. Anderson, J.R., Boyle, C.F., Reiser, B.J.: Intelligent tutoring systems. Science **228**(4698), 456–462 (1985)
5. Romero, C., Ventura, S.: Educational data mining: a review of the state of the art. IEEE Trans. Syst. Man Cybern. Part C Appl. Rev. **40**(6), 601–618 (2010)
6. Lindsey, R.V., Khajah, M., Mozer, M.C.: Automatic discovery of cognitive skills to improve the prediction of student learning. In: Advances in Neural Information Processing Systems, pp. 1386–1394 (2014)
7. Ritter, S., Anderson, J.R., Koedinger, K.R., et al.: Cognitive tutor: applied research in mathematics education. Psychon. Bull. Rev. **14**(2), 249–255 (2007)
8. Qiu, J., et al.: Modeling and predicting learning behavior in MOOCs. In: Proceedings of the Ninth ACM International Conference on Web Search and Data Mining (WSDM 2016), pp. 93–102. ACM, New York (2016). https://doi.org/10.1145/2835776.2835842
9. Li, X., Wang, T., Wang, H.: Exploring N-gram features in clickstream data for MOOC learning achievement prediction. In: Bao, Z., Trajcevski, G., Chang, L., Hua, W. (eds.) DASFAA 2017. LNCS, vol. 10179, pp. 328–339. Springer, Cham (2017). https://doi.org/10.1007/978-3-319-55705-2_26
10. Huang, Z., et al.: Question difficulty prediction for READING problems in standard tests. In: The 31st AAAI Conference on Artificial Intelligence (AAAI 2017), San Francisco, California, USA, 4–9 February 2017, pp. 1352–1359 (2017)
11. Su, Y., Liu, Q., Liu, Q., et al.: Exercise-enhanced sequential modeling for student performance prediction. In: Thirty-Second AAAI Conference on Artificial Intelligence (2018)
12. Wu, K., Tang, J., Zhang, C.: Where have you been? inferring career trajectory from academic social network. In: Twenty-Seventh International Joint Conference on Artificial Intelligence. IJCAI (2018)
13. Lian, D., Ye, Y., Liu, Q., Zhu, W., Xie, X., Xiong, H.: Mutual reinforcement of academic performance prediction and library book recommendation. In: 16th IEEE International Conference on Data Mining (ICDM 2016), Barcelona, Spain, November 2016, pp. 1023–1028 (2016)
14. Yao, H., Nie, M., Su, H., Xia, H., Lian, D.: Predicting academic performance via semi-supervised learning with constructed campus social network. In: The 22nd International Conference on Database Systems for Advanced Applications (DASFAA 2017), Suzhou, China, March 2017, pp. 581–593 (2017)
15. Nie, M., Yang, L., Sun, J.: Advanced forecasting of career choices for college students based on campus big data. Front. Comput. Sci. **12**(3), 494–503 (2018)
16. Cao, Y., Gao, J., Lian, D., et al.: Orderness predicts academic performance: behavioral analysis on campus lifestyle. J. R. Soc. Interface **15**(146), 20180210 (2017)
17. Bairui, T., et al.: Design of early warning system for student's poor academic performance based on SVM improved by KFCM. Res. Explor. Lab. **38**(05), 112–115 + 228 (2019)

18. Dinovi, K.N.: Using a risk prediction model with first-year college students: Early intervention to support academic achievement. Dissertations & Theses - Gradworks (2011)
19. Hunt, E.B.. Concept learning: an information processing problem. Am. J. Psychol. **77**(2) (1962)
20. Quinlan, J.R.: Induction of decision trees. Mach. Learn. **1**(1), 81–106 (1986)
21. Quinlan, J.R.: C4. 5: Programs for Machine Learning. Elsevier, Amsterdam (2014)
22. Liu, X.-Y., Li, Q.-Q., Zhou, Z.-H.: Learning imbalanced multi-class data with optimal dichotomy weights. In: Proceedings of the 13th IEEE International Conference on Data Mining (ICDM 2013), Dallas, TX, pp. 478–487 (2013)
23. Liu, C., Wang, W., Wang, M., et al.: An efficient instance selection algorithm to reconstruct training set for support vector machine. Knowl.-Based Syst. **116**, 58–73 (2017)
24. Fawcett, T.: An introduction to ROC analysis. Pattern Recogn. Lett. **27**(8), 861–874 (2006)
25. Davis, J., Goadrich, M.: The relationship between Precision-Recall and ROC curves. In: Proceedings of the 23rd International Conference on Machine Learning. ACM, pp. 233–240 (2006)
26. Caizi, X., Wang, X.-Y., Xu, J., Jing, L.-P.: Sample adaptive classifier for imbalanced data. Comput. Sci. **46**(1), 94–99 (2019)
27. Ali, A., Shamsuddin, S.M., Ralescu, A.L.: Classification with class imbalance problem: a review. Int. J. Adv. Soft Comput. Appl. **7**(3), 176–204 (2015)
28. Johnson, J.M., Khoshgoftaar, T.M.: Survey on deep learning with class imbalance. J. Big Data **6**(1), 27 (2019)

Intelligent Detection of Large-Scale KPI Streams Anomaly Based on Transfer Learning

XiaoYan Duan, NingJiang Chen$^{(\boxtimes)}$, and YongSheng Xie

Guangxi University, Nanning 53000, China
chnj@gxu.edu.cn

Abstract. In the complex and variable SDDC (Software Defined Data Center) environment, in order to ensure that applications and services are undisturbed, it needs to closely monitor various KPI (Key Performance Indicators, such as CPU utilization, number of online users, request response delays, etc.) streams of resources and services. However, the frequent launch and update of applications, which produces new KPIs or change the data characteristics of KPIs monitored, so that the original anomaly detection model may become unavailable. So we propose ADT-SHL (Anomaly Detection through Transferring the Shared-hidden-layers model). It first clusters historical KPIs by similarity, and then all KPIs in each cluster are through Shared-hidden-layers method to train VAE (Variational Auto-Encoder) anomaly detection model so as to reconstruct KPIs with corresponding characteristics. When a new KPI is generated, ADT-SHL classify it into similar cluster, and finally model of the cluster is transferred and fine-tuned to detect the new KPI anomalies. This process is rapid, accurate and without manual tuning or labeling for new KPI. and its F-scores range 0.69 to 0.959 for the studied KPIs from two cloud vendors and others, ADT-SHL is only lower a state-of-art supervised method under all labelings by 3.05% and greatly outperforming a state-of-art unsupervised method by 67.46%, and compared with them, ADT-SHL reduce model training time by 94% on average.

Keywords: KPI anomaly detection · Transfer learning · Shared-hidden-layers · KPI similarity · Data center

1 Introduction

With the maturity of SDDC (Software Defined Data Center) technology [1], more and more enterprises are moving services into the cloud to obtain good performance and security. To ensure the reliability and stability of resources, applications and services in the cloud platform, operators will monitor their KPI streams in real time (Key Performance Indicators, such as CPU utilization, request response delay, number of visits per second, etc.). Anomalies on the KPIs (e.g., a peak or a dip) may indicate insufficient resources, server failures, external attacks, or network overload [2], affect the user experience, finally affect revenue of cloud vendors, etc. Therefore, it is necessary to quickly and accurately detect whether the services are abnormal, so as to timely repair and ensure the quality and reliability of the services.

© Springer Nature Singapore Pte Ltd. 2019
H. Jin et al. (Eds.): BigData 2019, CCIS 1120, pp. 366–379, 2019.
https://doi.org/10.1007/978-981-15-1899-7_26

SDDC has a large number of servers and abundant products, and causing the new KPIs appears due to the following three reasons. First, new products are frequently launched into cloud, which will lead to produce new KPI streams. Second, with DevOps, blockchain, machine learning and microservices popular, software changes become more frequent [3], including software upgrades, expansions, shrinkages, migrations, and configuration updates to meet the needs of deploying new function, fixing bugs, and improving system performance, such as capacity expansion is to deploy the service on more servers, but the total request count is stable, the number of requests on a single server will be significantly reduced. Third, the enterprises take corresponding strategies according to market changes, and this will result in changes in the data characteristics of the monitored KPIs. How to quickly adapt to large-scale new characteristic or newly generated KPIs anomaly detection has become a common but urgent problem.

Although there have been many studies on KPI anomaly detection, including traditional statistical algorithms [3, 4], Supervised learning algorithms [2, 5], Unsupervised learning algorithms [6, 7], but they can't solve the above scenario well. So we propose ADT-SHL, a large-scale KPIs anomaly intelligent detection strategy based on Shared-hidden-layers model transferring. The idea of ADT-SHL is based on the following three observations. (1) Many KPIs are similar because of potential associations (such as the server traffic of the load balancing cluster and the response time of the two applications), so similar anomaly detection algorithms and parameters can be used for the similar KPIs, ROCKA [8] cluster and classify KPIs into clusters, but a cluster use the same anomaly detection algorithm and parameters, which makes the accuracy of detection slightly lower. (2) Transfer learning for depth neural network is that train the base network on the source dataset, and then the learned characteristics or model (network weights) are transferred to the second network and fine-tuned on the target dataset, which reduce the time of training target model [9]. (3) The application of VAE (Variational Auto-Encoder) in KPI anomaly detection has achieved remarkable results [6, 7], which gives a reasonable reference for us to choose VAE as the basic algorithm for anomaly detection, but a improved VAE can better solve the problem of large-scale new KPIs anomaly detection.

Based on the above three observations, ADT-SHL firstly clusters history KPIs based on the similarity, then the KPIs of a cluster jointly train the Shared-hidden-layers VAE anomaly detection model, reconstructed KPIs with corresponding characteristics, and then classify the new KPI into a similar cluster, finally, the model of similar cluster is intelligently transferred, fine-tuned and used to detect anomaly of the KPI. The whole process is very fast, without manual algorithms or parameters tuning, and labeling for any new KPI, and solve the problem of low accuracy or needing a large amount of history data of unsupervised learning algorithms, and these are verified in Sect. 4.

The contributions of this paper are summarized as follows:

- For the problem that new KPI has almost no label, the generation time is short, and the amount of data is small. The shared-hidden-layers model of existing KPI is transferred, fine-tunned and used to detect anomaly of the new KPI, without manual algorithm selection, parameters tuning, or label for any new KPI, and reduce training time of the model and the dependence on history data.

- For negative transfer causing performance loss (the transferred model is not suitable for anomaly detection of this KPI, a normal point of this KPI should be judged as anomalous in another KPI), we first fill missing points and extract baseline of history KPIs, classify into clusters by similarity, and then select the transferred model according to the distance with centroid KPI of each cluster.
- All KPIs of a cluster use the shared-hidden-layers way to train VAE detection model from both temporal and spatial dimension, use missing point injection and improved ELBO for training, MCMC imputation for detection, convolutional neural networks for encoding and decoding, and finally reconstruct all KPIs with their own characteristics. So as to simultaneously detect multiple KPIs anomaly.

2   Working Background

2.1   Clustering of KPI Streams

SDDC has a large number of KPIs to monitor and constantly generates new KPIs, luckily, and many KPIs are similar due to their implicit associations. [10] summarizes a number of approaches about time series clustering, such as raw-data-based, feature-based and model-based approaches, most of which are good at clustering idealized and smooth data and are infeasible for KPIs clustering due to a large number of anomalous and missing points.

In this work, we use DTW(Dynamic Time Warping) [10] to calculate the similarity of KPIs, which is capable of calculating similarities between time series of different lengths and shapes, and use well-known DBA(DTW Barycenter Averaging) to extract centroid KPI of every cluster, because it has more faster convergence speed [8] and does not consider the correlation between different dimensions, that is very suitable for one dimensional experimental data like (time, value).

2.2   Anomaly Detection Algorithms for KPI Streams

Anomaly detection algorithms for KPI deal with the problem of recognizing unexpected data points from normal mode, over the years, traditional statistical anomaly detection algorithms have been proposed to calculate the anomaly score [3, 4], these algorithms usually need the operators manually select anomaly detection algorithm and tune parameters for each KPI to improve detection accuracy, which is infeasible for the large-scale new KPIs owing to huge labor costs. In order to avoid the trouble of selecting algorithms and tuning parameters, some researchers put forward that using supervised learning algorithms detects anomaly. For example, ADS [2] and Opprentice [5], they need to collect the feedback information of operators as labels and use the anomaly scores of traditional detectors as characteristics to train the anomaly classifiers, In fact, KPIs are almost unlabeled, and when large-scale KPIs appear, labeling them will cost significant labor and time. In order to reduce the dependence on label data, researchers begin to use unsupervised learning anomaly detection algorithms, which dose not manually select detection algorithm, tune parameters and label for each KPI [6, 7], but they have lower accuracy [11] and require a large amount of training data. For example, the Donut [6] requires up to 6 months of training, which can't meet the fast detection of Kpis anomaly. The new data generation time is short and the amount of data is small and almost unlabeled, it is difficult to provide users with reliable and efficient services by the above methods. and the comparison between these algorithms and ADT-SHL is shown in Table 1.

Table 1. Comparison of KPI anomaly detection algorithms

Algorithms	Select algorithm and tune paramters	History data	Label data	Labor costs	Time consuming
Traditional statistics	Need	Some	No	Higher	Longer
Supervised learning	No	Rely on	Highly rely on	Higher	Long
Unsupervse learning	Some	Highly rely on	No	Some	Longer
ADT-SHL	No	Lighter	No	No	Shorter

2.3   Background of Variational Auto-Encoder

VAE is mainly composed by coding network and decoding network [12], respectively.

$$z \sim \text{Encoder}(x) = q_\phi(z|x), \bar{x} \sim \text{Decoder}(x) = p_\theta(x|z) \tag{1}$$

Where $q_\phi(z|x)$ is called a probability coder, $p_\theta(x|z)$ is called a probability decoder. The loss function $\mathcal{L}_{VAE}$ of the VAE is the sum between expected log likelihood (reconstruction loss function) $\mathcal{L}_{recon\_loss}$ and the KL (Kullback-Leibler) divergence $\mathcal{L}_{latent\_loss}$ which measures the fit degree of the latent variable in the unit Gaussian distribution.

$$\mathcal{L}_{VAE\_loss} = -E_{q_\phi(z|x)}\left[\log\frac{p_\theta(x|z)p_\theta(z)}{q_\phi(z|x)}\right] = \mathcal{L}_{recon\_loss} + \mathcal{L}_{latent\_loss} \tag{2}$$

In Eq. (2), where

$$\mathcal{L}_{recon\_loss} = -E_{q_\phi(z|x)}[\log p_\theta(x|z)] \tag{3}$$

$$\mathcal{L}_{latent\_loss} = \text{KL}[q_\phi(z|x)||p_\theta(x|z)] = E_{q_\phi(z|x)}(\log q_\phi(z|x) - \log p_\theta(z)) \tag{4}$$

3 Anomaly Detection Method Through Transferring Shared-Hidden-Layers Model

We propose the ADT-SHL method, which rapidly detect large-scale KPIs anomaly by transferring and fine-tuning the shared-hidden-layers VAE model of similar cluster. Figure 1 shows the overall framework of ADT-SHL, Firstly ADT-SHL preprocess the historical KPIs, including fill missing points, extract baseline and standardize for similarity judgment, then cluster the preprocessed KPIs according to the similarity and calculate centroid KPI of every cluster (Sect. 3.1). Secondly, all KPIs of a cluster together train the shared-hidden-layers VAE model, and if KPIs are not the same cluster, they do not train together to avoid negative transferring (Sect. 3.2). Thirdly, when a new KPI generates, ADT-SHL classify it into the most similar cluster by calculating the distance with each centroid KPI, and the anomaly detection model of the similar cluster is transferred and fine-tuned, gradually learn the characteristics of new KPI itself, so that the fine-tuned model can rapidly detect anomaly (Sect. 3.3), Finally, we use test data for detection (Sect. 3.4). This process no need to manually select algorithm, tune parameters and label for the newly generated KPIs, which can also reduce the training time of the anomaly detection model (detailed in Sect. 4.3) and the dependence on history data.

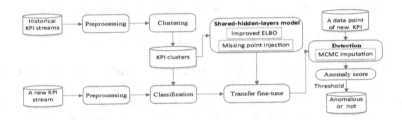

Fig. 1. The framework of ADT-SHL

3.1 Data Preprocessing and Clustering

Anomalous points on KPI usually have different characteristics from normal points. But missing points are different from the anomalous points, which is mainly caused by the monitoring system not working properly or collecting data, and it will increase the difficulty in training and detection of model, so we will fill missing points by using the linear interpolation method [8], and extract the baseline by using the moving average method with a small window of length W on KPI, for each data point x_t of KPI, the corresponding point on the baseline is represented as $x_t^{'} = \frac{1}{W} \sum_{i=1}^{W} x_{t-i+1}$. After standardizating, use DTW to calculate the similarity of KPIs. By the transitivity, if KPI a is similar to KPI b, and KPI b is similar to KPI c, the KPI a, b, and c can be divided into the same cluster, and reduce the similarity calculation between a and c. Use DBA to extract centroid KPI of every cluster, ADT-SHL calculates the centroid KPI only by iterating Eq. (5) until it converges, where D = $\{X_1, X_2, \cdots, X_n\}$ is a KPI cluster, X_i is a KPI, and $\bar{X} = medoid(D)$ is the average series of initial cluster D.

$$\bar{X} = DBA(\bar{X}, D) \tag{5}$$

However, when new KPIs generate, in order to avoid to calculate similarity with all historical KPIs, the calculation with the centroid KPI represented by each cluster is selected. the new KPI will classify the cluster that distance with the centroid KPI is the smallest and satisfy the specific threshold of the DTW algorithm. Otherwise, they become a cluster alone, and the centroid of the cluster is itself. At the same time, in the clustering process, the KPIs of the same type and different shapes can be clustered to different clusters, if the KPI types are different. they can not be divided into the same cluster, and the same cluster can share the parameters of the hidden layers of the anomaly detection model.

3.2   Shared-Hidden-Layers VAE Detection Model Training

The overall network structure of ADT-SHL is illustrated in Fig. 2. In this structure, the input layers and hidden layers extract common features, and KPIs of the same cluster share the parameters of input and hidden layers, but the output layer is not shared and used to learn the unique characteristics of every KPI. This network can be transferred and fine-tuned for another KPI. Using temporal Conv to extract the local features of KPI itself, and spatial Conv to find similar KPI fragments, because the KPIs use the time sliding window as input, the KPIs in a cluster does not mean that each segment of KPIs is similar. And using two Fully Connected Layers calculate the weights of each local feature, so as to obtain the mean(u) and variance vector (σ^2): $u = W_u^T f_\emptyset(x) + b_u$ and $\sigma = \log\left[\exp\left(W_\sigma^T f_\emptyset(x) + b_\sigma\right) + 1\right]$,among them, W_u, W_σ are weight matrix, b_u, b_σ are bias vector, and then combine with Gaussian noise to sample, obtain the latent variable z, and finally reconstruct each KPI stream in a cluster by Deconv, and the output KPI has its own characteristics.

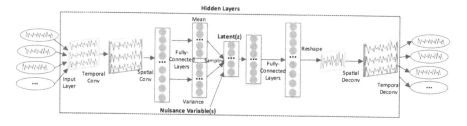

Fig. 2. Network structure of ADT-SHL

VAE is an unsupervised generation algorithm that the target dimension is equal to the input dimension. ADT-SHL works by learning the normal mode, uses time window $x_{t-W+1},\cdots,x_t$ as an input x, the output dimension is W, the moving step is 1, replacing the missing data with other synthetic values may loss performance and difficultly generate enough good normal data, so reset the filled missing points to zero. And in order to let the hidden layers find deeper characteristics and avoid simple learning, the normal points are randomly destroyed by a certain ratio λ, that is $x_t = 0$, and each time window is fixed λW destroyed points, as if they were missing points, and they will be restored after training is over. Since there are many missing points, when give an anomalous KPI, the detection model can be reconstructed well due to long-term contact with anomalous points.

Equation (2) minimizing loss function $\mathcal{L}_{VAE\_loss}$ is difficult to solve, in turn, we find the maximum of the evidence lower bound(ELBO, Eq. (6)).

$$
\begin{aligned}
\mathcal{L}(\theta,\emptyset;x) &= -\mathcal{L}_{VAE\_loss} = -\mathcal{L}_{recon\_loss} - \mathcal{L}_{latent\_loss} \\
&= logp_\theta(x) - \mathrm{KL}[q_\emptyset(z|x)\|p_\theta(x|z)] \\
&= E_{q_\emptyset(z|x)}[logp_\theta(x|z) + logp_\theta(z) - logq_\emptyset(z|x)]
\end{aligned}
\tag{6}
$$

ADT-SHL works by learns the normal mode, so we should ignore the existence of the anomalous mode as much as possible during training process, the ELBO loss function is improved by Eq. 7. where α_i is an index, $\alpha_i = 1$ when x_i is not a anomalous or missing point, otherwise $\alpha_i = 0$, $\sum_{i=1}^{W} a_i/W$ represents the proportion of normal points in x. Due to the priori of x is assumed to be Gaussian Distribution, its likelihood $logp_\theta(x|z)$ can be rewritten as $\sum_{i=1}^{W} a_i logp_\theta(x_i|z)$.

$$
\mathcal{L}(\theta,\emptyset;x) = E_{q_\emptyset(z|x)}\left[\sum_{i=1}^{W} \alpha_i \log p_\theta(x_i|z) + \sum_{i=1}^{W} \frac{\alpha_i}{W} * \log p_\theta(z) - logq_\emptyset(z|x)\right]
\tag{7}
$$

At this time, there is:

$$
\mathcal{L}_{recon\_loss} = E_{q_\emptyset(z|x)}\left(\sum_{i=1}^{W} \alpha_i \log p_\theta(x_i|z)\right)
\tag{8}
$$

$$
\mathcal{L}_{latent\_loss} = E_{q_\emptyset(z|x)}\left(logq_\emptyset(z|x) - \sum_{i=1}^{W} a_i/W * \log p_\theta(z)\right)
\tag{9}
$$

Algorithm 1 Training the shared-hidden-layers VAE anomaly detection model

---

Input: A array D that is a KPI cluster divided by DTW
Output: Shared-hidden-layers VAE anomaly detection model of the cluster
1. $\emptyset, \theta \leftarrow$ initalize network parameters;
2.*array_size* $\leftarrow$ Size(D);   // number of KPIs in the cluster
3.**repeat**
4. $x \leftarrow x_{t-w+1}, x_{t-w}, \cdots, x_t$ from D;
      // Encode x
5. $\mathcal{L}_{latent\_loss} \leftarrow E_{q_\theta(z|x)}\left(\log q_\theta(z|x) - \sum_{i=1}^{W} a_i/W * \log p_\theta(z)\right)$;
6.   **If**(*array_size*$\neq$1){ //not Dropout when the cluste has multiple KPIs
7.   intermediate_encode_data$\leftarrow$*ReLU(BN(W$\otimes$x + b))*;
8.   **else**
9.   intermediate_encode_data$\leftarrow$ *Dropout(ReLU(BN(W$\otimes$x + b)))*;
10. $\mu, \sigma \leftarrow$Fully-Connected Layers on intermediate_encode_data; //u is mean,and σ is variance
11. $N(0,I)\leftarrow\mu, \sigma$ and noise s generate standard normal distribution;
12. $z\leftarrow$ samples from prior $N(0, I)$;
      // Decode the latent variable z and reconstruct x
13. $\mathcal{L}_{recon\_loss} \leftarrow E_{q_\theta(z|x)}(\sum_{i=1}^{W} a_i \log p_\theta(x_i|z)$); // Equation (8)
14.   intermediate_decode_data$\leftarrow$Fully-Connected Layers on z and ReLU operation;
15.   **If**(*array_size*$\neq$1)
16. $\tilde{x} \leftarrow$*ReLU(BN(W'$\otimes$intermediate_decode_data + b))*;
17.   **else**
18. $\tilde{x} \leftarrow$ *Dropout(ReLU(BN(W'$\otimes$intermediate_decode_data + b)))*;
        //update parameters according to gradients using Adam
19. $\emptyset \overset{+}{\leftarrow} -\nabla_\emptyset(\mathcal{L}_{latent_{loss}} + \mathcal{L}_{recon\_loss})$;
20. $\theta \overset{+}{\leftarrow} -\nabla_\theta(\mathcal{L}_{recon\_loss})$;
21.**Until** $\mathcal{L}(\theta, \emptyset; x)$ get the max value
22.**return** Shared-hidden-layers VAE anomaly detection model of the cluster

---

As shown in Fig. 2, 1D CNN is selected as the coding and decoding network of VAE, and compared with the MLP(multi-layer perceptron, the VAE default coding and decoding network), which is equivalent to CNN where convolution kernel size and each layer inputs size is same. The basic blocks of ADT-SHL network are the combination of the convolutional layer $y = W \otimes x + b$, the batch normalization layer $s = BN(y)$ and the ReLU activation layer $h = ReLU(s)$. We use Dropout to avoid overfitting when the cluster only has a KPI, but don't use it when have multiple KPIs, and use batch normalization to accelerate convergence and improve generalization [13], ReLU is used to join non-Linear element during training so that can learn more features. KPIs of the same cluster share weights and together train VAE anomaly detection model due to the higher similarities, and training at the same time avoid that the model is over-fitting a KPI stream. The whole training process is shown in Algorithm 1, the complexity of model training is $O(n^2)$, n is the KPI number of a cluster, this model will reconstruct all KPIs of the cluster, with each KPI's own characteristics.

3.3 Model Adaption for New KPIs

When a new KPI generates, the KPI is first preprocessed, and calculate the distance with all centroid KPIs by using DTW. If the distance is within the threshold and is the smallest, the KPI is classified into similar cluster. The shared-hidden-layers model of the cluster is transferred, fine-tuned and used as the anomaly detection model of the KPI. The first few layers of ADT-SHL learn relatively generalized characteristics, and as the layers deepen, the unique characteristics of each KPI are learned. If the distance is not within the threshold range, the new KPI is separately classified into a cluster, and the anomaly detection model is trained from the beginning by using the proposed way in Sect. 3.2, and the model is used to detect anomaly in accordance with Sect. 3.4.

In most cases, the monitored KPIs are no labels, when transfer and fine-tune, there are three adjustment strategies for selection that is based on whether target domain has labels. The first is Supervised-Finetune, which uses label information to backpropagate and fine-tune ADT-SHL network, the second is for the unlabeled target domain, use Adam's reverse fine-tuning the entire network and is called as Unsupervised-Finetune, and the third fine-tune the learning depth feature's decoding layers and is called as Unsupervised-Decode, among them, the second and third are unsupervised adjustments. In the process of finding the optimal value of hyperparameters, if the learning rate is too small, the convergence will be too slow, but too large learning rate can lead to divergence. So we use an adaptive learning rate scheduler to calculate the ideal learning rate for the loss function, and use it into the inverse fine-tuning in the target domain, the final convergence model is used for anomaly detection of this KPI.

3.4   Detection

During KPI anomaly detection, the anomalous and missing points of a test window may make reconstruction inaccurate, and missing points are known but anomalous points aren't known at the beginning, ADT-SHL use MCMC [6] to interpolate missing points, reducing the bias of missing points incoming. After the test window x is placed into the training model, the missing points information of the reconstructed window is used instead of the missing points in x, and the non-missing points keep unchanged, this process is repeated L times, an iterative of MCMC interpolation is as shown in Fig. 3, where $x_{missing}$ represents the missing points in the KPI test window x, the non-missing points in x are represented by x_{others}, if we select Unsupervised-Decode strategy to fine-tune, x is reconstructed using $z \to p_\theta(x|z)$, and the reconstructed x is represented by $\bar{x}$, otherwise x is reconstructed by $q_\phi(z|x) \to z \to p_\theta(x|z)$, and we calculate the anomaly score by $\mathcal{L}_{recon\_loss} = -E_{q_\phi(z|x)}[\log p_\theta(x|z)]$.

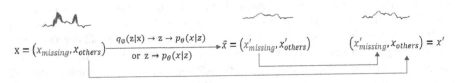

$$x = \left(x_{missing}, x_{others}\right) \xrightarrow[\text{or } z \to p_\theta(x|z)]{q_\phi(z|x) \to z \to p_\theta(x|z)} \bar{x} = \left(x'_{missing}, x'_{others}\right) \qquad \left(x'_{missing}, x'_{others}\right) = x'$$

Fig. 3.   An iterative of MCMC interpolation

After MCMC iteration, we only calculate anomaly score of the last point, such as x_t in $x_{t-W+1},\dots,x_t$, and each time window only calculates the last point, then the score is compared with setted threshold, if the score is greater than the threshold, x_t is judged as an anomalous point, otherwise, x_t is nomal. When this window has been detected, moving a unit to continue loop detection until end.

4   Evaluation

KPI anomaly detection can be regarded as a two-class classification problem, a data point is classified into a normal point or anomalous point, and the performance of the anomaly detection algorithms can be evaluated by using the indicators of two-class (precision, recall, and their comprehensive metric F-score), besides, we use model training time to verify the adaptation effect of transfer learning for KPI anomaly detection.

In order to evaluate the performance of ADT-SHL, we carried out a large number of experiments from two cloud vendors and other real data. the experiment datasets are first introduced, and then compare and verify the performance from the two aspects which include KPI anomaly detection and rapid adaptation. In all experiments, z dimension is set to 100, the Spatial Conv kernel of size E = 64, the Temporal Conv kernel of size C = 120, and the fully connected layer in the encoding and decoding network have 100 nodes. The window size in the baseline extraction is 1, and the time window is set to 120 points during model training, because if the time is too small, the detection model can not capture the normal mode. but if it is too large to easily cause over-fitting. The normal data destroyed ratio is $\lambda = 0.01$, iteration number of MCMC and DBA are fixed to 10, ADT-SHL is implemented by TensorFlow and use an Adam optimizer with an initial learning rate of 0.001, and use an adaptive learning rate for fine-tuning.

4.1   Datasets

In order to verify whether ADT-SHL can be used in the data center, we select public data from two cloud vendors, Yahoo and Donut as our experiments data. The details are shown in the Table 2, among them, Alibaba Cluster Data [14] contains five types of KPIs from 4034 servers and 71476 containers, Baidu [15] contains service indicators (CPU utilization, network bytes, disk read bytes, etc.) and business indicators (online advertising click rate, traffic etc.), Yahoo [16] contains more missing data so that we can verify missing points filling and MCMC interpolation technology, randomly select 75% of each as historical data, and the other 25% are newly generated KPIs, the last column in Table 2 indicates the cluster number of each dataset. In order to evaluate anomaly detection performance of ADT-SHL on a KPI, we also use Donut two public and different types of KPIs [6] which are called KPI A and KPI B. 60% for training, 40% for detection.

4.2   KPI Anomaly Detection Performance Evaluation

To evaluate the performance of ADT-SHL in KPIs anomaly detection, compare it with state-of-art supervised learning algorithm Opprentice [5] (base Random Forest and outperform most traditional statistical algorithms) and state-of-art unsupervised

Table 2. KPIs dataset statistics

Data Source	Total KPIs	KPI types	History KPIs	New KPIs	Clusters
Alibaba Cluster Data [14]	377550	10	283163	94387	1276
Baidu [15]	668	18	501	167	49
Yahoo [16]	367	24	276	91	40
Donut [6]	2(A, B)	2	No	2	2

learning algorithm Donut [6] (the first VAE-based anomaly detection algorithm with theoretical explanation). Donut uses the same network parameters as ADT-SHL, Opprentice maximum depth is 6 and component is setted to 200. The F-score evaluation is carried out on KPI A and KPI B. We select KPI A and KPI B as experimental data because those data are fully compliant with the Donut application scenario, and full labelings are also suitable for Opprentice evaluation.

As seen from Fig. 4, we compare the F-Score values of Donut, ADT-SHL, and Opprentice at KPI A and B, the overall F-Score values in B are higher than A. When A has a 100% label, the performance of ADT-SHL is slightly lower than that of Opprentice but the difference is not much, and is 0.03 higher than Donut. When B has a 100% label, it is again verified that the performance of ADT-SHL and Opprentice is similar, and is 0.034 higher than Donut. ADT-SHL learns from the time and space dimensions and uses fully connected layers to count learned characteristics, it can learns deeper features than Donut. However, we also found that ADT-SHL converges slower than Donut because of the complicated network structure.

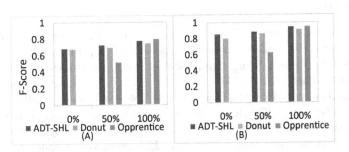

Fig. 4. F-score comparison on KPI A and B. 0%, 50%, 100% indicates the label ratio of training, at 0%, Opprentice can't detect anomaly.

4.3 Rapid Adaptation Performance Evaluation

In order to further verify that ADT-SHL can quickly detect anomaly of a new KPI, we select a semi-supervised KPI anomaly detection algorithm ADS [2] and a combinational algorithm of DTW + DBA + Donut for comparation, among them, the ADS first use ROCKA [8] for KPIs clustering, then extract the cluster centroid KPI and label it, when a new KPI arrives, classfy and train anomaly detection model together with centroid KPI. DTW + DBA + Donut first use DTW for clustering and DBA for centroid extraction, then use Donut to train model of centroid KPI, and manually adjust the trained models for anomaly detection of newly generated KPIs. We select all new KPIs with similar clusters as experimental target data, and Sect. 3.3 adjustment strategy is automatically selected, all models calculate the average F-score under the optimal threshold, we start timing after extracting centroid KPI of each cluster, and average F-score values of all models in different time periods are shown in Table 3.

Table 3. Average F-Score% in different time periods of Baidu, Alibaba Cluster Data, and Yahoo datasets (The best F-Score for each time period is indicated by bold)

Time	Model	Dataset		
		Baidu	Alibaba Cluster Data	Yahoo
10-min	ADT-SHL	**48.34**	**46.59**	**50.67**
	ADS	40.54	39.78	42.56
	DTW + DBA + Donut	32.77	28.85	33.45
30-min	ADT-SHL	**56.67**	**49.89**	**54.7**
	ADS	47.78	43.86	50.98
	DTW + DBA + Donut	37.45	36.49	45.67
1-h	ADT-SHL	**63.46**	**57.69**	63.78
	ADS	60.87	54.78	**61.85**
	DTW + DBA + Donut	46.68	42.32	52.5
1.5-h	ADT-SHL	65.89	60.53	68.76
	ADS	**69.48**	**64.88**	**70.9**
	DTW + DBA + Donut	48.23	46.34	53.64
2-h	ADT-SHL	**75.56**	**72.72**	**76.36**
	ADS	70.68	68.71	73.78
	DTW + DBA + Donut	55.83	50.92	59.48

As we can see from Table 3, in the first 10 min and 30 min, ADT-SHL shows a better F-score, but ADS takes a lot of time to label the centroid KPI. After 1-h, ADT-SHL is similar to ADS, ADS performance is best at 1.5-h, our model has learned the characteristics of new KPI through fine-tuning convergence at 2-h, and the performance is gradually restored. But in whole process, DTW + DBA + Donut is always weak due to slow convergence. ADT-SHL can have higher performance within 1 h of transferring and fine-tuning, it can reconstruct KPI with its own characteristics, and the performance is greatly improved.

In all the above experiments, although the Opprentice can perform slightly better than ADT-SHL by 3.05% when KPIs are all labeled, but it is infeasible for the large-scale new KPIs anomaly detection owing to a lot of labeling time and huge human cost. ADT-SHL learns the characteristics of the new KPI by fine-tuning convergence, and its adaptation performance is better and faster than semi-supervised algorithm ADS and the combinational algorithm of DTW + DBA + Donut, which shows that ADT-SHL can quickly and accurately detect KPI anomaly under complex and variable SDDC environment, and ensure the reliability of services, but the convergence of ADT-SHL is a little bit slower.

5 Conclusion

In this paper, we present ADT-SHL, a KPI anomaly detection algorithm by intelligently transferring the shared-hidden-layers VAE model, which can quickly and accurately detect anomaly of large-scale KPIs, without manual algorithm selection, parameters tuning or labeling for any KPI, and can cope with the complex and volatile SDDC environment to ensure the normal operation of various applications and services. ADT-SHL firstly clusters historical KPIs by similarity, then all KPIs of a cluster use the shared-hidden-layers way to train VAE detection model, when a new KPI generates, the model of similar cluster is transferred and fine-tuned for anomaly detection of the KPI. our extensive experiments show that ADT-SHL obtain F-score values from 0.69 to 0.95 on 283940 historical KPIs and 94647 new KPIs, almost the same as a state-of-art supervised algorithm Opprentice and greatly outperforming a state-of-art unsupervised algorithm Donut. In the future work, we will further optimize the backpropagate process of ADT-SHL by using learned characteristics in the GAN discriminator as basis for the VAE reconstruction KPIs.

Acknowledgment. This work was is supported by the National Natural Science Foundation of China (61762008), the Natural Science Foundation Project of Guangxi (2017GXNSFAA198141) and Key R&D project of Guangxi (No. Guike AB17195014).

References

1. Software Defined Data Center. https://www.vmware.com/cn/solutions/software-defined-datacenter.html. Accessed 23 Jun 2018
2. Bu, J., Liu, Y., Zhang, S., et al.: Rapid deployment of anomaly detection models for large number of emerging KPI streams. In: 2018 IEEE 36th International Performance Computing and Communications Conference (IPCCC), pp. 1–8. IEEE. Orlando (2018)
3. Chen, Y., Mahajan, R., Sridharan, B., et al.: A provider-side view of web search response time. In: ACM SIGCOMM Computer Communication Review, vol. 43, no. 4, pp. 243–254. ACM, New York (2013)
4. Choffnes, D.R., Bustamante, F.E., Ge, Z.: Crowdsourcing service-level network event monitoring. ACM SIGCOMM Comput. Commun. Rev. 41(4), 387–398 (2011)
5. Liu, D., Zhao, Y., Xu, H., et al.: Opprentice: towards practical and automatic anomaly detection through machine learning. In: Proceedings of the 2015 Internet Measurement Conference, pp. 211–224. ACM, Tokyo (2015)
6. Xu, H., Chen, W., Zhao, N., et al.: Unsupervised anomaly detection via variational auto-encoder for seasonal KPIs in web applications. In: Proceedings of the 2018 World Wide Web Conference on World Wide Web. International World Wide Web Conferences Steering Committee, Lyon, France, pp. 187–196 (2018)
7. Li, Z., Chen, W., Pei, D.: Robust and unsupervised KPI anomaly detection based on conditional variational autoencoder. In: 2018 IEEE 37th International Performance Computing and Communications Conference (IPCCC), pp. 1–9. IEEE, Orlando (2018)

8. Li, Z., Zhao, Y., Liu, R., et al.: Robust and rapid clustering of kpis for large-scale anomaly detection. In: 2018 IEEE/ACM 26th International Symposium on Quality of Service (IWQoS), pp. 1–10. IEEE, Banff (2018)
9. Csurka, G.: Domain adaptation for visual applications: a comprehensive survey. arXiv preprint arXiv:1702.05374 (2017)
10. Petitjean, F., Forestier, G., Webb, G.I., et al.: Dynamic time warping averaging of time series allows faster and more accurate classification. In: 2014 IEEE International Conference on Data Mining, pp. 470–479. IEEE, Shenzhen (2014)
11. Zhang, Y.L., Li, L., Zhou, J., et al.: Anomaly detection with partially observed anomalies. In: Companion of the Web Conference 2018 on The Web Conference 2018. International World Wide Web Conferences Steering Committee, Lyon, France, pp. 639–646 (2018)
12. Diederik, P., Welling, M.: Auto-encoding variational Bayes. In: Proceedings of the International Conference on Learning Representations (2013)
13. Ioffe, S., Szegedy, C.: Batch normalization: accelerating deep network training by reducing internal covariate shift. In: International Conference on International Conference on Machine Learning, Lille, France, vol. 37, pp. 448–456 (2015)
14. Alibaba Cluster Data. https://github.com/alibaba/clusterdata/tree/master/cluster-trace-v2018. Accessed 10 May 2019
15. Baidu Public Data. https://github.com/baidu/Curve. Accessed 10 Apr 2019
16. Yahoo Public Data. https://webscope.sandbox.yahoo.com/. Accessed 15 Jun 2019

Latent Feature Representation for Cohesive Community Detection Based on Convolutional Auto-Encoder

Chun Li, Wenfeng Shi, and Lin Shang$^{(\boxtimes)}$

State Key Laboratory for Novel Software Technology,
Nanjing University, Nanjing 210023, China
lichun@smail.nju.edu.cn, ffeng1993@gmail.com, shanglin@nju.edu.cn

Abstract. It is important to identify community structures for characterizing and understanding complex systems. Community detection models, like stochastic models and modularity maximization models, lack of the ability of nonlinear mapping, which leads to unsatisfactory performance when detecting communities of complex real-world networks. To address this issue, we propose a nonlinear method based on Convolutional Auto-Encoder (ConvAE) to improve the cohesiveness of community detection. We combine the convolution neural network and auto-encoder to improve the ability of nonlinear mapping of the proposed model. Moreover, to better characterize relations between nodes, we redefine the similarity between nodes for preprocessing the input data. We conduct extensive experiments on both the synthetic networks and real-world networks, and the results demonstrate the effectiveness of the proposed method and the superior performance over traditional methods and other deep learning based methods.

Keywords: Community detection · Latent feature representation · Auto-Encoder · Convolutional Neural Network

1 Introduction

Identifying community structure in networks has attracted significant attention in the past few years because of its wide applications on various fields, such as finding protein complexes in biological networks and obtaining disciplinary groups in collaborative networks. As an important part of community structure identification, community detection aims to identify groups of interacting vertices in a network depending on the properties of network structure. Since networks in real-world contain a large amount of nodes and the relationships between these nodes are intricate, it is inefficient to perform clustering algorithm directly on initial representations, hence the key issue of community detection is

Supported by organization x.

H. Jin et al. (Eds.): BigData 2019, CCIS 1120, pp. 380–394, 2019.
https://doi.org/10.1007/978-981-15-1899-7_27

obtaining efficient latent feature representations of the networks. Once the representations exactly reflect the cohesiveness of the network, clustering algorithms are expected to get properly communities in this latent space.

In the past few decades, various algorithms have been proposed to tackle the issue of community detection. Existing community detection methods can be roughly divided into two categories, modularity maximization methods and stochastic models. The former attempts to improve the modularity function, which is introduced by Newman [24]. And the latter focuses on deriving generative models. Essentially, both of them are mapping a network to a feature representation in latent space, where different communities are more separable for clustering algorithms. However, since eigenvalue decomposition (EVD) and nonnegative matrix factorization (NMF) [27] adopted by the modularity maximization methods and the stochastic models are linear mapping, these models only work well on condition that a linear mapping is constructed from the original feature space to the ideal latent feature space. Unfortunately, there are plenty of nonlinear properties in networks. For instance, the connection between two nodes is nonlinear. So the performance of current community detection methods are severely restricted when applied to real-world problems.

Recently, some researchers utilize neural networks to extract the latent feature representation of complex network, such as the deep nonlinear reconstruction (DNR) [21] and the community detection algorithm based on deep sparse Auto-Encoder (CoDDA) [16]. Both of them benefit from the nonlinear provided by neural network and achieve better performance compared to their linear competitors. However, both the DNR and CoDDA are using full connected layers where numerous parameters are needed to learn, hence the learning process is difficult.

Deep learning is popular in the past few years. In deep learning based models, the output of each layer serves as the input of the successive layer. Multiple processing layers in neural networks are performed to learn representations of data to form the abstraction of different levels. To take advantage of the local connectivity and weight sharing of convolution, we combine convolution and auto-encoder and propose an algorithm which adopt the Convolutional Auto-Encoder to detect the community in social network.

The contributions of this work can be summarized in three folds.

- We propose an unsupervised deep learning based algorithm called ConvAE to obtain better latent feature representation of network which is beneficial to community detection.
- We redefine the similarity of two nodes for preprocessing the input data.
- Abundant experiments are conducted on different datasets and the experimental results demonstrate the superior performance compared to the previous arts.

The rest parts of this paper are organized as follows. we will introduce the feature representation of social network and the deep learning including Auto-Encoder and Convolutional Neural Network in Sect. 2. The ConvAE we proposed and the preprocessing for the input data will be elaborated in Sect. 3. The Sect. 4

will show the experimental results to validate the exactitude and efficiency of our method. In Sect. 5, we will conclude this paper.

2   Related Work

Given a undirected and unweighted graph $G = (V, E)$ which is demonstrated by an adjacency matrix A, $V = (v_1, v_2, ..., v_N)$ represents N vertices and $E = \{e_{ij}\}$ represents edges. Each edge in E connects two vertices in V. The matrix $A = \{a_{ij}\} \in R_+^{N \times N}$ is a nonnegative symmetric matrix, where $a_{ij} = 1$ if the vertex i connects the vertex j, or $a_{ij} = 0$ if vertex i doesn't connects vertex j. Also, for all $1 \leq i \leq N$, $a_{ii} = 0$. We define $s_i = \sum_{j=1}^{N} a_{ij}$ as the degree of vertex i.

2.1   Latent Feature Representation of Community

Most algorithms of community detection focus on finding an effective representation of network.

Stochastic model [14] adopts NMF to map A into latent space H,

$$A \approx \hat{A} = HH^T \tag{1}$$

where $H = \{h_{ij}\} \in R^{N \times K}$ is a nonnegative membership matrix and h_{ij} represents the probability that node i is a member of community j. Thus, each row of H can be represented as a community membership distribution of the corresponding vertex. And the loss function can be defined to minimize the distance between A and HH^T. The distance function can be square of Frobenius norm [27,29], which is,

$$L(A, HH^T) = ||A - HH^T||_F^2 \tag{2}$$

Varied stochastic models, such as nonnegative matrix tri-factorization [33] and stochastic block model [19], have been proposed to find more effective representations. Especially, the mapping of stochastic block model is,

$$A = ZBZ^T \tag{3}$$

where $Z \in \{0, 1\}^{N \times K}$ is a fixed matrix that has exactly a single 1 in each row and at least one 1 in each column. $B \in [0, 1]^{K \times K}$ is a full rank and symmetric matrix. Z denotes which community the node pertains to, e.g. $Z_{ik} = 1$ represents that the node i is a member of community k. The matrix B consists of the probability of edges within and between communities.

The modularity maximization method employs the singular value decomposition (SVD) to mapping the input to latent representation for reconstruct the modularity matrix. The modularity can be computed by,

$$Q = \frac{1}{2m} \sum_{i,j \in V} (a_{ij} - \frac{s_i s_j}{2m}) \delta(c_i, c_j) \tag{4}$$

where m is the number of edges, and $\delta(c_i, c_j) = 1$ if node i and j belongs to the same community, otherwise $\delta(c_i, c_j) = 0$. By using matrix $B \in R^{N \times N}$ where $b_{ij} = a_{ij} - \frac{s_i s_j}{2m}$, we can obtain,

$$Q = Tr(H^T B H), s.t. Tr(H^T H) = N \tag{5}$$

where $Tr(.)$ represents the trace of a matrix. Based on Rayleigh Quotient [26], the largest K eigenvectors of the B are what we intend to find. To get them, there exist orthogonal decomposition $B = U \Lambda U^T$, where Λ is diagonal matrix with the eigenvalue of B as the diagonal elements and $UU^T = I$. Thus, above community detection algorithms can be regarded as mapping a network to a latent feature representation.

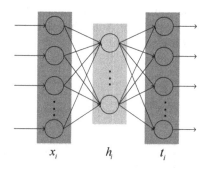

Fig. 1. The structure of an Auto-Encoder (AE).

2.2 Auto-Encoder

Auto-Encoder is a method used to learn efficient data codings in an unsupervised manner [9]. As is shown in Fig. 1, a Auto-Encoder consists of an input layer, an output layer and several hidden layers. Note that the number of nodes in the output layer is the same as which in the input layer so as to reconstruct the input of Auto-Encoder. Also, a Auto-Encoder contains two parts, the encoder and the decoder.

In Auto-Encoder, the encoder part maps the original data X to a latent embedding $H = [h_{ij}] \in R^{d \times N}$ where $d < N$. And the h_i, which is the ith column of H, represents vertex i in latent space,

$$h_i = f(x_i) = s_f(W_H x_i + d_H) \tag{6}$$

where $W_H \in R^{d \times N}$ is the weight matrix and $d_H \in R^{d \times 1}$ is the bias of encoder layer. Both of them are learnable. s_f is the activation function. After h_i is obtained, the decoder layer takes the latent representation h_i as input and reconstruct x_i following equation (7).

$$t_i = g(h_i) = s_g(W_T h_i + d_T). \tag{7}$$

where t_i is the result of reconstruction, $W_T \in R^{N \times d}$ is the weight matrix and $d_T \in R^{N \times 1}$ is the bias of decoder layer. Similarly, s_g is the activation function like s_f. To reconstruct the original data X accurately, it is natural to minimize the difference between X and T following the equation (8).

$$
\begin{aligned}
\arg \min \mathcal{L}(X, T) &= \arg \min \|X - T\|^2 \\
&= \arg \min \sum_{i=1}^{N} \|x_i - t_i\|^2 \\
&= \arg \min \sum_{i=1}^{N} \|x_i - s_g(W_T(s_f(W_H x_i + d_H)) + d_T)\|^2
\end{aligned}
\tag{8}
$$

where $\mathcal{L}(X, T)$ is the distance function which denotes the nuance between X and T. Once training process is over, the Auto-Encoder is able to extract efficient latent feature representation H of input.

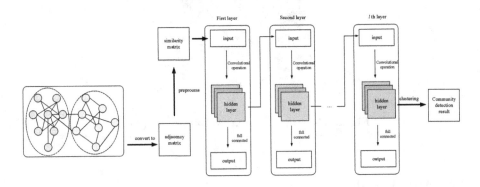

Fig. 2. The framework of ConvAE

3   Latent Feature Representation for Cohesive Community Detection Based on Convolutional Auto-Encoder

The framework of our method is shown in Fig. 2. Firstly, the network G is converted into its adjacency matrix A. We further preprocess the adjacency matrix to get the similarity matrix X which serves as the input of our models. Secondly, we train ConvAE layer by layer. After the training process finished, we regard the output of the encoder of lth layer as the latent feature representation of the network G. Finally, we use the latent feature representation and adopt clustering algorithm to find communities of network G.

3.1 The Similarity of Nodes

The adjacency matrix A are widely used to measure the connectivity of two nodes, but it lacks the ability to describe the similarity of two unconnected nodes [16]. Hence some information which is vital for community representation may be lost if we take the adjacency matrix as input directly. So we introduce a new similarity matrix of network G for better community representation.

The similarity of nodes is related with the distance between two nodes and the number of common adjacent nodes. In general, the shorter the distance between two nodes is, the higher the similarity is. And vice versa. Thus, if two nodes are close enough and share many adjacent nodes, there is a good probability that they belong to the same community.

The proposed similarity of node i and node j is defined as,

$$sim(i,j) = \alpha e^{1-dis(i,j)} + \gamma(1 - e^{-com(i,j)}) \tag{9}$$

where $dis(i, j)$ is the length of the shortest path between node i and node j, $com(i, j)$ represents the number of common adjacent nodes of node i and node j. α and γ are the weight factors which decide the importance of two parts.

We use the new similarity measure to build the input matrix X, where $x_{ij} = sim(i,j)$.

3.2 Structure of ConvAE

Motivated by the strong ability of convolutional neural network in feature extraction, the fully connected layer in encoder of Auto-Encoder is replaced by a convolutional layer in our model. For the input data $X \in R^{N \times N}$, we use the convolutional layer to extract the nonlinear features of network G. The convolutional layer has K kernels $\mathcal{W} = \{W_1, W_2, ..., W_K\}$ and biases $\mathcal{B} = \{b_1, b_2, ..., b_K\}$. Computing with the kernel weight W_k and bias b_k, we can extract a feature map H_k,

$$H_k = \sigma(W_k * X + b_k) \tag{10}$$

where $\sigma(.)$ is an activation function, like sigmoid function or ReLU. W_k is k^{th} convolution kernel of which the size is $m \times m$. In order to avoid the decrease of dimension of h_i^k, we employ padding when applying the convolution operation. After getting K feature maps $\mathcal{H} = \{H_1, H_2, ..., H_K\}$ where $H_k \in R^{N \times N}$ and $h_i^k \in R^{N \times 1}$, we connect all the feature maps to build a matrix T which shares the same size with the input as follows:

$$t_i = \sigma(\sum_{k=1}^{K} W_H^k h_i^k + d_H) \tag{11}$$

where $W_H^k \in R^{N \times N}$ and $d_H \in R^{N \times 1}$ are the kth weight matrix and the bias of output layer respectively. Similarly, $\sigma(.)$ is an activation function, like sigmoid function or ReLU.

Similar to Auto-Encoder, we also aim at learning a latent nonlinear representation $\mathcal{H}$ which can appropriately reconstruct the original input X. Thus, we minimize the reconstruction error between X and T by optimizing the parameter $\phi = \{\mathcal{W}, \mathcal{B}, W_H, d_H\}$,

$$
\begin{aligned}
\phi &= \arg\min_{\phi} L_{\phi}(X, T) = \arg\min_{\phi} \sum_{i=1}^{N} L_{\phi}(x_i - t_i) \\
&= \arg\min_{\phi} \sum_{i=1}^{N} L_{\phi}(x_i - \sigma(\sum_{k=1}^{K} W_H^k h_i^k + d_H)) \\
&= \arg\min_{\phi} \sum_{i=1}^{N} L_{\phi}(x_i - \sigma(\sum_{k=1}^{K} W_H^k \sigma(W_k^{m \times m} * x^{m \times m} + b_k) \\
&\quad + d_H))
\end{aligned}
\tag{12}
$$

where $L_{\phi}(X, T)$ represents the distance function which measures the reconstruction error. Specifically, we apply Euclidean function as the distance function.

As can be seen, there are a lot of parameters in K hidden layers, so we add sparsity restriction on our model. We adopt KL divergence as the sparsity restriction function and use L2 norm as the regularization to avoid overfitting as well as reducing the complexity of model.

The KL divergence can be defined as,

$$
\frac{1}{K} \sum_{k=1}^{K} KL(p \| \frac{1}{N} \sum_{i=1}^{N} h_i^k)
\tag{13}
$$

where p is a constant close to 0, $p_k = \frac{1}{N} \sum_{i=1}^{N} h_i^k$ is the average activation of k_{th} hidden layers. $KL(p \| p_k)$ is the relative entropy from p to p_k. The equation of $KL(p \| p_k)$ is,

$$
KL(p \| p_k) = p \log \frac{p}{p_k} + (1 - p) \log \frac{1 - p}{1 - p_k}
\tag{14}
$$

We also add L2 norm at the cost function. The final cost function is,

$$
\begin{aligned}
\arg\min_{\phi} (\sum_{i=1}^{N} L_{\phi}(x_i - \sigma(\sum_{k=1}^{K} W_H^k \sigma(W_k^{m \times m} * x^{m \times m} + b_k) \\
+ d_H)) + \beta \frac{1}{K} \sum_{k=1}^{K} KL(p \| \frac{1}{N} \sum_{i=1}^{N} h_i^k) + \lambda(\|\mathcal{W}\|_2 + \|W_H\|_2)
\end{aligned}
\tag{15}
$$

while β and λ are hyperparameters which control the relative importance of the sparsity restriction and the penalty of model complexity compared to the reconstruction error, respectively.

All the above are just one layer of our model. However, with the increase of the number of layers, the amount of learnable parameters also grows exponentially, which makes adjusting these parameters inefficient. So we follow the strategy that has been applied on Stacked Auto-Encoder [5] and train our model layer by layer. Specifically, for the first layer, the input is $X^{(1)}$, the K feature maps are $\mathcal{H}^{(1)} = \{H_1^{(1)}, H_2^{(1)}, ..., H_K^{(1)}\}$, and the output is $T^{(1)}$. After we have trained the first layer, we sum up the K feature maps of the first layer and average them out to obtain a matrix $X^{(2)} = \frac{H_1^{(1)} + H_2^{(1)} + ... + H_K^{(1)}}{K}$ with the same size of the input of the first layer $X^{(1)}$. For the second layer, the input is $X^{(2)}$ and we can get matrix $X^{(3)}$ in the same way. When the training process of the last layer finished, we can obtain matrix $X^{(l)}$ which is the final feature representation of the input network. To measure the quality of the feature representation, we adopt clustering algorithm K-Means to detect communities by using the feature representation since the community detection can be regarded as a type of clustering problem.

4 Experiments

In this section, firstly, we introduce the real-world datasets and experiments setup. Then we evaluate our method with K-Means clustering algorithm on these synthetic and real-world networks respectively. Finally, we apply our feature representations on different clustering algorithms to identify communities, and evaluate them by the results of community detection.

4.1 Description of Datasets

To validate the effectiveness of our proposed method, we choose two types of real-world datasets for our experiments. One with known community structures and the other with unknown community structures. The former is described as follows:

Football: a network recording football games of Division IA colleges in regular season Fall 2000 [24].

Karate: a social network represents the friendships of members in karate club at a US university in 1970s [31].

Polblogs: a directed network of hyperlinks between weblogs on US politics in 2005 which is recorded by Adamic and Glance [20].

Dolphins: a social network which represents the frequent associations of 62 dolphins in New Zealand.

Polbooks: a network in which 105 politics books are collected by Kerbs.

School Friendship: a social network which is based on self-reporting from the students in high school and the partition of the network is roughly corresponds to the different grade of students in the survey [17].

The statistics of above networks are shown in Table 1.

Similarly, Table 2 shows the statistics of the latter.

Table 1. Characteristics of some real-world network with ground-truth

Dataset	Nodes	Edges	Communities
Football	115	613	12
Karate Club	34	78	2
Polblogs	1490	16718	2
Dolphins	62	159	2
Polbooks	105	441	3
School Friendship6	68	220	6
School Friendship7	68	220	7

Table 2. Characteristics of some real-network without ground-truth

Dataset	Nodes	Edges	Communities
Lesmis	77	254	6
Jazz	198	2742	4
Adjnoun	112	425	7
Neural	297	2148	2

4.2   Experiments Setup

Comparison Methods. We compare our method with other community detection approaches which can be separated into two categories based on adopting deep learning or not. The deep learning based methods consist of DNR_L2 [21], DNR_CE [21] and CoDDA [16] and the remaining algorithms include spectral algorithm (SP) [22], the Fast-Newman (FN) algorithm [23], the external optimization algorithm (EO) [15], modeling with node degree preservation (MNDP) [11], Karrer's method [4], Ball's method [3] and BNMTF method [33].

DNR and CoDDA apply the Auto-Encoder to extract the features of the network. The DNR takes modularity matrix as the input while CoDDA adopts the similarity matrix as the input. DNR_2 denotes DNR with L2 norm and DNR_CE denotes DNR with the cross-entropy distance. SP and FN are proposed by Newman to detect the community, which are based on modularity. MNDP uses random-graph null model to find community structure. Karrer's method adopts likelihood probability as objective function. Ball's method is designed for link partitions. BNMTF uses squared loss objective function and adopts NMF for optimization.

Parameter Setup. In our experiments, the number of layers of our ConvAE model is set to 3 and the size of the convolution kernel is 2×2. We use padding when applying the convolution operation. The number of feature maps k is set to 3. The activation function σ is ReLU. We set the constant p to 0.1 and the hyperparameter β to 2.0, λ to 0.0001.

Evaluation Metrics. Towards datasets without ground-truth communities, we employ modularity function [22] to evaluate the performance of various methods on these datasets. For datasets with ground-truth communities, we employ Normalized Mutual Information (NMI) index [12] to evaluate the performance.

Modularity: the popular modularity Q is defined as equation (4). The larger the modularity is, the greater possibility the community structure we detect is the true structure.

NMI index: The Normalized Mutual Information calculates the similarity between the known communities of the network A and the predicted ones B.

$$NMI(A,B) = \frac{-2\sum_{i=1}^{C_A}\sum_{j=1}^{C_B} C_{ij}log(C_{ij}N/C_iC_j)}{\sum_{i=1}^{C_A} C_ilog(C_i/N) + \sum_{j=1}^{C_B} C_jlog(C_j/N)} \tag{16}$$

where C_A, C_B is the number of communities in partition A and partition B, respectively. C_i and C_j is the sum of elements of C in row i and row j. If $A = B$, then *NMI(A, B)* = 1; if $A <> B$, then *NMI(A, B)* = 0. The larger the NMI is, the closer the community structure we detect to true structure is.

4.3   Experimental Results

Validation on Synthetic Networks. We use the Girvan-Newman [24] networks as one part of the synthetic datasets. At the same time, we generate community networks with Lanchichinetti-Fortunato-Radicchi benchmark [2], LFR networks. These two types of synthetic datasets are artificial networks that resemble real-world networks to evaluate the methods.

Each GN network contains more than one hundred nodes and four communities. Each node has two probability: one is the possibility that current node connected to the nodes in the same community, and another presents the possibility that current node connected to the nodes in different community.

The NMIs of different algorithms on GN networks are shown in Fig. 3. The values on x-axis denotes the average number of inter-community edges per vertex, and the values on y-axis represents the result of evaluation function NMI.

Figure 3 shows the curve of our model as well as other methods, which include DNR, FN, SP, EO and CoDDA. As is shown in Fig. 3, the NMI of our method is superior to other competing methods. Especially, when the average of inter-community edges per vertex is smaller than 7, the NMI of ConvAE is 1, which means our method performs perfectly in these situations.

The advantage of LFR over other generating methods is that it accounts for the heterogeneity in the distribution of both the node degrees and community sizes. We set the number of nodes N to 1000, the average degree $avek$ to 20, the maximum degree $maxk$ to 25, the exponent for the degree distribution to -2, the exponent for the community size distribution to -1, the minimum for the community sizes $minc$ to 20 and the maximum for the community sizes $minc$ to 100. We set parameter μ from 0.6 to 0.8 so each node has $(1-\mu)^{avek}$ intra community links on average. The results are shown in Fig. 4. The values on x-axis

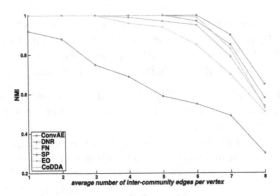

Fig. 3. The results of different algorithms for community detection on GN networks

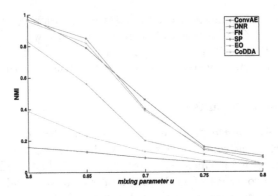

Fig. 4. The results of different algorithms for community detection on LFR networks

denote the mixing parameter μ, and the values on y-axis represent the result of evaluation function NMI.

As shown in Fig. 4, our algorithm has given good results on LFR networks. We can see that the curve of our method is above other curves at the majority situations, which means the NMI of our method is larger than others besides $\mu = 0.65$. Even though in situation where $\mu = 0.65$, the performance of our method is also competitive.

Validation on Real-World Networks with Known Community Structures. We present results of our proposed method on seven popular real-world networks with ground-truth. We use the NMI to evaluate our method and other methods, including SP, FN, EO, CoDDA, DNR_L2 and DNR_CE. The results of community detection are presented on Table 3.

As Table 3 showing, our method outperforms the majority of other algorithms. Especially, on the network Karate, our NMI is 1.000, which means the true community structure of network is accurately identified by our method. Moreover, our method performs best on the largest network polblogs, which

Table 3. NMI on real-network with ground-truth

Dataset	SP	FN	EO	CoDDA	DNR_L2	DNR_CE	ConvAE
Football	0.334	0.698	0.885	0.920	0.927	0.914	**0.929**
Karate Club	**1.000**	0.692	0.587	0.731	**1.000**	**1.000**	**1.000**
Polblogs	0.511	0.499	0.501	0.493	0.389	0.517	**0.689**
Dolphins	0.753	0.572	0.579	0.753	**0.889**	0.818	**0.889**
Polbooks	0.561	0.531	0.557	0.497	0.552	0.582	**0.651**
School Friendship6	0.418	0.727	**0.952**	0.893	0.888	0.924	0.931
School Friendship7	0.477	0.762	0.910	0.904	0.907	**0.932**	0.929

Table 4. Modularity on real-network without ground-truth

Dataset	MNDP	Karrer	Ball	BNMTF	CoDDA	DNR_L2	ConvAE
Lesmis	0.543	0.458	0.521	0.549	0.516	0.609	**0.716**
Jazz	0.438	0.369	0.435	0.435	0.542	0.574	**0.606**
Adjnoun	0.271	−0.104	0.259	0.263	0.442	0.594	**0.718**
Neural	0.381	0.262	0.364	0.369	0.387	0.406	**0.613**

owes to the usage of KL divergence and L2 norm which restricts the sparsity and model complexity.

Validation on Real-World Networks with Unknown Community Structures. In Lesmis [8], Jazz [25], Adjnoun [22] and Neural [10] where no ground-truth albel for each node in the networks are provided. We use the modularity to evaluate the networks with unknown community structure. The statistics of these datasets are shown in Table 2. And the results shown in Table 4 indicate our method is better than other algorithms. Especially, modularity of our method is bigger than other methods on all four networks by a large margin.

Also, Tables 3 and 4 manifest that our method is more effective compared to other methods on the real-world networks no matter the structure of network is known or not.

From all the experiments above, we can see that the deep learning methods (including DNR, CoDDA and our method) can detect the community structure more precisely than other methods. It is because the deep learning based methods have strong power to exact the nonlinear feature than conventional methods.

Table 5. ConvAE on different clustering algorithms

Dataset	K-means	Affinity Propagation	Spectral	MiniBatchKMeans	Agglomerative
Polblogs	0.689	0.437	0.512	0.614	0.587
Polbooks	0.651	0.485	0.591	0.632	0.573

Among the deep learning methods, our method also outperforms the rest two methods(DNR, CoDDA), which can be seen in Tables 3 and 4. This is mainly because the weight sharing of the convolutional layer is based on the local similarity which also exists in the social network. Since the community is formed by those who have similar interests or properties, the convolution operation is naturally appropriate to address community detection problem, which results in the superior performance of the proposed method.

Validation on Community Representation. Besides K-means, we also apply other clustering algorithms to detect community in networks, such as Affinity Propagation Clustering, SpectralClustering, MiniBatchKMeans and AgglomerativeClustering. We can see the comparing result on Table 5.

In Table 5, we can see the majority of clustering algorithms achieves good results which is better than the results of most algorithms in Table 3. Even though the result is the worst within Table 5, it is not terrible in fact. For example, the worst result on polblogs is 0.437 which is achieved by Affinity Propagation Clustering. However, DNR_L2 can just get 0.389 shown in Table 3. The reason that Affinity Propagation Clustering behave badly may be that the parameter of Affinity Propagation Clustering is set to default and the clustering number can not be set. From this experiment, we can see that our method can extract more cohesive latent feature representation.

5 Conclusion

In this work, we present a method which combines convolutional neural network and auto-encoder for getting a cohesive feature representation of network and use the representation to identify the community in social network. Motivated by the strong representation ability of convolutional neural network, we apply the convolutional operation to our model, by which our method combines the advantages of both auto-encoder and convolutional operation. To avoid the drawback of adjacency matrix, we propose a new method to preprocess the input data and obtain similarity matrix. Finally, we validate our method through extensive experiments and compare our method with other community detection algorithms, including deep learning methods, on synthetic and real-world networks. Furthermore, we compare the NMI metric of community detection with different clustering methods. All of the experiments indicate our method can extract nonlinear features more effectively and obtain more cohesive representations of networks.

Acknowledgments. This work is supported by the Natural Science Foundation of China (No. 61672276), Natural Science Foundation of Jiangsu Province of China (No. BK20161406).

References

1. Krizhevsky, A., Sutskever, I., Hinton, G.: ImageNet classification with deep convolutional neural networks. In: Proceedings of the Advances in Neural Information Processing Systems, pp. 1097–1105 (2012)
2. Lancichinetti, A., Fortunato, S., Radicchi, F.: Benchmark graphs for testing community detection algorithms. Phys. Rev. E **78**(4), 046110 (2008)
3. Ball, B., Karrer, B., Newman, M.E.J.: Efficient and principled method for detecting communities in networks. Phys. Rev. E **84**(2), 036103 (2011)
4. Karrer, B., Newman, M.E.J.: Stochastic blockmodels and community structure in networks. Phys. Rev. E **83**(1), 016107 (2011)
5. Schölkopf, B., Platt, J., Hofmann, T.: Greedy layer-wise training of deep networks. In: International Conference on Neural Information Processing Systems, pp. 153–160 (2006)
6. Ansótegui, C., Giráldez-Cru, J., Levy, J.: The community structure of SAT formulas. In: Cimatti, A., Sebastiani, R. (eds.) SAT 2012. LNCS, vol. 7317, pp. 410–423. Springer, Heidelberg (2012). https://doi.org/10.1007/978-3-642-31612-8_31
7. Kemp, C., Tenenbaum, J.B., Griffiths, T.L., Yamada, T., Ueda, N.: Learning systems of concepts with an infinite relational mode. In: National Conference on Artificial Intelligence, pp. 381–388 (2006)
8. Knuth, D.E.: The Stanford GraphBase: A Platform for Combinatorial Computing, pp. 41–43. ACM, Quezon City (1993)
9. Liou, C.Y., Cheng, W.C., Liou, J.W., Liou, D.R.: Autoencoder for words. Neurocomputing **139**(139), 84–96 (2014)
10. Watts, D.J., Strogatz, S.H.: Collective dynamics of 'small-world' networks. Nature **393**(6684), 440–442 (1998)
11. Jin, D., Chen, Z., He, D., Zhang, W.: Modeling with node degree preservation can accurately find communities. Math. Biosci. **269**, 117–129 (2015)
12. Wu, F., Huberman, B.A.: Finding communities in linear time: a physics approach. Eur. Phys. J. B **38**(2), 331–338 (2004)
13. Bourlard, H., Kamp, Y.: Auto-association by multilayer perceptrons and singular value decomposition. Biol. Cybern. **59**(4–5), 291–294 (1988)
14. Psorakis, I., Roberts, S., Ebden, M., Sheldon, B.: Overlapping community detection using Bayesian non-negative matrix factorization. Phys. Rev. E **83**(2), 066114 (2011)
15. Duch, J., Arenas, A.: Community detection in complex networks using extremal optimization. Phys. Rev. E **72**(2), 027104 (2005)
16. Shang, J.W., Wang, C.K., Xin, X., Ying, X.: Community detection algorithm based on deep sparse auto-encoder. Ruan Jian Xue Bao/J. Softw. **28**(3), 648–662 (2017). (in Chinese)
17. Xie, J., Kelley, S., Szymanski, B.K.: Overlapping community detection in networks: the state of the art and comparative study. ACM Comput. Surv. **45**(4), 1–37 (2013)
18. Nowicki, K., Snijders, T.A.B.: Estimation and prediction for stochastic blockstructures. Publ. Am. Stat. Assoc. **96**(455), 1077–1087 (2001)
19. Rohe, K., Chatterjee, S., Yu, B.: Spectral clustering and the high-dimensional stochastic blockmodel. Ann. Stat. **39**(4), 1878–1915 (2010)
20. Adamic, L., Glance, N.: The political blogosphere and the 2004 US election: divided they blog. In: Proceedings of the 3rd International Workshop on Link Discovery, vol. 62, no. 1, pp. 36–43. ACM (2005)

21. Yang, L., Cao, X., He, D., Wang, C., Wang, X.: Modularity based community detection with deep learning. In: International Joint Conference on Artificial Intelligence, pp. 2252–2258 (2016)
22. Newman, M.E.J.: Modularity and community structure in networks. Proc. Natl. Acad. Sci. **103**(23), 8577–8582 (2006)
23. Newman, M.E.J.: Fast algorithm for detecting community structure in networks. Phys. Rev. E **69**(6), 066133 (2004)
24. Girvan, M., Newman, M.E.J.: Community structure in social and biological networks. Proc. Natl. Acad. Sci. U.S.A. **99**(12), 7821 (2002)
25. Gleiser, P.M., Danon, L.: Community structure in Jazz. Adv. Complex Syst. **6**(04), 565 (2003)
26. Horn, R.A., Johnson, C.R.: Matrix Analysis, emphGraduate Texts in Mathematics, pp. 176–180. Cambridge University Press, Cambridge (1990)
27. Wang, R.S., Zhang, S., Wang, Y., Zhang, X.S., Chen, L.: Clustering complex networks and biological networks by nonnegative matrix factorization with various similarity measures. Neurocomputing **72**(1), 134–141 (2008)
28. Fortunato, S., Castellano, C.: Community structure in graphs. In: Meyers, R. (ed.) Computational Complexity, pp. 490–512. Springer, New York (2012). https://doi.org/10.1007/978-1-4614-1800-9
29. Zhang, S., Wang, R.S., Zhang, X.S.: Uncovering fuzzy community structure in complex networks. Phys. Rev. E **76**(4), 046103 (2007)
30. Raghavan, U.N., Albert, R., Kumara, S.: Near linear time algorithm to detect community structures in large-scale networks. Phys. Rev. E **76**(2), 036106 (2007)
31. Zachary, W.: An information flow model for conflict and fission in small groups. J. Anthropol. Res. **33**(4), 452–473 (1977)
32. Lecun, Y., Bottou, L., Bengio, Y., Haffner, P.: Gradient-based learning applied to document recognition. Proc. IEEE **86**(11), 2278–2324 (1998)
33. Zhang, Y., Yeung, D.: Overlapping community detection via bounded nonnegative matrix tri-factorization. In: Proceedings of the 18th ACM SIGKDD International Conference on Knowledge Discovery and Data Mining, pp. 606–614 (2012)

Research on Monitoring Method of Ethylene Oxide Process by Improving C4.5 Algorithm

Xuehui Jing$^{(\boxtimes)}$, Hongwei Zhao, Shuai Zhang, and Ying Ruan

Shenyang University, Shen Yang, Liaoning, China
690152489@qq.com

Abstract. Aiming at the serious safety hazards and environmental pollution problems in the process of extracting ethylene epoxidation to produce ethylene oxide, a chemical process monitoring method based on improved decision tree C4.5 is proposed. Compared with the previous chemical process monitoring methods, this paper is based on the data characteristics of the extraction process. And the information gain rate is obtained by various characteristics in the process of preparing ethylene oxide, and appropriate node branches are selected to classify the impact effects of the accident. On this basis, analysis of the extracted accident classification, forming a safety production evaluation and environmental risk assessment model for the production of ethylene oxide. The experimental results verify that the chemical process monitoring method of C4.5 is improved is useful for the chemical hazard prevention is made in advance for the chemical process, and effectively reduces the risk factor of the chemical process.

Keywords: C4.5 algorithm · Risk assessment · Ethylene oxide · Decision tree

1 Introduction

With the continuous development of computing networks, various kinds of information continue to proliferate, the amount of information collected increases, and a variety of data form the ocean of information. Based on this, a new technology is proposed to process massive data and extract valuable information, namely data mining technology [1]. Data mining technology using a combination of multiple algorithms is a multi-disciplinary crossover technology [2].

The chemical industry has always been closely related to our production and life. However, due to its unique industrial nature, environmental pollution and even dangerous incidents are inevitable [3]. Ethylene oxide (EO) is also known as oxidized ethane. Because it is a highly toxic gas, also known as cycloalkane, it is the simplest cyclic ether. The hazards in the production of ethylene oxide are mainly present in the oxidation and purification of ethylene oxide. During the oxidation reaction, the device contains a large amount of combustible gas. It may contain a mixture of ethylene and methane, both of which are susceptible to oxidative exothermic reaction with oxygen and present a risk of explosion. Since the ethylene oxidation reaction is exothermic, especially when the ethylene oxide is completely oxidized to carbon dioxide. If this heat is not released in time, the reaction may be uncontrolled. The large amount of heat

© Springer Nature Singapore Pte Ltd. 2019
H. Jin et al. (Eds.): BigData 2019, CCIS 1120, pp. 395–405, 2019.
https://doi.org/10.1007/978-981-15-1899-7_28

accumulated may result in loss of assets, threats to human health or a safe environment [4]. Combustion reactions that occur before a formal reaction, sudden drops in temperature, and flammable gases in circulation can cause an explosion. During the purification of ethylene oxide, the higher purity ethylene oxide may undergo a flare reaction without manual regulation. And the high purity ethylene oxide is sensitive. It is susceptible to various impurities and increases the unsafe factors in the reaction, causing fire and explosion hazards. There are many hidden dangers in safety, and it is also an urgent problem to be solved in the current production of ethylene oxide.

The original decision tree C4.5 algorithm can extract and analyze the data effectively and accurately, and the result is more intuitive. However, it is applicable to various fields in various disciplines and lacks certain pertinence [5]. Although the data processing results are clear at a glance, easy to understand. However, we cannot accurately extract the key information we need, and the accuracy is not enough. The data set is calculated according to the order of comparison, and the number of repetitions is large, the algorithm takes a long time and is low efficiency [6]. The existing improved C4.5 algorithm is mainly aimed at improving the efficiency, solving the overfitting problem, and paying less attention to the accuracy rate. In order to reduce the impact of these shortcomings, domestic and foreign experts have proposed a variety of improved algorithms, applied to various industries.

This paper applies the theoretical knowledge and mining methods of data mining, and combines the special industry nature of the chemical industry to conduct research. The decision tree C4.5 algorithm for data mining is mainly improved. Select characteristic parameters suitable for the chemical industry and modify them on the original algorithm. In view of the above characteristics, data can be extracted according to the characteristics of incidents and substances in the past, and the rules can be found, the experience can be summed up, and preventive measures should be taken in advance. Taking the production of ethylene oxide as an example, the application of data mining technology in its safety and environmental protection data is divided into the following five parts. Select the appropriate raw data set from the existing incident data. Collect valuable data, preprocess the collected data, use the decision tree algorithm for data mining process, and analyze the data [7]. Using the above five parts to analyze the extracted ethylene oxide safety production accident classification data. A safety production evaluation and environmental risk assessment model for the production of ethylene oxide is obtained. It is effective to verify the chemical process monitoring method that improves the C4.5 algorithm. Compared with the improved algorithm in [8, 9], it is more specific and suitable for chemical industry safety risk assessment and environmental pollution assessment. The analytical results obtained are in line with the actual situation. For the chemical industry production process, take precautions against chemical hazards in advance, reduce the risk factor of the chemical process. Having certain practical significance (Fig. 1).

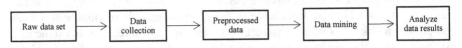

Fig. 1. Data mining process [10]

2 Application of Improved C4.5 Algorithm in Preparation of Ethylene Oxide

2.1 C4.5 Decision Tree Algorithm

Decision tree algorithm is one of the commonly used algorithms of data mining technology. It divides the complex decision process into a series of simple choices and can explain the whole process intuitively [11]. Common decision tree algorithms are as follows: ID3, C4.5, C5.0, CART, etc. [12].

First select a root node to start training. Then select the corresponding features and classify the samples according to the selected features. If the samples belong to the same class, then the node becomes a leaf node. If not the same class, select the current node with the largest impurity (Indicates the class distribution balance of the samples falling on the current node) [13], select new features, and classify according to the new features. Until the same feature attribute sample is assigned to the same class, or no attribute can provide the sample to continue partitioning. Finally, the decision tree pruning process is used to prun the decision tree to remove overfitting and complex samples.

The relevant calculation formula is as follows:

1. Information entropy: used to describe the concept of node uncertainty, and to obtain nodes with large impurity [13].
   The information entropy expression is:

$$Entropy(t) = -\sum_{c=1}^{c} p\left(\frac{c}{t}\right) \log_2 p\left(\frac{c}{t}\right) \tag{1}$$

The data set has a total of C categories;
$p(c/t)$: The frequency of occurrence of a class c sample in a t-node.

2. Information gain: the decrease of information entropy compared with the parent node, and the split scheme with the largest information gain.
   The information gain expression is:

$$E_k = \sum_{k=1}^{K} \frac{n_k}{n} Entropy(t_k) \tag{2}$$

$$InfoGain = Entropy(t_0) - E_k \tag{3}$$

There are K child nodes;

$Entropy(t_0)$: Indicates the information entropy of the parent node;
$Entropy(t_k)$: Indicates the information entropy of the child node;
t_0: Parent node;
t_k: Child node;
n: The number of samples of the parent node t_0;
n_k: The number of samples of the child node t_k.

3. Information gain rate: The concept proposed on the basis of information entropy can effectively solve the over-fitting problem when decision tree branches.
Split information:

$$SplitInfo = -\sum_{k=1}^{K} \frac{n_k}{n} log_2 p\left(\frac{n_k}{n}\right) \tag{4}$$

$p\left(\frac{n_k}{n}\right)$: Ratio of child nodes to parent node samples.
Information gain:

$$InfoGainRatio = \frac{InfoGain}{SplitInfo} \tag{5}$$

$$InfoGR = \frac{Entropy(t_0) - \sum_{k=1}^{K} \frac{n_k}{n} Entropy(t_k)}{-\sum_{k=1}^{K} \frac{n_k}{n} log_2 \left(\frac{n_k}{n}\right)} \tag{6}$$

2.2   Improved Decision Tree Algorithm

This paper modifies the information gain algorithm on the basis of the original. According to the characteristics of the hidden dangers in the chemical production, the following modifications are proposed.

In the original C4.5 algorithm, the information gain is calculated in this way. First calculate the amount of information that divides the subset. Then, by using the information entropy of the parents, the average value of the information entropy of each child node is directly subtracted. The algorithm is too simple and single. This paper combines the most theoretical ideas of expectation. This method can effectively solve the problem of incomplete data loss in the chemical industry safety assessment. The parameters are changed according to the characteristics of the data model. Making a change in the idea of first calculating the average and then calculating the impairment. The idea of improved the algorithm is first to find the calculating the impairment and then to find the calculating the average. This speeds up the convergence while solving the lack of data. At the same time, the data coverage is wide and the results are more accurate.

The improved information gain expression is:

$$InfoGain = \sqrt{\sum_{k=1}^{K} \frac{(Entropy(t_0) - n_k Entropy(t_k))^2}{n^2}} \tag{7}$$

The Improved information gain rate expression is:

$$InfoGR = \frac{\sqrt{\sum\limits_{k=1}^{K} \frac{(Entropy(t_0) - n_k Entropy(t_k))^2}{n^2}}}{-\sum\limits_{k=1}^{K} \frac{n_k}{n} \log_2\left(\frac{n_k}{n}\right)} \tag{8}$$

After many calculations, the results show that the improved information gain from the original information gain is more differentiated than the original information gain. The classification is more detailed and accurate. The improved algorithm is better than the original algorithm in classification information gain, the calculation result is more detailed, and the error rate is significantly reduced. It is suitable for the calculation of safety accident classification methods in chemical industry.

2.3 C4.5 in Ethylene Oxide Safety and Environmental Protection Application

According to the decision tree algorithm, all possible incidents are selected as the original data set, and valuable incident data is collected in the original data set to clean the data. Then select data with valuable informations. The simulation results of incident prediction during the safe production of ethylene oxide were obtained. The range of incident damage only simulates the safe production area of ethylene oxide in the entire chemical plant, and does not account for damage to other areas. The simulation is based on the principles of integrity and relevance. The leaf nodes in the figure indicate the most serious consequences. In the incident part of the ethylene tank area, the incident occurred in the center of the ethylene tank, and the incident within 60 m radius is divided according to the distance from the location to the ethylene tank.

Ethylene oxide incidents prediction simulation process:

1. Collecting incident as the root node classification, calculate the following information about the number of incident in the next level of child nodes.
2. Using Eq. (1) to calculate the incident information entropy of each child node.
3. Using Eq. (6) to calculate its information gain.
4. At this time, use Eqs. (4) and (5) to derive from Eq. (6) computes information gain rate.
5. Calculate the information gain rate according to multiple feature classification methods, and whether it is related to the environment has the largest information gain rate, so it is divided into two sub-nodes: environmental pollution and process incident.
6. Still using the above process, selecting multiple feature comparisons, and sequentially branching the lower child nodes by calculating the information gain rate.
7. Until there is only one incident result in the child node, stop the build.
8. Get the final result.

The decision tree is shown below (Fig. 2):

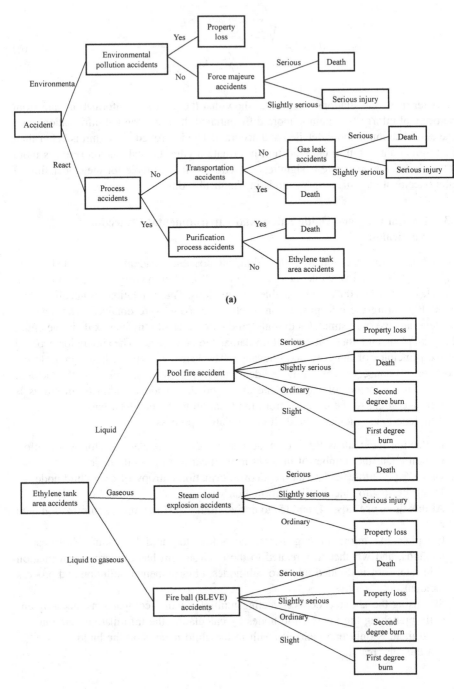

Fig. 2. Ethylene oxide incidents prediction simulation results [14]

3 Experiment Analysis

Based on the decision tree C4.5 algorithm, the process of producing ethylene oxide from a chemical plant in Shaanxi was selected, and the results of the main safety production accident prediction were data sets. Using the python programming language. By importing the numpy library, the sklearn library, and the matplotlib library. Then import the Decision Tree Classifierhuanshu function from the tree module in the sklearn library and Standard Scaler function from preprocessing data processing module. Finally, use plt. show() to get the decision tree model.

3.1 Ethylene Oxide Incident Prediction Simulation

3.1.1 Ethylene Oxide Incident Prediction Simulation Results
From the selection of all possible incident raw data sets, through environmental and reaction processes, it is divided into environmental pollution incident and process incident. If the environmental pollution incident is only environmental pollution, it will be divided into property losses without causing personal injury or death. There are some natural factors of force majeure in environmental pollution incident, such as typhoons, earthquakes, hail, etc. Classified according to the severity of the impact of the incident, seriousness causing death, slightly seriousness causing serious injury.

Process incidents are divided into two aspects: the incident of ethylene oxide purification process and the auxiliary process incident of preparing ethylene oxide. Auxiliary process incidents include transportation incidents and gas leaks. Ethylene oxide will release heat during transportation. If this heat is not released in time, the reaction may be uncontrolled, and the accumulated large amount of heat may cause loss of assets, personnel or environmental damage [15]. Personnel supervise the transportation process, and if there is an incident, there will be casualties. Ethylene oxide is also known as oxidized ethane, and because it is a highly toxic gas, it is also called methane [16]. When the amount of ethylene oxide leakage is small, it will have a serious impact on the surrounding air, causing serious injuries. When ethylene oxide leaks more, it involves a wide area and reaches a certain concentration, which has extremely bad consequences, causing serious impact and directly leading to death [17].

In the process of purifying ethylene oxide, the device contains a large amount of combustible gas and oxygen, which may contain a mixture of ethylene and methane [18], both of which are susceptible to oxidation and exothermic reaction with oxygen, and there is a risk of explosion, such as an incident will inevitably result in casualties. If it is not an ethylene oxide purification process incident, it will be classified into an ethylene tank incident and continue to apply the decision tree algorithm classification.

3.1.2 Ethylene Tank Area Incident Prediction Simulation Results
According to the type characteristics of incidents, the explosion incidents in the ethylene tank area can be divided into three types of incidents: pool fire incident, steam cloud explosion incident and fire ball (BLEVE) damage. Distance from location to incident site, pool fire incidents are divided into four levels of damage, property damage, death, second degree burns and first degree burns. The steam cloud explosion

incident is divided into three levels: death, serious injury and property loss. Fireball incident is the same as pool fire incident [19].

In summary, leaf nodes can be divided into death, property damage, first degree burns, second degree burns and serious injuries. It can be clearly seen that the various possible incidents collected are divided into several different damage levels according to their characteristics. Clearly react to possible casualties and property damage in the event of an incident.

3.2   Comparative Analysis of Examples

Through experiments, the results are shown in Fig. 3. Based on the results of the same risk accident, the accuracy of the improved C4.5 algorithm is better than the original C4.5 algorithm.

Collecting the types and quantities of accidents that may occur during the safe production of ethylene epoxidation to produce ethylene oxide. The classification information extracted by the improved C4.5 algorithm is better than the original C4.5 algorithm. It can be clearly seen from the experimental results chart that the average accuracy is improved about 0.04. This also shows that there are more accurate methods to analyze and evaluate accidents in chemical production, and can effectively propose countermeasures.

It can be seen from Table 1 that the improved algorithm is more detailed and accurate for event analysis, and the classification performance is higher than C4.5. The classification error rate has dropped from 0.12 to 0.10, with a significant increase.

The simulation results of the improve decision tree reflect that the decision tree has a maximum depth of 6, and contains 18 leaf nodes. There were 7 incidents that occurred in death, accounting for 39% of all incidents, which was the highest probability of occurrence in all incidents. Incidents in transportation incidents and purification processes must directly lead to death, accounting for 29% of deaths. When five incidents are seriously affected, it may result in death, accounting for 71% of deaths.

There were 4 incidents in which property damage occurred, accounting for 22% of all incidents. Mainly in the ethylene tank incident, accounting for 75%, mainly in slightly serious characteristics. The result of environmental pollution is property damage, and there is no other impact on casualties, It is not very serious.

There were 3 incidents that were seriously injured, accounting for 17% of all incidents. Occurs in force majeure incidents, toxic gas leaks and steam cloud explosions. There were no obvious correlations between the three types of incidents, but they all showed a 'slightly serious' characteristic.

There are 2 kinds of incidents in second degree burns, accounting for 11% of all incidents, which is ordinary impact. There were 2 types of incidents in first degree burns, accounting for 11% of all incidents, which was a slight impact. The percentage of incidents in the ethylene tank area was 18%, accounting for 25% of the pool fire incidents, and 25% of the fireball incidents. The probability of occurrence among the various incidents was the least.

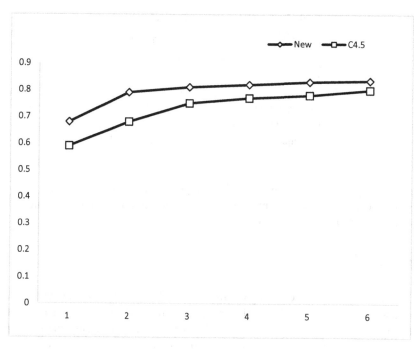

(a) Ethylene oxide accident

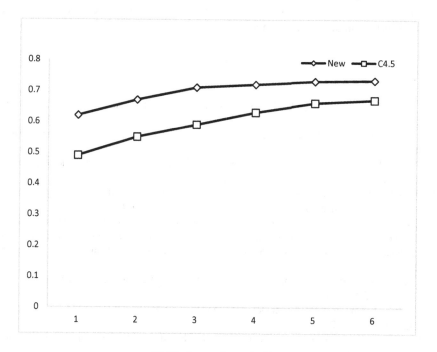

(b) Ethylene tank area incident

Fig. 3. Experimental results. X-axis is number of experiments. Y-axis is accuracy.

Table 1. Example comparative analysis table

Classification	Number		Proportion	
	Improve	C4.5	Improve	C4.5
Death	7	6	39%	33%
Property loss	4	5	22%	28%
Second degree burn	3	4	17%	22%
First degree burn	2	0	11%	0
Serious injury	2	3	11%	17%

In summary, the ethylene oxide production process leads to more deaths, resulting in fewer incidents of second degree burns and first degree burns. If an incident occurs, it will lead to serious consequences.

4 Conclusion

The improve decision tree C4.5 algorithm of data mining is used to extract the safety hazard and environmental pollution information in the production process of ethylene oxide, and the decision tree model for the incident prediction of ethylene oxide production process is obtained. This model classifies various possible incidents according to their own characteristics, and combines the characteristics of ethylene oxide production process, the incidents of death, property loss, second degree burn, first degree burn and serious injury were classified.

References

1. Cai, M., Zhang, W., Wang, H.: Overview of data mining in big data era. Value Eng. **38**(05), 155–157 (2019)
2. Zhao, Y., Shang, M.: Data mining as a feature of interdisciplinary subjects. Times Finance **40**(09), 263–264 (2017)
3. Yin, Z.: Research on green chemical technology in chemical engineering and technology. Shandong Ind. Technol. **38**(10), 69 (2019)
4. Chen, L., Liu, H., Zou, R., Ge, C.: Simulation and system analysis of ethylene oxide process. J. Beijing Univ. Chem. Technol. (Nat. Sci. Ed.) **44**(04), 20–26 (2017)
5. Liu, Y.: C4.5 crossover algorithm based on mapreduce in university big data analysis. In: Proceedings of the 2nd International Seminar on Artificial Intelligence, Networking and Information Technology (ANIT 2018), p. 7. Boss Academic Exchange Center, Shanghai Hao Culture Communication (2018)
6. Liu, R.: C4.5 performance analysis of decision tree classification algorithm. Inf. Syst. Eng. (01), 153–154 (2019)
7. Yu, Q.: Data mining process model and innovative application. Electron. Technol. Softw. Eng. 7(01), 173–174 (2018)
8. Liu, H., Su, Y.: Mobile phone client application based on C4.5 algorithm. Comput. Dig. Eng. **47**(08), 2090–2093 (2019)

9. Nie, B., Luo, J., Du, J., et al.: Improved algorithm of C4.5 decision tree on the arithmetic average optimal selection classification attribute. In: 2017 IEEE International Conference on Bioinformatics and Biomedicine (BIBM). IEEE (2017)
10. Luo, X.: Application of data mining technology in hospital information system. Electron. Technol. Softw. Eng. **7**(03), 204–205 (2018)
11. Shao, Y.: A review of typical algorithms for decision trees. Comput. Knowl. Technol. **14** (08), 175–177 (2018)
12. Zhang, Y., Cao, J.: Decision tree algorithm for big data analysis. Comput. Sci. **43**(S1), 374–379+383 (2016)
13. Gaoyan, O., Zhanxing, Z., Bin, D., et al.: Data Science Guidance, pp. 81, 84–86, 1st edn. Higher Education Press, Beijing (2018)
14. Peng, C., Wen, Yu., Li, C.: Medical big data based on decision tree algorithm. Inf. Technol. Inf. **43**(09), 70–74 (2018)
15. Zhang, Z.: Design points and process paradigm of safety facilities for ethylene oxide tanks. Shanghai Chem. Ind. **48**(11), 30–34 (2018)
16. Savaris, M., Carvalho, G.A., et al.: Chemical and thermal evaluation of commercial and medical grade PEEK sterilization by ethylene oxide. Mater. Res. **19**(04), 807–811 (2016)
17. Wang, Y., Cao, J., Jiang, Y.: Occupational acute ethylene oxide poisoning incident and typical case analysis. Occup. Health Emerg. Rescue **36**(01), 78–80 (2019)
18. Yu, H.: Process and safety design of ethylene oxide tank area. J. Daily Chem. **39**(11), 40–43 (2016)
19. Huang, H., Zhang, L., et al.: Python Data Analysis and Application, 1st edn, pp. 167–198. People's Posts and Telecommunications Press, Beijing (2018)

Web API Recommendation with Features Ensemble and Learning-to-Rank

Hua Zhao[1], Jing Wang[2], Qimin Zhou[1], Xin Wang[1], and Hao Wu[1(✉)] (iD)

[1] School of Information Science and Engineering,
Yunnan University, Kunming, China
haowu@ynu.edu.cn

[2] National Pilot School of Software, Yunnan University, Kunming 650500, China

Abstract. In recent years, various methods against service ecosystem have been proposed to address the requirements on recommendation of Web APIs. However, how to effectively combine trivial features of mashups and APIs to improve the recommendation effectiveness remains to be explored. Therefore, we propose a Web API recommendation method using features ensemble and learning-to-rank. Based on available usage data of mashups and Web APIs, textual features, nearest neighbor features, API-specific features, tag features of APIs are extracted to estimate the relevance between the mashup requirement and the candidates of APIs in a regression model, and then a learning-to-rank approach is used to optimize the model. Experimental results show our proposed method is superior to some state-of-the-art methods in the performance of recommendation.

Keywords: Web API recommendation · Features ensemble · Learning-to-rank · Top-N recommendation

1 Introduction

With the continuous growth of the Web API resource, the ecosystem and economic form centered on APIs is gradually emerging [1]. ProgrammableWeb, the largest API online registry, has indexed more than 20,000 APIs and over 7,000 mashups. Web APIs, as the most widely used and powerful components, have the advantages of easy access, development, combination and personalization [2] and have been widely used by many enterprises. Faced with the rapid development of information society and the emergence of a large number of additional requirements, developers usually need to integrate multiple services to meet complex requirements. Mashups [3] can integrate existing APIs with various functions to add value to these services. This makes mashup services a new kind of Web application and the development direction of service composition.

However, it has become increasingly challenging for a significant number of developers to quickly choose the right Web API from the large number of candidates. Therefore, it is important to help developers choose Web APIs that

© Springer Nature Singapore Pte Ltd. 2019
H. Jin et al. (Eds.): BigData 2019, CCIS 1120, pp. 406–419, 2019.
https://doi.org/10.1007/978-981-15-1899-7_29

meet development needs in less time. To solve this problem, researchers have done a lot of works to improve the performance of Web API recommendation. These works can be roughly divided into content-based Web API recommendation [4–8], recommendation based on collaborative filtering [9–13], and combined recommendation [14–17]. Although these works fulfill API recommendation to some extent, there are also some disadvantages. Firstly, some researchers only recommend APIs from single meta-information such as text or tags or popularity, resulting in sparse data. For example, recommendations based on textual similarity or LDA models do recommend APIs not well when the textual descriptions of mashups or APIs are short. In previous work [18], we combined the random walk algorithm to implement the recommendation of mashups on knowledge graph. However, most developers cannot really label or categorize a mashup until it is generated. Furthermore, Web APIs may have different importance to different mashups, some APIs are important to mashups but get low-ranking leading to a wrong choice. In addition, the random walk approach does not utilize textual information and other information should also be taken into account to improve the performance of the API recommendation.

Motivated by these factors, we propose an API recommendation method based on features ensemble and learning-to-rank. Trivial features are extracted by content and structural information of mashups and APIs, and then merged into multi-dimensional features. Through regression analysis, a pairwise algorithm in learning-to-rank [19] is used to implement the recommendations of APIs. The main contributions of this paper are as follows:

- We define and extract a series of features to make an ensemble model and estimate the relevance between mashups and APIs dataset. We use a pairwise learning-to-rank approach to optimize the weights in features ensemble model.
- Experiments on the ProgrammableWeb dataset show that our proposed method can effectively improve the performance of recommendations in comparison with some state-of-the-art methods.

The rest of this paper is organized as below: Sect. 2 describes feature definition and generation methods of mashups and APIs. Section 3 introduces the feature selection. Section 4 introduces learning-to-rank based recommendation method. Section 5 gives the experiment and result analysis. Section 6 reviews related work. Finally, we draw conclusions and discuss future work in Sect. 7.

2   Feature Definition and Generation

We mainly obtain features of mashups and APIs in various dimensions from two aspects of content and structure. Firstly, we use the content information to extract the corresponding features. Then, the remaining features are extracted using structural information. To extract valid content information, there needs some preprocessing steps of textual description of mashups and APIs.

- Convert the textual description of mashups and APIs to lowercase and turn it into a list of words.

- Remove stop words and special symbols from the list.
- Extract the stem of a word and make it into a standard form.
- Count the number of occurrence of each word, and remove those words that appear less.

To accurately represent each feature, we list some common symbols in Table 1.

Table 1. Notation list.

Meaning	Notation
Collection of textual descriptions of APIs and mashups	D
Single textual description of the API/mashup	d, d_s, d_a
Words in the text description of an API/ a mashup	w
Tags in API	t
A single mashup	s, u
A single API	a

2.1   Text Features

Text features focus on exploiting the textual descriptions of Mashups/APIs to generate their vector representations using different weighting schema, such as TF, TFIDF and LDA, and then use *cosine* similarity to estimate the relevance between mashups and APIs.

Term Frequency Feature (f-txt-vsm-tf): The term frequency mainly counts the number of word occurrence in the text. We use $tf(w, d_s)$ and $tf(w, d_a)$ to express the term frequency of a word w in the text of a mashup s and the text of an API a, respectively. Then, we generate vector respresentations for Mashups and APIs.

TFIDF Feature (f-txt-vsm-tfidf): TF-IDF considers not only the frequency of words appearing in the text, but also the importance of words in the corpus. The *tfidf* multiplies the aforementioned frequency by relative distinctiveness of the word in D. The distinctiveness (idf) is measured by $\log \frac{|D|}{n_w}$, where $|D|$ is the total number of mashups and APIs, n_w is the number of texts of mashups or APIs in which the word w occurred [22]. Then, we generate vector respresentations for Mashups and APIs.

LDA Feature (f-txt-vsm-lda): Latent Dirichlet Allocation (LDA) [23] is a probabilistic model, which can carry out semantic topics among texts in a large scale corpus. We calculate the topical representation of mashups and APIs using their textual information. Then, the *cosine* similarity is used to calculate the relevance of both to get the topical feature.

2.2 Nearest Neighbor Feature

We consider the idea of mashup-based collaborative filtering to generate this feature. Firstly, we retrieve a nearest neighbor set, $N(s)$, for target mashup s. The relevance of target mashup s and candidate API a can be estimated by:

$$cf(s,a) = \sum_{u \in N(s)} sim(s,u) \cdot rel(u,a) \qquad (1)$$

where $sim(s,u)$ represents the *cosine* similarity of s and u described in text, $rel(u,a)$ is the relatedness between u and a. For brevity, if a is invoked by mashup u, $rel(u,a) = 1/m$ where m is the total number of Web APIs invoked by u; otherwise, $rel(u,a) = 0$. In the similarity calculation, TFIDF-based and LDA-based vector representation of mashups are both used, so two kinds of the nearest neighbor features can be obtained: *f-mashup-cf-tfidf* and *f-mashup-cf-lda*.

2.3 API-Specific Features

Popularity of API (f-api-pop): We count the number of times each API participated in mashups in the dataset. The more the API participated, the more popular it is and the better it matches with mashups. Based on this observation, the feature of API popularity in mashups can be obtained as $\log(1 + count(a))$, where $count(a)$ represents the number of times a participates in mashups.

Rating of API (f-api-rating): The rating score reflects the API's reputation in user community. The higher the score is, the greater the probability that it will be adopted by users. We use rating information of APIs by normalizing its original score, which ranges generally from 0.0 to 5.0.

Creation Time of Mashup/API (f-api-creattime): Each API has a life cycle, which has a subtle relationship with whether it can be adopted by users. For example, APIs that are too old or too young are not necessarily preferred by users. Based on this assumption, we consider the interval of creation time between API and mashup as an additional feature.

2.4 Tag Features of APIs

Tags are widely used in organizing and sharing information resources, and it can be regarded as the accurate summarization of API functions, carrying richer and more accurate information. So, we also consider to enhance the API recommendation effect by using the tag feature.

Relevance of API Tags to Mashup (f-tag-api-bm25) : BM25 [24] is an algorithm based on probabilistic retrieval model to estimate the relevance between the query term and texts. We use all tags of the API as the query terms and estimate its relevance to mashup texts as follows:

$$bm25(a, s) = \sum_t \frac{idf(t) \cdot tf(t, d_s) \cdot (k+1)}{tf(t, d_s) + k \cdot (1 - b + b \cdot (|d_s|/avg(|d_s|)))} \tag{2}$$

where $k = 2$, $b = 0.75$ are default tuning parameters, and avg is the average length of mashups texts.

3 Feature Selection

Various features contribute differently to the recommendation results, and some of them may correlate with each other. It is necessary to select the most important features to reduce the model complexity. We use *backward stepwise regression* to select features. At first, a general predictor is established to predict the relevance of target mashup to APIs and implement the recommendation of APIs, as shown as follows:

$$r = \delta_2(\mathbf{h}^T \delta_1(\mathbf{W}x + \mathbf{b})) \tag{3}$$

where r is the relevance score, $x = \{f_1, f_2, ...f_n\}$ are the features vector, δ_1 is the *ReLU* function, δ_2 is the *sigmoid* function, $\mathbf{h}^T$ is a $n \times 1$ vector of regression coefficient (also the parameters of feature weight), $\mathbf{W}$ is a $n \times n$ matrix, $\mathbf{b}$ is a $n \times 1$ vector.

We start the test with all available features, deleting one feature (the most useless one) at a time as the prediction model progresses. This process will be repeated until the retained features are all significant to the prediction model. The impact of each feature on the model is shown in Table 2, where the Recall@20 is used to measure the impact of each feature in the model, "*None*" represents all the features in the model, and the *underline* represents the elimination of this feature in this selection.

As can be seen from Table 2, after **f-txt-vsm-tf** is removed in the first round, the recommendation performance by Recall@20 decreases from 80.55% to 76.26%, but it increases from 80.55% to 82.44% after **f-api-pop** is removed, so **f-api-pop** is removed. In the second round, the performance increases from 82.44% to 82.53% after removing **f-api-rating**, while the performance decreases after the others are removed, so **f-api-rating** is removed. In the third round, the performance is improved from 82.53% to 82.99% after removing **f-api-creattime**, so this feature is useless. In the fourth round, the feature selection is terminated, as removing anyone of the remaining features will significantly reduce the model performance. Finally, we have six features: **f-txt-vsm-tf**, **f-txt-vsm-tfidf**, **f-txt-vsm-lda**, **f-mashup-cf-tfidf**, **f-mashup-cf-lda**, and **f-tag-api-bm25** to make a regression-based ensemble model.

Table 2. Feature selection.

Candidate features	1st selection	2nd selection	3rd selection	4th selection
None	0.8055	—	—	—
f-txt-vsm-tf	0.7626	0.8145	0.8198	0.8203
f-txt-vsm-tfidf	0.7551	0.7776	0.7896	0.7686
f-txt-vsm-lda	0.7403	0.8189	0.8222	0.8213
f-mashup-cf-tfidf	0.7500	0.7981	0.8005	0.8017
f-mashup-cf-lda	0.7626	0.8168	0.8225	0.8199
f-api-pop	<u>0.8244</u>	—	—	—
f-api-rating	0.7638	<u>0.8253</u>	—	—
f-api-creattime	0.7744	0.8129	<u>0.8299</u>	—
f-tag-api-bm25	0.7645	0.8186	0.8214	0.8212

4   Learning-to-Rank Based Recommendation Method

We combine learning-to-rank algorithm with the selected features to implement API recommendation. RankNet [19] is an effective pairwise method in learning-to-rank and uses a probabilistic loss function to learn a ranking function. It can be described as follows: given a mashup requirement description, there are a set of API pairs $<a_i, a_j>$ which have different relevance r_i and r_j to the target mashup, and the parameters are trained using neural networks for the features vector x_i/x_j w.r.t a_i/a_j. If there exist $r_i > r_j$, a_i should precede to a_j, namely $a_i \succeq a_j$, and thus the prediction probability of a_i ranking higher than a_j which can be defined as formula 4.

$$p_{ij} = p(a_i \succeq a_j) = \frac{1}{1 + e^{-(r_i - r_j)}} \tag{4}$$

The true probability of a_i ranked higher than a_j is: $\hat{p}_{ij} = 0.5(1 + R_{ij})$. $R_{ij} \in \{-1, 0, 1\}$ are the true preferences of a_i and a_j, indicating that a_i is *lower, equal, higher* ranked than a_j, respectively. The loss function of the true probability and predicted probability is defined as:

$$\begin{aligned}
\mathbb{L} &= -\hat{p}_{ij} \log p_{ij} - (1 - \hat{p}_{ij}) \log(1 - p_{ij}) + L_2(\theta) \\
&= 0.5 \cdot (1 - R_{ij})(r_i - r_j) + \log(1 + e^{-(r_i - r_j)}) + L_2(\theta)
\end{aligned} \tag{5}$$

To minimize the loss function $\mathbb{L}$ and optimize the parameters of weights to achieve API recommendation, we build a hidden layer neural network and adopt Stochastic Gradient Descent (SGD) [25] to learn model parameters. Also, a L_2 regularization is used to prevent overfitting of parameters $\theta = \{\mathbf{h}, \mathbf{W}, \mathbf{b}\}$: $L_2(\theta) = \|\mathbf{h}\|_2^2 + \|\mathbf{W}\|_2^2 + \|\mathbf{b}\|_2^2$.

5   Experimental Results and Analysis

5.1   Dataset

In our experiments, we use the linked Web API dataset [26] of which semantics and structure of mashups and APIs retrieved from ProgrammableWeb.com. There are 11339 Web APIs, 7415 mashups, 3286 tags and other information.

According to experimental requirements, more than 10,000 APIs not invoked by mashups and 93 mashups without APIs record are deleted, and the dataset is divided into two stages according to the time line. The test set is the latter stage of the dataset. Since there are relatively small mashups data for 2013 and 2014, the data for those two years are combined into one test set. The statistics of the dataset are shown in Table 3.

Table 3. Statistics information of the dataset.

Statistics	Dataset1	Dataset2
Mashup in the training dataset	6271	6785
Mashup in the test dataset	514	537
Web APIs	1000	1186

5.2   Evaluation Metrics

For Web API recommendation, it is necessary to evaluate the accuracy and diversity of recommendation. Therefore, three commonly used metrics, including Recall, NDCG and Hamming distance, which are applied to evaluate the effects of ranking and recommendation diversity in the information retrieval area. In the experiments, the larger the three metrics are, the better the performance is.

Recall is the ratio of the actual APIs in the top-N API recommendation list to the actual APIs used by the mashup based on mashup requirements. It can be defined as:

$$Recall@N = \frac{|actual APIs \cap Top - N APIs|}{|actual APIs|} \tag{6}$$

NDCG gives different weight to each API in the list of recommend top-N APIs, and the higher weight the former API is. One of the commonly used description is:

$$DCG@N = \sum_{i=1}^{n} \frac{2^{rel(i)} - 1}{\log_2(i+1)}, IDCG@N = \sum_{i=1}^{c} \frac{1}{\log_2(i+1)} \tag{7}$$

where $rel(i)$ is a binary value indicating whether the target mashup uses candidate APIs. If true, $rel(i) = 1$, otherwise, $rel(i) = 0$. c is the number of APIs that are truly used by the target mashup in the top-N candidate list. nDCG@N is implemented by standardizing DCG@N (nDCG@N = (DCG@N)/(IDCG@N)) to evaluate the recommendation accuracy.

Hamming distance is a "diversity" measure of recommending results [18]. For the target mashup pair $<s, u>$ can be defined as:

$$HD@N = 1 - \frac{Overlap_{su}@N}{N} \qquad (8)$$

where $Overlap_{su}@N$ represents the number of overlaps of the API in the two top-N recommendation lists for s and u. We use the average hamming distance of all target mashup pairs to represent the hamming distance of the entire recommendation system.

5.3 Evaluation Methods

In this section, our proposed method is compared with some representative methods mentioned in related works.

- **Vector space model (VSM)**. VSM mainly uses TFIDF technology to evaluate the importance of words in the text. Specifically, TFIDF technology is used to represent texts of target mashups and APIs as vectors, then the similarity is calculated and top-N similar APIs to the target mashup are recommended.
- **Collaborative filtering model (CF)**. This method is similar to item-based collaborative filtering, where mashups is equivalent to items. The most similar N mashups to the target mashup is found from the training set. The API used by these mashups is then recommended to the target mashup. The recommendation rate between the target mashup s and candidate APIs are expressed in formula 1.
- **Service profile recommendation model (SPR)**. The service profile recommendation model [5] uses mashup textual descriptions and structures to discover important lexical features of APIs, and bridge the vocabulary gap between mashup developers and service providers. It uses the Author-Topic Model (ATM) [27] to jointly model the mashup textual description and component services for refactoring the service profile.
- **Logistic Regression (LR)**. Logistic regression is mainly used to calculate the probability of occurrence in binary classification. Multi-dimensional features of the target mashup and APIs are used to predict the relevance of the APIs and the target mashup, then the recommendation list is generated to realize the API recommendation.
- **Random Forest (RF)**. Random Forest is an algorithm that integrates multiple trees by using the idea of ensemble learning. The weight of multi-dimensional features is learned in binary classification. Finally, the learned weights are used to predict APIs for the target mashup, then the API recommendation list is generated to realize the API recommendation.
- **RankNet (RN)**. The proposed method combines the regression of multi-dimensional features learning-to-rank method to calculate the recommendation rate of each pair of target mashup and API, then recommendation list is generated by sorting these pairs according to their recommendation rate.

5.4   Experimental Analysis

Recall, NDCG and hamming distance are used to compare the performance of the above methods. We calculate five API recommendation lists for target mashup: top-1, top-5, top-10, top-15, and top-20. Figures 1 and 2 show the recommendation performance of selected methods against two datasets.

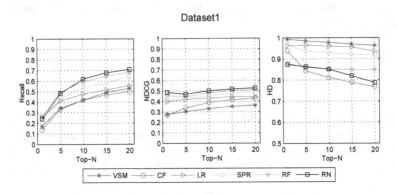

Fig. 1. Comparison of method indicators in dataset 1.

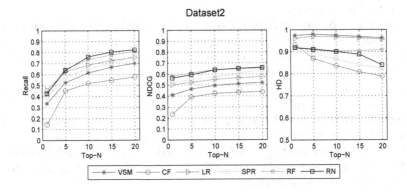

Fig. 2. Comparison of method indicators in dataset 2.

In term of ranking performance, method which only uses text information such as VSM and CF are obviously not as robust as those methods with feature combinations. As combining text information and API co-occurrence patterns, SPR and LR achieve superior performance than VSM and CF.

Regarding to the ensemble-based methods, RN is significantly better than LR and RF in both of Recall and NDCG. RN respectively improves RF by 1.7%–3.9% in terms of Recall@20 and 0.7%–3.2% in NDCG@20. Compared with

Table 4. Case studies of API recommendations.

Target mashup	Methods	Top 3 Recommendations of APIs		
Quizlio (2012)	VSM	FreebieSMS	Quizlet Flashcards	GoMoText SMS Gateway
	CF	Clockwork SMS	Google Maps	True Knowledge
	SPR	Google Maps	Twilio SMS	GoMoText SMS Gateway
	LR	Twilio SMS	Quizlet Flashcards	GoMoText SMS Gateway
	RF	Hoiio SMS	Clockwork SMS	Google Maps
	RN	Twilio	Quizlet Flashcards	Google Maps
	―	Twilio	Twilio SMS	Quizlet Flashcards
LyricStatus (2012)	VSM	ChartLyrics Lyric	LyrDB	Lyricsfly
	CF	Twilio	Lyricsfly	Facebook
	SPR	Facebook	Lyricsfly	ChartLyrics Lyric
	LR	LyricFind	ChartLyrics Lyric	Lyricsfly
	RF	Lyricsfly	Facebook	Facebook Graph
	RN	Lyricsfly	YouTube	Facebook
	―	YouTube	Facebook Graph	Facebook
Twenue (2013)	VSM	Tweet Press	Tweet Scan	Foursquare
	CF	Twitter	Yahoo Geocoding	411Sync
	SPR	Bing Maps	Google Maps	OpenStreetMap
	LR	Scribble Maps	Google Maps	Fwix
	RF	Twitter	Google Maps	Facebook
	RN	Twitter	Google Maps	Foursquare
	―	Twitter	Google Maps	Foursquare
Holidayen (2014)	VSM	Yahoo Travel	Cleartrip	Travel Booking Network
	CF	Google Maps	Flickr	Facebook
	SPR	Yahoo Travel	HotelsCombined	sletoh.com
	LR	Yahoo Travel	Tripit	Cleartrip
	RF	Google Maps	Yahoo Travel	Flickr
	RN	Google Maps	Flickr	Twitter
	―	Viator	Google Maps	Flickr

LR, there is an improvement of 9.4%–27.7% in Recall@20 and 14.9%–19.0% in NDCG@20. RF filters the candidate APIs and makes decisions on whether to recommend them. LR makes probability prediction on the candidate APIs and recommends APIs with high probability for target mashup. Neither of them optimizes the model parameters based on ranking preference, so both of them are inferior to RN in terms of ranking performance.

With respect to the diversity metric, there also exists an apparent accuracy–diversity dilemma from the aggregate perspective that increasing recommendation diversity most often compromises recommendation accuracy, and vice versa. Our experimental results are in good agreement with this phenomenon. In term of HD@10, both perform better than that of CF or SPR, and act more robust.

5.5   Case Study

To qualitatively compare selected methods, we choose four target mashups randomly from two datasets as: **Quizlio, LyricStatus, Twenue, Holidayen**. The results are shown in Table 4 (" – " represents the APIs truly used by the mashup).

For the first case, "**Quizlio**" is a platform that test with Quizlet memory cards, and its tag information includes "Flashcard", "Quiz", "SMS", and "Test".

On the recommendation list, CF and RF fail to recommend any relevant APIs, VSM and SPR can exactly recommend one, both LR and RN can hit two. For the second case, "**LyricStatus**" is a platform for sharing favorite lyrics with Facebook friends, and its tag information includes "Lyrics", "Music", "Quote", "Social" and "Status". On the recommendation list, neither VSM or LR does recommend relevant APIs, CF and SPR recommends one, RF and RN hit two. For the third case, "**Twenue**" is a platform which uses Twitter and Google maps to find and share the location, its tag information includes "Geotagging", "Local", "Mapping", "Social" and "Tweets". On the recommendation list, VSM, CF, SPR and LR hit one, and RF hit two, RN can recover all. As for the last case, "**Holidayen**" is a Travel planning platform whose tag information includes "Travel" and "Trip Planner". On the recommendation list, VSM, SPR and LR do not recommend any relevant APIs, CF, RF and RN exactly recover two.

6   Related Works

In recent years, with the emergence of lightweight Web APIs, mashups and APIs recommendation has become more and more active, attracting wide attention from domestic and foreign peers and industry [20]. Different from traditional recommendation, an important feature of Web APIs is to participate in service combine to realize application value-added [21]. Tasks and goals of recommendation vary from in how to effectively recommend APIs to meet the need of mashup. Therefore, we summarize the related work of Web API recommendation from the following three aspects: content-based recommendation, recommendation based on collaborative filtering, and combined recommendation.

Firstly, content-based recommendation mainly uses related technologies (such as TFIDF, LDA model, connection prediction, etc.) to evaluate the relevance between mashups and APIs, and then recommends Web API that meet the needs of developers [4–8]. Likewise, different characteristics of mashups and Web APIs are utilized to implement recommendations, such as text descriptions, tags, popularity, network information, etc. Vector space models (such as TFIDF) are mainly used to analyze and calculate the similarity between target mashup and Web APIs [4]. In order to improve the accuracy of similarity measurement, some researchers propose probabilistic models to mine the potential relevance of API documents. Cao et al. [7] proposed a Web API recommendation for mashup development by integrating content and network-based service clustering. Using relationships between mashup services, they developed a two-level topic model to mine potential, useful, and novel topics, thereby improving the accuracy of service clustering. Considering that different attributes may have different contributions to service links when calculating the semantic distance between services. Bao et al. [8] developed a potential attribute modeling method to reveal the context-aware attribute distribution.

Secondly, recommendation based on collaborative filtering [9–13] mainly uses the usage records of mashups and Web APIs to predict and recommend top-N Web APIs. Cao et al. [9] proposed a mashup service recommendation method based on content similarity and collaborative filtering, which combined user interest value and score prediction value to rank and recommend mashup services. Jiang et al. [10] calculated the similarity of users or services by using the nearest neighbor based collaborative filtering method, and then predicted the service grade by gathering the scores of the former k-nearest users or services.

The current recommendation usually does not adopt a single recommendation mechanism and strategy, but combine several methods to achieve a better recommendation. Thus, combined recommendations [14–17] are generated. Rahman et al. [14] proposed a matrix factorization method based on integrated content and network-based service clustering. This not only enables recommendations within a short list, but also allows for potential relationships. Xia et al. [15] proposed a category-aware API clustering and distributed recommendation method, used the extended k-means to cluster Web APIs, and developed a distributed machine learning framework to predict service ranking.

As can be seen from the above work, most of these methods use a single meta-information, either using the information of text or tags or popularity. At present, there is little research on the integration of multiple features. Li et al. [17] proposed a Web API recommendation based on tags, topics, cooccurrence and popularity. The tags are first used to calculate the similarity between the RTM exported mashup and the topic information from the Web APIs. Then, the popularity of the Web APIs is built by combining the historical call time and the category information. Finally, the factorization machine is used to train multi-dimensional vectors to predict and recommend top-N Web APIs. At the same time, most of the recommend results are a list, and the ranking of the recommendation results is also important. However, the ranking of the recommendation results is not considered in the above work. Therefore, a Web API recommendation method based on features ensemble and learning-to-rank is proposed, considering the multi-information of content and structural features of mashups and Web APIs.

7 Conclusion

We have proposed a Web API recommendation method based on features ensemble and the principle of learning-to-rank, which takes the textual description features and structural features of mashups and APIs into account, and uses an effective strategy to optimize the feature weights in the regression-based recommendation model. Experimental results show that a better recommendation performance can be achieved by combining trivial features and optimizing ranking preferences of Web APIs. Combination methods also do better in dealing with the dilemma of accuracy and diversity of recommendations. What we should do next is to use deep neural networks to develop recommendation methods by deep understanding of the textual description and structural characteristics of mashups and APIs.

Acknowledgements. This work is supported by the National Natural Science Foundation of China (61562090, 61962061), partially supported by the Yunnan Provincial

Foundation for Leaders of Disciplines in Science and Technology, the Program for Excellent Young Talents of Yunnan University, the Project of Innovative Research Team of Yunnan Province (2018HC019).

References

1. Tan, W., Fan, Y., Ghoneim, A., et al.: From the service-oriented architecture to the web API economy. IEEE Internet Comput. **20**(4), 64–68 (2016)
2. Zhao, H.B. Prashant, D.: Towards automated RESTful web service composition. In: 2009 IEEE International Conference on Web Services, pp. 189–196. IEEE, USA (2009)
3. Liu, X., Hui, Y., Sun, W., et al.: Towards service composition based on Mashup. In: 2007 IEEE Congress on Services, pp. 332–339. IEEE, USA (2007)
4. Cao, B., Liu, J., Tang, M., et al.: Mashup service recommendation based on user interest and social network. In: 2013 IEEE 20th International Conference on Services, pp. 99–106. IEEE, USA (2013)
5. Zhong, Y., Fan, Y., Tan, W., et al.: Web service recommendation with reconstructed profile from mashup descriptions. IEEE Trans. Autom. Sci. Eng. **15**(2), 468–478 (2018)
6. Li, C., Zhang, R., Huai, J., et al.: A novel approach for API recommendation in mashup development. In: 2014 IEEE International Conference on Web Services, pp. 289–296. IEEE, USA (2014)
7. Cao, B., Liu, X., Rahman, M., et al.: Integrated content and network-based service clustering and web APIs recommendation for mashup development. IEEE Trans. Serv. Comput. **99**, 1 (2017)
8. Bao, Q., Gatlin, P., Maskey, M., et al.: A fine-grained API link prediction approach supporting mashup recommendation. In: 2017 IEEE International Conference on Web Services (ICWS), pp. 220–228. IEEE Computer Society, Honolulu (2017)
9. Cao, B., Tang, M., Huang, X.: CSCF: a mashup service recommendation approach based on content similarity and collaborative filtering. Int. J. Grid Distrib. Comput. **7**(2), 163–172 (2014)
10. Jiang, Y., Liu, J., Tang, M., et al.: An effective web service recommendation method based on personalized collaborative filtering. In: IEEE International Conference on Web Services, pp. 211–218. IEEE, Washington (2011)
11. Zheng, Z., Ma, H., Lyu, M.R., et al.: QoS-aware web service recommendation by collaborative filtering. IEEE Trans. Serv. Comput. **4**(2), 140–152 (2011)
12. Yao, L., Wang, X., Sheng, Q.Z., et al.: Mashup recommendation by regularizing matrix factorization with API co-invocations. IEEE Trans. Serv. Comput. **99**, 1 (2018)
13. Lo, W., Yin, J., Deng, S., et al.: collaborative web service Qos prediction with location-based regularization. In: 2012 IEEE 19th International Conference on Web Services, pp. 464–471. IEEE Computer Society, Honolulu (2012)
14. Rahman, M.M., Liu, X., Cao, B.: web API recommendation for mashup development using matrix factorization on integrated content and network-based service clustering. In: 2017 IEEE International Conference on Services Computing (SCC), pp. 225–232. IEEE Computer Society, Honolulu (2017)
15. Xia, B., Fan, Y., Tan, W., et al.: Category-aware API clustering and distributed recommendation for automatic mashup creation. IEEE Trans. Serv. Comput. **8**(5), 674–687 (2017)

16. Gao, W., Chen, L., Wu, J., et al.: Joint modeling users, services, mashups, and topics for service recommendation. In: 2016 IEEE International Conference on Web Services (ICWS), pp. 260–267. IEEE, San Francisco (2016)

17. Li, H., Liu, J., Cao, B., et al.: Integrating tag, topic, co-occurrence, and popularity to recommend web APIs for mashup creation. In: IEEE International Conference on Services Computing (SCC), pp. 84–91. IEEE, Honolulu (2017)

18. Wang, X., Wu, H., Hsu, C.H.: Mashup-oriented API recommendation via random walk on knowledge graph. IEEE Access **7**, 7651–7662 (2019)

19. Li, H.: A short introduction to learning to rank. IEICE Trans. Inf. Syst. **E94–D**(10), 1854–1862 (2011)

20. Wu, H., Yue, K., Li, B., et al.: Collaborative QoS prediction with context-sensitive matrix factorization. Future Gener. Comput. Syst. **82**, 669–678 (2018)

21. Huang, K., Fan, Y., Tan, W.: An empirical study of programmable web: a network analysis on a service-mashup system. In: 2012 IEEE 19th International Conference on Web Services (ICWS), pp. 552–559. IEEE, Honolulu (2012)

22. Wu, H., Pei, Y., Li, B., Kang, Z., Liu, X., Li, H.: Item recommendation in collaborative tagging systems via heuristic data fusion. Knowl.-Based Syst. **75**(1), 124–140 (2015)

23. Hoffman, M.D., Blei, D.M., Bach, F.R.: Online learning for latent dirichlet allocation. In: Advances in Neural Information Processing Systems, vol. 23, pp. 856–864. Natural and Synthetic, Canada (2010)

24. Robertson, S.E., Zaragoza, H.: The probabilistic relevance framework: BM25 and beyond. Found. Trends® Inf. Retr. **3**(4), 333–389 (2009)

25. Ketkar, N.: Stochastic gradient descent. In: Deep Learning with Python, pp. 113–132. Apress, Berkeley (2017)

26. Dojchinovski, M., Vitvar, T., Hoekstra, R.: Linked web apis dataset. Seman. Web **9**(3), 1–11 (2017)

27. Rosen-Zvi, M., Griffiths, T., Steyvers, M., et al.: The author-topic model for authors and documents. In: Proceedings of the 20th Conference in Uncertainty in Artificial Intelligence, pp. 487–494. DBLP, Canada (2004)

EagleQR: An Application in Accessing Printed Text for the Elderly and Low Vision People

Zhi Yu[1,2,3,4(✉)], Jiajun Bu[1,2,3,4], Chunbin Gu[1,2,3,4], Shuyi Song[1,2,3,4], and Liangcheng Li[1,2,3,4]

[1] Zhejiang Provincial Key Laboratory of Service Robot,
College of Computer Science, Zhejiang University, Hangzhou, China
{yuzhirenzhe,bjj,guchunbin,brendasoung,
liangcheng_li}@zju.edu.cn
[2] Alibaba-Zhejiang University Joint Institute of Frontier Technologies,
Hangzhou, China
[3] MOE Key Laboratory of Machine Perception, Beijing, China
[4] Ningbo Research Institute, Zhejiang University, Ningbo 315100, China

Abstract. Most people ignore the difficulties for the elderly and low vision people to access the printed text, which is of critical importance in their daily lives. To resolve this issue, an application named Voiceye is designed to get the audio text by scanning the QR code. But this application has the limitation on the length of the printed text and takes no account of information security. In this paper, we design an advanced application EagleQR that store encrypted URL instead of the whole printed text. On one hand, we convert the printed text into online version and then encrypt and compress its URL into the QR code on the server. On the other hand, we decompress and decrypt the data in the QR code and then get the audio text to the client. This application has significance to broaden the information access to the printed text for the elderly and low vision people.

Keywords: Accessible interaction · Elderly · Low-Vision people · QR code · Printed text

1 Introduction

Visually impaired people are defined as those with vision loss that is caused by lesions in eyes as a result of partly or fully negative transformation of ocular organs or optic nerves in brain. According to the latest data of W.H.O., there are 253 million visual impaired people globally, containing 36 million of blind people [1]. 81% of these people are over 50 years old [2] and most of them are illiterate. These people are unable to work, study, or perform daily tasks well as healthy people - some of them even lost all visual functions [3], but they also need to collect much information related to medical treatment, job opportunities and public welfare to support their lives.

In daily life, the special populations, especially the low vision people and the elderly, have more difficulties in accessing the information from electronic media and printed text than the healthy people, such as taking medicine, visiting the places of

© Springer Nature Singapore Pte Ltd. 2019
H. Jin et al. (Eds.): BigData 2019, CCIS 1120, pp. 420–430, 2019.
https://doi.org/10.1007/978-981-15-1899-7_30

interest, and so on. To break the barriers, many efforts have been made, including designing some assistive hardware devices, getting information by touching, and software products which can convert the text into audio. Most of these assistive technologies are focus on the electronic information and partial printed information. However, another kind of information that comes from the printed text should not be overloaded. The printed texts are everywhere and the contents are usually brief but of great importance for all. For instance, people will see the date of production when they buy food and will read the instructions before taking the medicine. Such examples are very common in daily life.

To solve this problem, Voiceye [4] in South Korea developed an application to assist the visual-impaired people in reading creatively. The application gets the printed text by scanning the QR code, into which the text on the A4 paper is compressed. Then the client can get the audio text. This application is of great significance to the breakthrough of interactive gap between printed text and the visual-impaired people.

However, there are still some problems existing in this application. (1) Because the capacity of QR codes is limited by their size and this application compresses the full text into the QR code, so Voiceye doesn't support the long printed texts. (2) There are many risks because of the impossibility of reading content directly, but this application takes no account of the factors of information security. (3) This application does not enable the visually impaired people to autonomously choose when to stop reading and how to resume the context correctly next time when they start reading, other than getting the audio text from the very beginning again.

Furthermore, we made a requirement research about this application at Changchun Union University in China. There were seven low-vision people participating in this investigation, including three males and four females, and all of them are looking forward to this application.

In this paper, we propose our application EagleQR that we encrypt and compress the URL of the text into the QR code on the server and then get the audio text by scanning the QR code on the client. To break the length limitation of the printed text, we store the URL instead of full printed texts. Considering the security when scanning the QR code, we encrypt and compress the URL into the QR code rather than store the real URL directly. Furthermore, we add the function of reading whenever you want to improve the user's experience.

The remaining of this paper is structured as follows: Sect. 2 provides the related work to the accessible interactive mode of the traditional presswork and some study about QR code. Section 3 introduces our platform design in detail and demonstrate its UIs and functions. Section 4 presents a preliminary pilot test. Section 5 concludes our work.

2 Related Work

To help the elderly and low-vision people get the printed text, several studies have been done. China Braille Press [5], which is a comprehensive culture publishing house serving 17 million blind people in China, has made great efforts. It has published a lot of Braille books and audiobooks. More than that, With the joint effects between us and

China Braille Press, a series of hardware products were developed such as Sunshine reader and Wenxing audiobooks, which can automatically translate text messages in multiple formats including TXT, WORD and HTML into voice message. In spite of the remarkable progress of such hardware, it is still only available for reading specific materials, restricting the discretionary options of users. AI-Khalifa introduced a barcode-based system to assist the blind and visually impaired people identify objects [6]. This system is forward-looking, but its design of storing pre-recorded audio file in the server will disable us to edit the file. And considering the large volume of audio files, it will be slow to transfer them from the server to the client. Voiceye in South Korea assisted the blind in reading by developing application creatively. The blind people can get the text on paper by utilizing this application which can scan the QR code that stores the corresponding content in advance on presswork. Then users can make the text display or broadcast in voice according to personal needs. This product is of great significance to the breakthrough of interactive channel between traditional presswork and blind people and wins the ICT Special Educational Needs Solution award, known as the Oscar Award of education software, at the BETT conference. However, there are still some imperfections in this application. Storing the full data of the e-text in a standard QR code will limit the article length and increase the difficulty of scanning. What's more, there are much risks when using the QR code. Krombholz et al. reported the security risks caused by QR code and showed that some malicious QR codes may be printed and replace ones which are benign [7]. Dabrowski et al. also described the phishing attacks and targeted exploits based on the QR code [8]. So We encrypt QR code to guarantee security in the process of scanning. But Voiceye takes little account of the security factors.

3 Application Design

In this section, we introduce the design of the application EagleQR in detail. We divided this section into three parts, architecture, UIs and functions on the server, UIs and functions on the client. In the section architecture, we introduce EagleQR in a holistic way. In the other two sections, we demonstrate the UIs and functions of this app in the view of the user.

3.1 Architecture

As Fig. 1 depicts, EagleQR can be divided into two parts: server (orange blocks) and client (gray blocks). The server part consists of text generation process and QR code generation process. The text generation process is responsible for getting online text and then the QR code generation can create the QR code by encrypting and compressing the URL. While the client is composed of three parts, preparation, URL acquisition and information acquisition. The function of preparation module is to carry on authentication. In the module of URL acquisition, the real URL can be acquired after decrypting and uncompressing the information in the QR code. The module of information acquisition can get the online text from the URL and then read the text by linking the assistive technology.

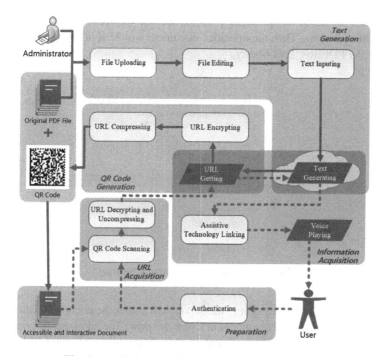

Fig. 1. Architecture of EagleQR (Color figure online)

On the server, this application starts with uploading a document by administrator. Then the online document will be split into several pages, thus the user can get the content from the one his/her just visited. And then the document will be converted into an editable version so that the administrator can improve the content into accessible text, such as adding alternative text for images. With URL extraction of the accessible text, we encrypt and compress its URL into a QR code under RSA [9] encryption algorithm and hexadecimal converting compression method.

On the client, a blind user should carry on authentication and log in the application. Then he/she can scan the corresponding QR code for obtaining accessible document. From the QR code we extract and parse the URL of the document by decrypting and decompressing technology, we can obtain its accessible content. Finally, the blind user can acquire the audio alternative information instead of unreadable printed text via assistive technology.

3.2 Server UIs and Functions

The key role of the server part is the administrator, which manages several modules mainly including document uploading, document editing and text input. The server UI provides the interactive platform for the administrator and the online documents.

As Fig. 2 shows, the administrator can only operate the system after logging in successfully. In the operation interface, the administrator has the permission to upload documents to the server by clicking the upload button, with some detailed information which shows the status of uploading process, such as the speed, time-cost and document size of uploading. Then the uploaded document will be split into several pages on the server via PDFBox Java library.

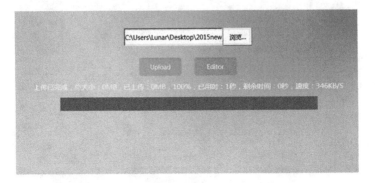

Fig. 2. Interface of document uploading

As Fig. 3 shows, after uploading, the server will gather the documents and for each document it shows the editable content in a new window. The administrator can modify some inaccessible and abnormal content in the editing box with a readable version. When all the edits are done, by clicking the Finish button the document will be generated with a new QR code placed in the top right corner of the page. In order to preventing potential security risk of QR code, RSA encryption algorithm is applied to encrypt the URL into the QR code.

Fig. 3. Text editing and QR code generating

3.3 Client UIs and Functions

The key role of the client part is the user, which participates in several modules mainly including authentication, QR code scanning, decoding, text generating and screen reader linking.

The users can use Scan QR Code function after logging in and the client will identify the real URL after a series of processes including Hexadecimal conversion, decompressing, inverse operation etc. Then the WebView will send a request to the server and load related data. It is noteworthy that the data will not be loaded correctly if user uses other scanning applications different without application. Figure 4 shows the different scanning results of two different scanning applications.

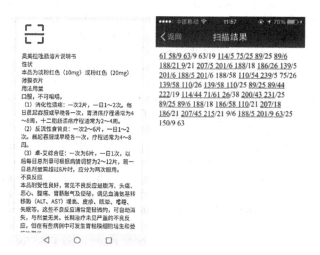

Fig. 4. Scanning results of EagleQR in this paper (left) and another application (right)

When the data is loaded successfully, the client will automatically link to the screen reader software called Voiceover, which is originally built in Apple devices, or other Text-to-Speech application. These can read the words displaying on the screen and transform the text into speech. Considering the fact that visual-impaired people have difficulty in turning pages in reading, gestures for page turning are of great importance to them. In order to differentiate our gestures from the standard gestures of Voiceover, we finally determined the following gestures: double-click with a single finger stands for paging down; Triple-click with a single finger stands for paging up. In addition, for avoiding conflict with the existed gesture in Voiceover, there is no definition for a two-fingered swipe in our gestures.

In practice, in view of less information stored by QR code, the fault tolerance rate in the process of scanning QR code is high that even if the scanning deviation appears, it is still verified quickly.

4 Experiment

In this section, we carry on two kinds of tests, simulation test and real user test. In the simulation test, we test the performance of this application. And collect the real user's feedback in another test.

4.1 Preliminary Pilot Test

We performed a pilot test with eleven low vision volunteers in order to improve the functions of EagleQR and the user experience. There were nine males and two females in different age group using different kinds of mobile phones.

Before the test, we prepared an article which is printed on an A4 sheet with a corresponding QR code on the top right corner. Around the QR code we add some rough flags to get easy perception. The participants are trained to know how to use the application before the test. During the test, all the participants are required to get the article via EagleQR independently as soon as possible under the same network. Meanwhile, we will record the duration of searching QR code for understanding the fluency of this operation. During the test, we record the duration between the start of the test and the time when they scanned the QR code successfully independently.

All the participants show their appreciation to this application and were satisfied with its safety, which was achieved by encrypting the QR code, and its convenience that allowing them to start reading from the page that they left off last time.

The results show that. all the subjects could finish the test within fifteen seconds and the average time is nine seconds. Meanwhile, from the participants' feedback, most of them suggested that the fault tolerance rate that is a parameter used to adjust the matching degree of scanning of the QR code should be increased. This suggestion was then adopted because less content in the QR code made it possible for us to achieve a higher fault tolerance rate. In following tests, we increase the fault tolerance rate by adjusting related parameters, which stands for the error allowed when scanning. The result demonstrates that increasing the fault tolerance rate makes it easier for the users to successfully scan the QR code (even scanning part of the QR code).

4.2 Real User Test

We invite 7 visual-impaired people whose vision is between level 1 and level 4 for this test. There are 3 males and 4 females. And they do not use any assistive tool in their life.

They are asked to use this application to get the printed text with different mobile phones in the same environment. They give some kinds of useful feedbacks about this application.

1. Rather than getting information by Braille, they prefer audio text.
2. This kind of access, which is based on QR code, to printed text responses faster than OCR and is easier to align to.
3. Sometimes OCR identification may lead to misprinting and wrong paragraph structures, while EagleQR is able to ensure the explicit paragraph structures.
4. Compared to OCR, EagleQR supports skipping in reading.

Fig. 5. Elderly people read traditional medicine specification (left) vs. use EagleQR instead (right)

Fig. 6. Elderly people read traditional introduction (top) vs. use EagleQR instead (down)

Totally, we can know that EagleQR have better performance on the speed and accuracy in recognition process, showing better users experience. After the test, they all agreed that it was a practical tool for the visual-impaired people to obtain information around them.

At the same time, we translate the OR code into WeChat QR code and use WeChat small program to achieve the function. Then we create 2 WeChat QR codes for talking medicine and visiting the places of interest. After that, we invite 2 elderly people to try this application in real as Figs. 5 and 6.

Through the actual test comparison, we found that for the elderly with weakened vision, because the printed text of the drug manual is small, even after wearing the reading glasses, the elderly can not clearly distinguish the above words. At the same time, the text on the attraction card is difficult to distinguish because of the long-term sun and rain. With our products, the experience of older people reading drug brochures and attraction cards has improved.

4.3   Compared with Similar Works

In the barrier-free interaction mode of traditional presswork, Korea's Voiceye has creatively developed mobile applications to help visually impaired people read. However, the mobile application still has more shortcomings: because all the information is compressed into the QR code, the amount of information is too large, the dot matrix in the QR code is very dense, and the printing quality must be very high to ensure good scanning results, so the cost is high and the information that can be saved is still limited. At the same time, due to the dense array of QR code and large amount of information, the mobile phone can be hardly accurately aligned with the two-dimensional code. To this end, Voiceye Co., Ltd. produces mobile phone scanning brackets, and strictly stipulates the position of the QR code printing to ensure better scanning results. However, this is not conducive to the visually impaired people to carry and use with them, and for the actual use environment of the visually impaired people, it does not perfectly match the optimal scanning position at any time.

Blind-assisted glasses are currently being developed. They have four main functions: ground detection, obstacle detection, indoor navigation and human recognition. The main functions are obstacle avoidance. OCR technology can also be used to achieve certain text recognition capabilities. However, due to the limitations of the OCR technology itself, the text recognition speed is slow and cannot be 100% accurate. Once the recognition text is wrong (for example, the amount of medicine taken is wrong), it may cause more inconvenience to the user, and the responsibility will be traced to the blind glasses itself. Therefore, the practicality is limited. The hardware cost is high, and the actual wear experience needs to be improved.

Based on the experience of summarizing existing products, this product develops a new two-dimensional code-assisted barrier-free interaction method. Since only the URL address of the text is stored, the load of the two-dimensional code is reduced and the accuracy of the scan code is improved. With speed. At the same time, as the information transmission party rather than the publisher, it avoids the responsibility of unnecessary information generation errors. And the use of encryption technology can better protect copyright and personal privacy. At the same time, no additional hardware costs are required, and the price is low, which is suitable for promotion.

5 Conclusion and Future Work

In this paper, we design an application supporting the interaction with the printed text for the visually impaired people. This application helped them overcome the difficulty in getting the printed text. The process was divided into several steps including uploading the content in the printed materials to the cloud, editing the uploaded text, the users scanning the QR code in the client, downloading the text from the cloud after scanning successfully and displaying audio version of the text. Compared to Voiceye, which stored the whole text into the QR code, this platform generated QR code that only stored URL of the text, so that the fault tolerance rate of QR code recognizing was greatly increased and the printing cost could be reduced. Also, the operating experience of the user could be optimized, and it supported the assistive reading technology of page up and page down. Further, we can connect server with some mainstream and provide the function of documents importing instead of uploading documents manually. And besides Text-to-Speech that we are using now, we try to use other voice software which supports the voice interaction with users to help elderly and low vision people to input voice more conveniently.

Acknowledgments. This work is supported by Alibaba-Zhejiang University Joint Institute of Frontier Technologies, The National Key R&D Program of China (No. 2018YFC2002603), Zhejiang Provincial Natural Science Foundation of China (No. LZ13F020001), the National Natural Science Foundation of China (No. 61972349, 61173185, 61173186) and the National Key Technology R&D Program of China (No. 2012BAI34B01, 2014BAK15B02).

References

1. WHO. World Health Organization. www.who.int/mediacentre/factsheets/fs282/en/
2. Bourne, R.R.A., Flaxman, S.R., Braithwaite, T., Cicinelli, M.V., Das, A., Jonas, J.B., et al.: Vision loss expert group magnitude, temporal trends, and projections of the global prevalence of blindness and distance and near vision impairment: a systematic review and meta-analysis. Lancet Glob. Health 5(9), e888–e897 (2017)
3. Meng, X., Liu, X.D.: Encoding study of 1 dimensional multiple system color bar code. J. Zhejiang Univ. **38**, 559–561 (2004)
4. VOICEYE. www.voiceye.com/kor/
5. CBA. China Braille Press. www.cbph.org.cn/
6. Al-Khalifa, H.S.: Utilizing QR code and mobile phones for blinds and visually impaired people. In: Miesenberger, K., Klaus, J., Zagler, W., Karshmer, A. (eds.) ICCHP 2008. LNCS, vol. 5105, pp. 1065–1069. Springer, Heidelberg (2008). https://doi.org/10.1007/978-3-540-70540-6_159
7. Krombholz, K., Frühwirt, P., Kieseberg, P., Kapsalis, I., Huber, M., Weippl, E.: QR code security: a survey of attacks and challenges for usable security. In: Tryfonas, T., Askoxylakis, I. (eds.) HAS 2014. LNCS, vol. 8533, pp. 79–90. Springer, Cham (2014). https://doi.org/10.1007/978-3-319-07620-1_8

8. Dabrowski, A., Krombholz, K., Ullrich, J., Weippl, E.R.: QR inception: barcode-in-barcode attacks. In: ACM CCS Workshop on Security and Privacy in Smartphones and Mobile Devices, pp. 3–10. ACM (2014)
9. Guruswami, V., Sudan, M.: Improved decoding of Reed-Solomon and algebraic-geometric codes. In: 1998 Proceedings of the 39th Annual Symposium on Foundations of Computer Science, pp. 28–37. IEEE (1998)

Author Index

Bi, Jianquan   312
Bo, Rongrong   73
Bu, Jiajun   420

Cai, Saihua   3
Cai, Tao   120
Cao, Lifeng   255
Cao, Wenzhi   209
Chan, Yeung   107
Chen, Bo   341
Chen, NingJiang   366
Chen, Wei   271
Chen, Yao   351
Chu, Deming   297

Dong, Wenyong   135
Dong, Yindong   148
Duan, XiaoYan   366

Feng, Jun   73

Gao, Shengxiang   89
Gao, Zhensheng   255
Gong, Cheng   73
Gu, Chunbin   420

He, Dongjian   325
He, Jinrong   325
He, Xiaoyu   194
He, Yonghou   341
Huang, Jin   351
Huang, Ren-ji   177

Jiang, Chengshun   240
Jin, Hai   57
Jing, Xuehui   395

Kang, Yilin   351
Kou, Zhijuan   57

Lai, Hua   89
Li, Bo   33
Li, Chun   380

Li, Chunbin   148
Li, Jian   159, 194
Li, Li   3
Li, Liangcheng   420
Li, Mo-Ci   177
Li, Xiaodong   73
Li, Yuanxi   341
Li, Zhi   351
Li, Zhuo-lin   159
Liang, Yu   46
Liao, Yulei   224
Lin, Jingjing   297
Liu, Bin   325
Liu, Jiao   194
Liu, Tianquan   120
Liu, Yawen   120
Lu, Xin   255

Mu, Huiyu   3

Niu, Dejiao   120

Ren, Fuji   148
Ruan, Ying   395

Shang, Lin   380
Shao, Yanling   135
Shi, Peishen   285
Shi, Wenfeng   380
Song, Shuyi   420
Song, Zhongshan   312
Sun, Chong   312
Sun, Jiayu   209
Sun, Ruizhi   3
Sun, Xiao   148

Tan, Liang   224
Tao, Li   341
Tong, Jianing   271

Wang, Chunqing   341
Wang, Hua   57

Wang, Jing   406
Wang, Ke   209
Wang, Meng   46
Wang, Puyu   285
Wang, Wenfa   325
Wang, Xin   406
Wang, Zhentong   351
Wei, Dajiang   351
Wei, Zhihua   271
Wu, Hao   406
Wu, Lisha   46
Wu, Wenjun   46

Xie, Xia   57, 209
Xie, YongSheng   366
Xu, ChenYang   16
Xu, Haibo   240
Xu, Yu   89

Yang, Chunming   33
Yang, DongJu   16
Yang, Wen   73
Ye, Qing   177
Yin, Fan   312
Yu, Zhengtao   89
Yu, Zhi   420

Yuan, Hao-chen   159
Yuan, Pingpeng   57

Zhang, Fan   297
Zhang, Hai   285
Zhang, Huan-xiang   159
Zhang, Hui   33
Zhang, Jia   33
Zhang, Lei   312
Zhang, Qianqian   3
Zhang, Shuai   395
Zhang, Xia Jie   107
Zhang, Xiao-Lin   159
Zhang, Xiaolin   194
Zhang, Zili   341
Zhao, Chen   89
Zhao, Hongwei   395
Zhao, Hua   406
Zhao, Xujian   33
Zheng, Xia   120
Zhou, Qimin   406
Zhou, Shijie   120
Zhou, Weiyu   312
Zhu, Jianlin   351
Zhu, Jing Hua   107
Zhu, Rongbo   351
Zhu, Zhanbing   255

Printed in the United States
By Bookmasters